Bill Rouse

Electronics Mathematics

Thomas Power

Delmar Publishers Inc.

This book is dedicated to my wife Starene

Cover photograph by Manfred Kage–Peter Arnold, Inc. The photograph shows a segment of a piezoelectrical surface wave filter. The geometrical structure of the chrome/gold electrode converts electrical into acoustic energy. Reflected light interference contrast after Normanski. Magnification 140×.

Delmar Staff
   Administrative editor: Mark Huth
   Project editor: Jonathan Plant

For information address Delmar Publishers Inc.,
2 Computer Drive West, Box 15–015,
Albany, New York 12212

Copyright © 1985 by Delmar Publishers Inc.

All rights reserved. No part of this work covered by the copyright hereon may be reproduced or used in any form or any means—graphic, electronic, or mechanical, including photocopying, recording, taping, or information storage and retrieval systems—without written permission of the publisher.

Printed in the United States of America
Published simultaneously in Canada by Nelson Canada,
a division of International Thomson Limited

10 9 8 7 6 5 4 3 2

Library of Congress Cataloging in Publication Data

Power, Thomas C.
   Electronics mathematics.

   Includes index.
   1. Electronics—Mathematics. I. Title.
TK7835.P69   1985   512'.1   84–7828
ISBN 0–8273–2410–3

Contents

Preface v
Flowchart vi
List of Abbreviations and Greek Symbols vii, viii

Part One: Review of Algebra

Chapter 1 Introduction and Signed Numbers 2
Chapter 2 Addition and Subtraction of Algebraic Expressions 9
Chapter 3 Multiplication and Division of Binomials and Polynomials 25
Chapter 4 Factoring 37

Part Two: Mathematics for DC and AC Circuits

Chapter 5   Powers of Ten 51
Chapter 6   Metric Units with Applications 64
Chapter 7   Equations 79
Chapter 8   Fractions 92
Chapter 9   Fractional Equations 106
Chapter 10 Right-Angle Trigonometry 128
Chapter 11 Angles 146
Chapter 12 Simultaneous Equations and Determinants 164
Chapter 13 Networks 184
Chapter 14 Thevenin's Theorem with Applications 196
Chapter 15 The Bridge 213
Chapter 16 Principles of Vector Algebra 229
Chapter 17 Sine Waves 254
Chapter 18 Phasors 277
Chapter 19 Phasor Algebra 293
Chapter 20 AC Series Circuits 307
Chapter 21 AC Parallel Circuits 331
Chapter 22 AC Series-Parallel Circuits and Resonance 356
Chapter 23 AC Circuit Analysis 374

Part Three: Computer Mathematics, Logarithms, and Quadratic Equations

Chapter 24	Math for the Computer	403
Chapter 25	Principles of Boolean Algebra	426
Chapter 26	Logarithms	456
Chapter 27	Applications of Logarithms	471
Chapter 28	Quadratic Equations	495

Appendix	513
Answers to Odd-Numbered Exercises	520
Index	534

Preface

It has been said, "What the world does not need is another electronics math text." However, this is more than just another conventional textbook—it is a learning tool for electronics students which uses a class-tested workbook format. For each problem in the book, space is provided for computing the answer. This special format not only allows students to use text explanations as reference, but also provides them with a convenient method for keeping solved problems on hand. Because of this the textbook can serve as a valuable resource for students during their educational program.

The book is intended for electronics technology students at community colleges, vocational schools, and industry training programs. The material included requires the user to have taken at least one year of high school algebra. Part One provides a review of selected algebra topics required for further study of this text. While some material in Part Two may also be considered algebra review, it is presented with applications in electronics. The chapters have been arranged to parallel as closely as possible the typical beginning courses in electronics theory, and the popular textbooks available for such courses. *Electronics Mathematics* is designed to be studied concurrently with dc and ac theory courses. At the same time, it offers the instructor the opportunity to introduce selected topics as needed, such as ac circuit analysis, quadratics, computer math, and logarithms (Part Three). A chapter dependency flowchart is provided on the following page to show how each chapter affects the study of others. It is intended to show the level of math required for each chapter rather than suggest a particular course outline.

Topics for this text have been selected with the electronics technology student in mind. Mathematical explanations are kept to a minimum, while numerous worked-out examples and problems are included throughout. Each chapter ends with a set of evaluation problems which provide an effective review of the material covered.

Much of this textbook has been class-tested by four instructors at Santa Rosa Junior College: Gordon Shimizu, Ed Sikes, Dave Herrington, and Frank Pugh. The author extends his thanks to them, and also to the reviewers of the manuscript: Ted Harris (Albuquerque Technical-Vocational Institute), Dr. Vijay Joshi (Durham Technical College), Len Mrachek (Hennepin Technical Center), Darrell Anderson (Milwaukee Area Technical College) and Harry Waller (San Antonio College).

FLOWCHART SHOWING CHAPTER DEPENDENCY — This chart shows how each chapter relates to the others. It can be helpful in coordinating electronics theory and electronics mathematics.

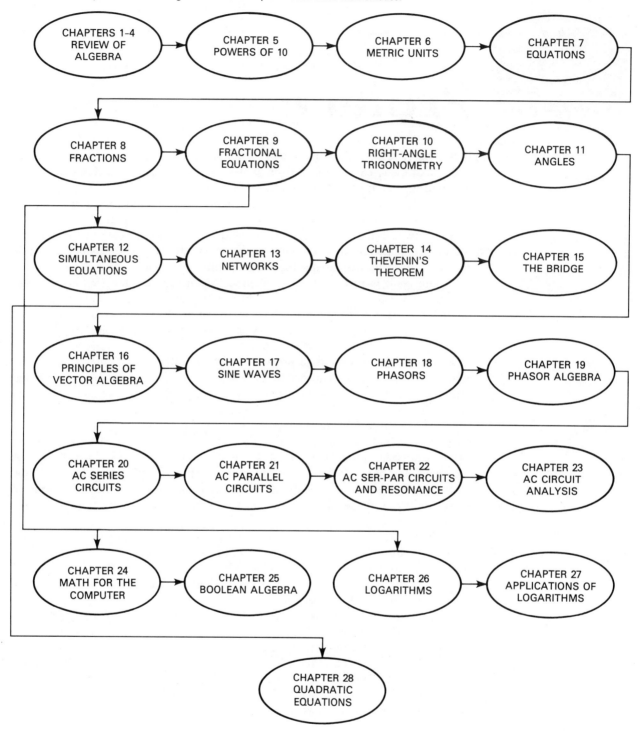

LIST OF ABBREVIATIONS

alternating current	ac	microsiemens	µS
ampere	A	microvolt	µV
angle whose		microwatt	µW
tangent is	arctan, $\tan^{-1}$	maximum	max
average	ave	mil	mil
centimeter	cm	milliampere	mA
cosine	cos	millihenry	mH
coulomb	C	millijoule	mJ
cycles per second	cycles/s, cps	milliohm	mΩ
decibel	dB	millisecond	msec
decibel milliwatt	dBm	millisiemens	mS
degree	deg	millivolt	mV
direct current	dc	milliwatt	mW
farad	F	minimum	min
foot	ft	minute	min
giga hertz	gHz	nanoampere	nA
gram	gm	nanofarad	nF
henry	H	nanosecond	ns
hertz	Hz	ohm	Ω
hour	hr	picofarad	pF
inch	in.	power factor	PF
kilohertz	kHz	radian	rad
kilohm	kΩ	revolutions per	
kilovolt	kV	minute	rev/min, rpm
kilovolt-ampere	kVA	revolutions per	
kilowatt	kW	second	rev/sec, rps
kilowatt-hour	kWh	root-mean-square	rms
maximum	max	second	s
megahertz	MHz	siemens	S
megawatt	MW	sine	sin
megohm	MΩ	tangent	tan
meter	m	volt	V
microampere	µA	volt-ampere	VA
microfarad	µF	volt-ampere	
microhenry	µH	reactance	VAR
microsecond	µs	watt	W

GREEK SYMBOLS

	Capital	Lower-case	Commonly Used to Designate
Alpha	A	α	Angles, area, coefficients
Beta	B	β	Angles, flux density, coefficients
Gamma	Γ	γ	Conductivity, specific gravity
Delta	Δ	δ	Variation, density
Epsilon	E	ϵ	Base of natural logarithms
Zeta	Z	ζ	Impedance, coefficients, coordinates
Eta	H	η	Hysteresis coefficient
Theta	Θ	θ	Temperature, phase angle
Iota	I	ι	
Kappa	K	κ	Dielectric constant, susceptibility
Lambda	Λ	λ	Wavelength
Mu	M	μ	Micro, amplification factor, permeability
Nu	N	ν	Reluctivity
Xi	Ξ	ξ	
Omicron	O	o	
Pi	Π	π	Ratio of circumference to diameter = 3.1416
Rho	P	ρ	Resistivity
Sigma	Σ	σ	Summation
Tau	T	τ	Time constant
Upsilon	Y	υ	
Phi	Φ	ϕ	Magnetic flux, angles
Chi	X	χ	
Psi	Ψ	ψ	Dielectric flux, phase difference
Omega	Ω	ω	Capital, ohms; lower case, angular velocity

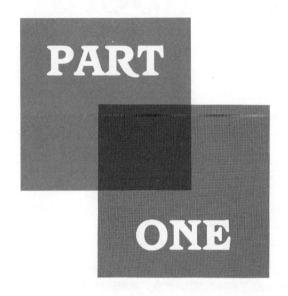# PART ONE Review of Algebra

CHAPTER 1
Introduction and Signed Numbers

Objectives

After completing this chapter you will be able to:
- Identify positive and negative numbers
- Add and subtract signed numbers
- Identify absolute quantities
- Multiply and divide signed numbers

Arithmetic is defined as the science of computing by positive real numbers, and it reaches certain limitations in problem solving. Algebra can be defined as a branch of mathematics that uses positive and negative numbers and letters to express relationships between quantities. Arithmetic has limitations in problem solving that can be expanded by the use of algebra.

Equations in electronics not only represent a method of solving problems but also define the theory associated with a circuit. The equation $V = IR$ actually represents the relationship of voltage, current, and resistance in an electric circuit. The equation $R_T = R_1 + R_2 + R_3$ defines a series circuit and, more specifically, a series circuit containing three resistive components.

1–1 EQUATIONS. Equations can be considered a shorthand notation for a group of words that describe an algebraic process, a theory of operation, or a descriptive figure.

EXERCISE 1–1   Write an equation for each description.

1. The number a when divided by the number b equals the number d.

2. Three times the number n divided by 8 is equal to c.

3. The area A of a triangle is defined as the altitude a times the base b divided by 2.

4. The number Z minus the number w is equal to the number p divided by 3.

5. The distance B added to point A is equal to one-half times D.

   $1/2 D = B + A$

6. The voltage v produced by a generator is equal to 6.3 times the number a less the number L.

   $V = 6.3a - L$

7. Four-tenths of the voltage V_{max} is equal to the voltage v.

   $V = .4 V_{may}$

8. The total length l of aluminum used to fabricate a chassis is equal to twice its length plus twice its width plus 2 times b_a.

   $L = 2L + 2W + 2ba$

9. The time t required to produce one cycle of alternating current is equal to the number 1 divided by f.

   $t = \frac{1}{f}$

10. Quarts q of volume can be converted to the metric unit of liters L by dividing the number of quarts by 0.946.

    $L = \frac{q}{.946}$

1–2 SIGNED NUMBERS. Signs of numbers can be explained by use of a *number line* as shown in Figure 1–1. An arbitrary point, usually the center point, is chosen and considered to be zero. All numbers to the right of that point are considered positive numbers (+), and those to the left are negative (−).

The *sign of a number* does not determine a value, but shows the direction taken from the zero point or point of origin. The number +3 has the same value as the number −3. Each number has three units between 0 and 3, but negative numbers are to the left and positive numbers are to the right. When no sign is affixed to a number, it is considered positive.

A practical example of these numbers is a thermometer for measuring temperature. In reading a thermometer, +15° is 15° above zero, and −15° is 15° below zero. They represent 15° difference between 0 and 15, but each represents a different point or temperature.

Algebra requires the use of positive and negative numbers. A system of operations and rules has been established.

Addition To add two or more numbers of like signs, find their sum; the sign of the answer will be the common sign.

$$\begin{array}{rrr} +3 & -3 & +6 \\ +4 & -4 & +8 \\ \hline +7 & -7 & +14 \\ & & +28 \end{array}$$

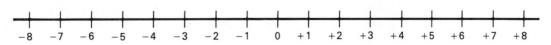

FIGURE 1–1 Number line

In algebraic problems this operation usually occurs with the numbers presented in a horizontal position.

$$3 + 4 = 7 \qquad -3 - 4 = -7 \qquad 6 + 8 + 14 = 28$$

To add a positive and a negative number, find the difference between the numbers; the sign of the answer will be the sign of the larger number.

$$\begin{array}{r}3\\-4\\\hline -1\end{array} \text{ (or) } 3 - 4 = -1 \qquad \begin{array}{r}-3\\+4\\\hline 1\end{array} \text{ (or) } -3 + 4 = 1$$

$$6 - 8 + 14 = 12$$
$$10 - 12 + 6 - 5 - 7 = -8$$

Subtraction To subtract one number from another, change the sign of the subtrahend (bottom number), algebraically add, and follow the rules for addition.

(sign of operation, subtraction)

$$(-) \quad \begin{array}{r}+7 = +7 \text{ (minuend)}\\+2 = -2 \text{ (subtrahend)}\\\hline +5 \text{ (remainder)}\end{array} \qquad \begin{array}{l}7 - (+2) = 7 - 2 = 5\\ \text{(after the sign is changed it}\\ \text{becomes an addition problem)}\end{array}$$

$$(-) \quad \begin{array}{r}+20 = +20\\-8 = +8\\\hline +28\end{array} \text{ (or)} \qquad 20 - (-8) = 20 + 8 = 28$$

In subtraction, when the problem is presented horizontally, a double-sign conflict occurs, as in the case above, where $20 - (-8) = 28$. Change the sign within the parentheses and drop the sign of operation.

$$-14 - (4) = -14 - 4 = -18$$
$$-15 - (-2) = -15 + 2 = -13$$

EXERCISE 1–2

Addition: Evaluate the following problems.

1.	26	2.	15.6	3.	−36	4.	−0.75
	−8		−9		8.5		−0.35
	18		*6.6*		*−27.5*		*−1.1*

5.	17	6.	96	7.	136.5	8.	−31
	−8		32		−82		−5
	−6		−40		76.8		−17
	15		−56		50		−3.6
	18		*32*		*181.3*		*−56.6*

9.	360	10.	−47	11.	−18.5	12.	80
	120		−62		7.6		−33
	560		−82		−13.4		107
	−2700		27		19.2		−154
	−1660		*−164*		*−5.1*		*0*

13. 6 − 7 + 4 − 3 = __0__

14. 16 − 12 + 3 − 8 = __−1__

15. −37 − 102 − 3 + 10 = __−132__

16. 63 + 72 + 18 − 6 = __147__

17. 8 − 4 + 7 − 6 + 9 − 4 + 7 = __17__

18. −13 − 7 − 12 − 4 − 6 − 10 = __−52__

19. −140 − 62 + 32 + 100 + 10 + 60 = __0__

20. +1 − 3 − 1 + 4 − 7 + 8 + 2 = __4__

Subtraction:

21. 17   22. −365   23. 1040
    − 8   − 40   − 820
    ―――   ―――   ―――――
     25   −325   1860

24. −16.4   25. 38.7   26. −7
     − 8.2   −50   6
     ―――――   ―――   ―――
     −8.2   88.7   −13

27. (68) − (19) = __49__   28. (−102) − (−17) = __−85__

29. (45) − (−15) = __60__   30. (−76) − (18) = __−94__

1–3 ABSOLUTE VALUE. An absolute value of a number is its value without regard to sign. In other words, +9 volts is as valid as −9 volts. Each unit contains 9 volts, and a volt is a volt regardless of sign. The absolute value can be indicated by vertical lines, such as $|-9|$ or $|+9|$, which are absolute values and have the same numerical value.

$$\text{or} \quad |-9| = |+9|$$
$$\text{and} \quad 9 = 9 \quad \text{(neither has a sign)}$$

Example A

$$\begin{array}{r} -8V \\ (-) \quad +10V \\ \hline |-18V| \end{array}$$

EXERCISE 1–3 Determine the absolute value of each of the following.

1. $|-6V|$ = __6V__   2. $|-100V|$ = __100V__

3. $|0V|$ = __0V__   4. $|6.3V|$ = __6.3V__

5. $|-25°|$ = __25°__   6. $|212°|$ = __212°__

7. $|6.4 \text{ ft}|$ = __6.4'__   8. $|-0.5 \text{ ft}|$ = __.5'__

9. $|32°|$ = __32°__   10. $|-373°|$ = __373°__
11. $|-9V| - |-3V|$ = __6V__   12. $|-10°| - |2°|$ = __8°__
13. $|-10°| - |-2°|$ = __8°__   14. $|-68V| + |-12V|$ = __80V__
15. $|-100V| - |100V|$ = __0V__   16. $|180\text{ ft}| + |-16\text{ ft}|$ = __196'__

1–4 MULTIPLICATION AND DIVISION OF SIGNED NUMBERS.

Multiplication

Rule 1: The product of two numbers with like signs will result in an answer with a positive sign.

Rule 2: The product of two numbers with unlike signs will result in an answer with a negative sign.

Example B

$$(3)(5) = 15$$
$$(-3)(5) = -15$$
$$(-3)(-5) = 15$$
$$(2)(-4)(5) = -40$$
$$(-3)(-4)(2)(5) = 120$$

Division The rules for division of signed numbers are the same as those for multiplication. Like signs result in a positive answer and unlike signs result in a negative answer. Division in algebra is usually presented as a fractional value. A fraction is a sign of division, meaning: divide the numerator by the denominator.

Example C

$$(+8) \div (+2) = +4 \quad \text{or} \quad \frac{(8)}{(2)} = +4$$

$$(-8) \div (-2) = +4 \quad \text{or} \quad \frac{(-8)}{(-2)} = +4$$

$$(+8) \div (-2) = -4 \quad \text{or} \quad \frac{(+8)}{(-2)} = -4$$

$$(-8) \div (+2) = -4 \quad \text{or} \quad \frac{(+8)}{(-2)} = -4$$

An even number of negative signs results in a positive answer, and an odd number of negative signs will result in a negative answer, regardless of the number of negative signs in the operation.

Example D

three negative signs = negative answer

$$(-3)(-4)(-6)(+2) = -144$$

four negative signs = positive answer

$$(-2)(-3)(+4)(-5)(-1)(+3) = +360$$

It is usually agreed that when performing several operations in a problem, multiplication and division should be done before addition and subtraction.

*Chapter 1
Introduction and
Signed Numbers*

EXERCISE 1–4 *Multiplication:* Find the resultant products.

1. $(-7)(6) = $ __−42__
2. $(15)(-5) = $ __−75__
3. $(-3)(-7) = $ __21__
4. $(0.4)(-0.6) = $ __−.24__
5. $(-6)(-1)(-4) = $ __−24__
6. $(7)(12)(10)(4) = $ __3360__
7. $(4)(-5)(3)(6) = $ __−360__
8. $(-120)(-4)(5)(-2) = $ __−4800__
9. $(6)(-320)(-4) = $ __7680__
10. $(-8)(-1)(-3)(-7)(-2) = $ __−336__

Division: Algebraic division is generally expressed as a fraction. Find the resultant quotients.

11. $\dfrac{16}{4} = $ __4__
12. $\dfrac{-33}{11} = $ __−3__
13. $\dfrac{-160}{-40} = $ __4__
14. $\dfrac{121}{-11} = $ __−11__
15. $\dfrac{-16.4}{0.5} = $ __−32.8__
16. $\dfrac{-9.6}{-1.2} = $ __8__
17. $\dfrac{88.8}{-3.3} = $ __−26.909__
18. $\dfrac{-16000}{400} = $ __−40__
19. $\dfrac{-896}{-1242} = $ __.7214__
20. $\dfrac{-0.0036}{0.08} = $ __−.045__

Chain operation of signed numbers:

21. $\dfrac{(-2)(-3)}{-1} = $ __−6__
22. $\dfrac{(8)(-3)(3)}{24} = $ __−3__
23. $\dfrac{(240)(-1)}{(-4)(6)} = $ __10__
24. $\dfrac{(0.05)(-10)}{(-1)(5)} = $ __.1__
25. $\dfrac{(160)(-2)(3)(-1)}{(-4)(6)} = $ __−40__
26. $\dfrac{(-0.01)(80)(-6)}{(8)(-3)(-0.5)} = $ __.4__

8

Part 1
Review of Algebra

27. $\dfrac{(-2)(-4)(-5)(-6)}{(-1)(-7)(-10)(-3)} =$ _1.1428_

28. $\dfrac{(-1)(0.1)(0.001)(10)}{(100)(-0.001)(-0.5)} =$ _−.02_

EVALUATION EXERCISE   Combine terms:

1. $3 + 2 - 4 + 2 - 5 =$ _−2_
2. $-4 - 7 - 6 + 3 - 7 =$ _−21_
3. $184 - 16 + 37 - 892 =$ _−687_
4. $5.5 + 1.7 - 0.8 - 1.2 + 18 =$ _23.2_
5. $0.09 - 0.4 + 0.75 - 0.65 =$ _−.21_
6. $(-3)(4) + 8 - 7 =$ _−11_
7. $(6)(-3) - (7)(-8) =$ _74_
8. $(10)(3) + (14)(-1) =$ _16_
9. $28 - 6(4) + (7)(2) =$ _18_
10. $8(2) + 7(3) - 8 =$ _29_
11. $(10)(-3) + 4(8) - 6 =$ _−4_
12. $\dfrac{8}{-2} + \dfrac{10}{5} - 3(4) =$ _−14_
13. $\dfrac{-18}{9} - 6(4) + \dfrac{15}{-3} =$ _−31_
14. $\dfrac{3}{4} + \dfrac{5}{-8} - \dfrac{2}{3} =$ _−.541_
15. $\dfrac{(12)(3)}{-6} + \dfrac{22}{11} - 8 =$ _−12_
16. $(|6|)(-3) =$ _−18_
17. $(|-2|)(|-4|) + (4)(-6) =$ _−16_
18. $\dfrac{|18|}{|3|} + \dfrac{|24|}{|-4|} - \dfrac{|39|}{|-13|} =$ _9_
19. $\dfrac{(14)(|-2|)}{-7} + |-6| =$ _2_
20. $|-0.5| + \dfrac{0.75}{|-0.5|} - (0.3)(-0.6) =$ _2.18_

CHAPTER 2
Addition and Subtraction of Algebraic Expressions

Objectives

After completing this chapter you will be able to:
- Identify algebraic expressions
- Identify literal numbers, terms, factors, coefficients, and exponents
- Classify expressions
- Add and subtract monomial expressions
- Multiply monomial expressions
- Divide monomial expressions
- Multiply monomial expressions by polynomial expressions
- Divide polynomial expressions by a monomial
- Simplify expressions by removing parentheses, brackets, and braces

2-1 DEFINITIONS. Before beginning the study of addition and subtraction of algebraic expressions, some general definitions should be agreed upon. The following definitions are common to algebra, and many can be expressed by illustrated example.

- *Literal or general number:* a letter of the alphabet used to represent unknown quantities; they are also called general numbers. They can be any letter of the alphabet, either upper or lower case ($a, b, c, d,$ etc.; $A, B, C, D,$ etc.; or symbols).

- *Algebraic expression:* literal or numeral numbers, or a combination of the two.

These are all examples of algebraic expressions:

$$3a \quad 3a + b \quad 4ab$$

$$2x^2 + y^3 - 7 \quad 4ab - 3a + 7$$

- *Term:* quantity separated from another term by either a plus or a minus sign, such as:

$a + b$	(both are terms)
$2a + 3b$	(two terms)
$2a + 3b - 8$	(three terms)
$\dfrac{ab}{c} - 7$	(two terms)

- *Factor:* quantity that is multiplied:

$2b$	(2 and b are both factors)
$3BC$	(3, B, and C are all factors)
BCD	(B, C, and D are all factors)
B^3	($B \cdot B \cdot B$ are three factors)
B^3C^2	($B \cdot B \cdot B \cdot C \cdot C$ are five factors)

- *Coefficient:* a factor, as in the expression $2bc$.

2 is a coefficient of bc
$2b$ is a coefficient of c
b is a coefficient of $2c$
c is a coefficient of $2b$

All of the above fit the definition of a coefficient; in most algebra work, however, the coefficient is referred to as the numeral number (that is, the coefficient of the expression $2bc$ is considered to be 2).

- *Classification of expressions:* categories of algebraic expressions such as these:

monomials	$3ab$	(one term)
binomials	$3ab + 8$	(two terms)
trinomials	$3ab + c - 8$	(three terms)
polynomials or multinomials	$3ab + c - 8 + 2y$	(usually more than three terms)

- *Symbol:* any notation used to represent an unknown quantity other than a numeral number.

- *Exponent:* the unit at which a number can be raised to; for instance:

4^2	2 is the exponent of 4; mathematically it is $4(4)$
3^4	4 is the exponent of 3; mathematically it is $3(3)(3)(3)$

$a^3 = a(a)(a)$

$x^2y^2 = x(x)(y)(y)$

2–2 ADDITION AND SUBTRACTION OF MONOMIALS. Only like things can be added or subtracted. What makes one expression like another is the likeness of their literal numbers; they must be alike in all aspects. The literal numbers must be identical, and the exponents must be to the same power. Rules of signed numbers are applied to the coefficients.

Chapter 2
Addition and
Subtraction of
Algebraic Expressions

Example A

Add:

$$\begin{array}{cccc} 2a & 2a^2 & 8ab^3 & 7a \\ \underline{4a} & \underline{4a^2} & \underline{-3ab^3} & \underline{3b^2} \\ 6a & 6a^2 & 5ab^3 & 7a + 3b^2 \\ & & & \text{or: } 3b^2 + 7a \end{array}$$

Subtract:
(change sign of subtrahend)

$$\begin{array}{ccc} 8a = & 8a & \\ \underline{6a} = & \underline{-6a} & \\ & 2a & \end{array} \qquad \begin{array}{ccc} 5a^2 = & 5a^2 \\ \underline{3a^2} = & \underline{-3a^2} \\ & 8a^2 \end{array}$$

$$\begin{array}{ccc} 6a^2b = & 6a^2b \\ \underline{10a^2b} = & \underline{-10a^2b} \\ & -4a^2b \end{array} \qquad \begin{array}{ccc} 9a = & 9a \\ \underline{3ab^2} = & \underline{-3ab^2} \\ & 9a - 3ab^2 \\ & \text{or: } -3ab^2 + 9a \end{array}$$

EXERCISE 2–1 Add or subtract as indicated.

Add:

1. $7a^2$
   $\underline{40a^2}$
   $47a^2$

2. $1.3ab$
   $\underline{6ab}$
   $7.3ab$

3. $1500b^2$
   $\underline{360b^2}$
   $1860b^2$

4. $18R$
   $\underline{-6R}$
   $12R$

5. $-11I^2$
   $\underline{-30I^2}$
   $-41I^2$

6. $650abc^3$
   $\underline{-46abc^3}$
   $604abc^3$

7. $136a^2b$
   $\underline{36ab^2}$
   $136a^2b + 36ab^2$

8. $-8.2I^2R$
   $\underline{-3.2IR}$
   $-8.2I^2R - 3.2IR$

9. $25iR$
   $\underline{10Ir}$
   $25iR + 10Ir$

10. $160V^2R$
    $\underline{-30V^2R}$
    $130V^2R$

11. $-9IR$
    $\underline{-50IR}$
    $-59IR$

12. $16ab$
    $\underline{-ab}$
    $15ab$

Subtract:

1. $18a$
   $\underline{6a}$
   $12a$

2. $17b^2$
   $\underline{-4b^2}$
   $21b^2$

3. $25I$
   $\underline{-150I}$
   $175I$

4. $-620R$
   $\underline{470R}$
   $-1090R$

5. $12.5V$
   $\underline{-6.3V}$
   $18.8V$

6. $-24iR$
   $\underline{-15iR}$
   $-9iR$

7. $16I^2$
   $\underline{4i^2}$
   $16I^2 - 4i^2$

8. $117IR$
   $\underline{-27I^2R}$
   $117IR + 27I^2R$

9. $32A$
   $\underline{-16A}$
   $47A$

10. $45i^2r$
    $\underline{-15i^2r}$
    $60i^2r$

11. $72Z$
    $\underline{-140Z}$
    $212Z$

12. $-120IZ$
    $\underline{30I^2Z}$
    $-120IZ - 30I^2Z$

2–3 HORIZONTAL ADDITION AND SUBTRACTION. Most equations are presented and solved by horizontal addition and subtraction. This is sometimes referred to as "collecting terms."

To add a quantity such as $3a - 6a + 7a - 2a + 12a$, first collect all positive values, then all negative values. Next, collect the two remaining positive and negative values.

Example B

$3a - 6a + 7a - 2a + 12a = 22a - 8a = \underline{14a}$ (answer)

$4a^2 + 7a^2 - 3b - 2a^2 - 2b = 11a^2 - 2a^2 - 5b = \underline{9a^2 - 5b}$ (answer)

$6ab - 3a^2b - 7ab - 2ab + 5a^2b =$

$\qquad 6ab - 9ab - 2a^2b = \underline{-3ab - 2a^2b}$ (answer)

EXERCISE 2–2 Collect like terms.

1. $6a + 3a - 4a + 2a = $ _7a_

2. $4i^2 + 16i - 3I^2 + 16i^2 - 8i = $ _$20i^2 - 3I^2 + 8i$_

3. $36R + 8R - 4R^2 - 3R + 2R^2 = $ _$41R - 2R^2$_

4. $18E - 11ir - 6E + 8ri = $ _$12E - 3iR$_

5. $15IX^2 - 12IX^2 + 3X - 4IR - 6 = $ _$3IX^2 + 3X - 4IR - 6$_

6. $42E^2I^4 - 17I^4E^2 + 6E^2 = $ _$25E^2I^4 + 6E^2$_

7. $32 - 17E - 6I^2 + 4E - 3I^2 - 8 = $ _$24 - 13E - 9I^2$_

8. $14a^2 + 6ab + 3a^2 - 2ab - 2a^2 - a = $ _$15a^2 + 4ab - a$_

9. $24B^2 - 7R - 11B^2 + 3R - 12B^2 + R + 3R - B^2 = $ _0_

10. $100 - 10IR + 32 - 8IR - 6R + 10i = $ _$132 + 10i - 6R - 18IR$_

11. $6v - 7v^2 - 3v - 9 + 3v - v^2 = $ _$-8v^2 - 9$_

12. $ab + b - ba + 4 - b + 6a^2b = $ _$4 + 6a^2b$_

2-4 MULTIPLICATION OF MONOMIALS.

The coefficients of monomials are multiplied following the rules for signed numbers. The exponents of like symbols are algebraicly added.

The expression $a^2(a)$ can be written

$$a \cdot a \cdot a \text{ or } a^3$$

Therefore, adding the exponents of the symbols is the same as multiplying the symbols.

Example C

$$(-6)a^2(3a^3) = -18a^5$$
(multiply coefficients, add exponents)

$$(3b^2)(-4b^3)(2) = -24b^5$$
(multiply coefficients, add exponents)

$$(-2a^2b)(-5ab^3c)(2abc) = 20a^4b^5c^2$$

Squaring, cubing, or raising a quantity to a power such as $(3a^2)^2$ can be performed by:

$$(3a^2)^2 = (3a^2)(3a^2) = 9a^4$$
(square coefficient, add exponents)

It can also be done by multiplying the exponent outside the parentheses by the exponent of the symbol:

$$(3a^2)^2 = 9a^4$$
(square coefficient, multiply exponents)

Example D

$$(4b^3)^2 = 16b^6$$

$$(3E^4)^3 = 27E^{12}$$

$$(a^2b^3)^4 = a^8b^{12}$$

$$(c^{-2})^3 = c^{-6}$$

$$(c^2)^{-3} = c^{-6}$$

Part 1
Review of Algebra

EXERCISE 2–3 Multiply the monomials:

1. $2v(6v^3)(4Vv) =$ _48Vv⁵_

2. $7I(3IR)(4I^2R^3) =$ _84I⁴R⁴_

3. $(3a^2)(4b^3a^2)(6abc)(a^4) =$ _72a⁹b⁴c_

4. $16i^2r(r)(-i^3)(3r) =$ _−48i⁵r³_

5. $10V^2(RV)(10R^2)(8) =$ _800V³R³_

6. $(ab)(a^2b^2)(6a)(b^4) =$ _6a⁴b⁷_

7. $(3I^2)^2(4IR) =$ _36I⁵R_

8. $(V^3R^2)^3(6R^3)(-V^2)^4 =$ _6V¹⁷R⁹_

9. $(a^2)^3(a^3)^2(a) =$ _a¹³_

10. $(4v^{-4})^3(6v^{-3})^2 =$ _2304/v¹⁸_

11. $(6F)^3(3F^{-3}) =$ _648_

12. $50H(-3HR)(R^3)^2 =$ _−150H²R⁷_

2–5 DIVISION OF MONOMIALS. A fraction is one method of expressing a division operation; a/b means to divide a by b. In algebra, division problems are generally expressed as a fraction. As in the chapter on the power of 10, division of exponents is the algebraic subtraction of exponents.

Example E

$$\text{divide} \longrightarrow \frac{6a^3}{3a^2} \overset{\text{subtract}}{=} 2a$$

$$\frac{8b}{2b} = 4$$

$$\frac{6I^2R^4}{3IR^2} = 2IR^2$$

$$\frac{12a^4b^3}{3a^{-2}b^{-3}} = 4a^6b^6$$

14

A general number can be moved from the numerator of a fraction to the denominator of a fraction, or vice versa, by changing the sign of the exponent. This allows fractions to be expressed with positive exponents, which are preferred as final answers.

Chapter 2
Addition and
Subtraction of
Algebraic Expressions

Example F

$$\frac{3}{a^{-2}} = 3a^2$$

$$\frac{4a^{-3}}{a} = \frac{4}{a^4}$$

$$\frac{3a^{-3}}{b^{-4}} = \frac{3b^4}{a^3}$$

EXERCISE 2–4 Perform the indicated division. Leave answers with positive exponents.

1. $\dfrac{16a^2b}{4ab} = $ _4a_

2. $\dfrac{10E^3B^2}{5BE} = $ _2E²B_

3. $\dfrac{IR^2}{I} = $ _R²_

4. $\dfrac{V^3v^2}{-2Vv} = $ _−V²v/2_

5. $\dfrac{16i^2rv}{4ir^2v} = $ _4i/r_

6. $\dfrac{120a^2b^{-4}}{60ab} = $ _2a/b⁵_

7. $\dfrac{H^5R^3}{3H^3R^5} = $ _.33H²/R²_

8. $\dfrac{1}{2}\left(\dfrac{a^2b^3}{-ab}\right) = $ _−.5ab²_

9. $\dfrac{64M}{4M^{-3}} = $ _4M⁴_

10. $\dfrac{44f^2}{22F^4} = $ _2f²/F⁴_

11. $\dfrac{18a^{-3}b}{9ab^{-3}} = $ _2b⁴/a⁴_

12. $\dfrac{-36EI}{4ei} = $ _−9EI/ei_

2–6 MULTIPLICATION OF A MONOMIAL BY A POLYNOMIAL.

Example G

$$6a(3a + 6) \quad \text{(multiply each term by } 6a\text{)}$$
$$= 18a^2 + 36a$$

$$4b(3a + 6b - 3c^2 - 2)$$
$$= 12ab + 24b^2 - 12bc^2 - 8b$$

EXERCISE 2–5 Perform the indicated multiplications and collect like terms.

1. $3v(8v^2 + 6R) = $ $24v^3 + 18vR$

2. $5I(6IR^2 - 3IR + 4) = $ $30I^2R^2 - 15I^2R + 20I$

3. $-6V^2(8vV - 6 + 3IR) = $ $-48vV^3 + 36V^2 - 18V^2IR$

4. $RV(3IR + 7V^2 - 3V + 6) = $ $3IR^2V + 7V^3R - 3V^2R + 6VR$

5. $8a^3b^2c(abc + a^2 + b + c^3) = $ $8a^4b^3c^2 + 8a^5b^2c + 8a^3b^3c + 8a^3b^2c^4$

6. $0.5E(30E - 6 + 42I - E) = $ $14.5E^2 - 3E + 21EI$

7. $-1(bc + 7 - ab + a^4) = $ $-bc - 7 + ab - a^4$

8. $\frac{1}{4}(I + 8I^2 - 20ir + v) = $ $.25I + 2I^2 - 5iR + .25v$

9. $6\left(\frac{3i}{2} + \frac{4R^2}{3} - E\right) = $ $9i + 8R^2 - 6E$

10. $-8\left(\frac{V^2}{4} + \frac{V^2}{8} - 8\right) = $ $-3V^2 + 64$

11. $4c\left(\frac{c}{4} - \frac{9}{2}V + 8\right) = $ $c^2 - 18cV + 32c$

12. $2a\left(\frac{10b}{5} - \frac{16c^2}{4} - 1\right) = $ $4ab - 8ac^2 - 2a$

2–7 DIVISION OF POLYNOMIALS BY A MONOMIAL. A fraction written as $\dfrac{6a^2 + 2a}{a}$ can also be expressed $\dfrac{6a^2}{a} + \dfrac{2a}{a}$. The numbers $6a^2$ and $2a$ are expressed over a common denominator a in the first expression. If the problem is expressed the second way, it can be divided as it is in the problems in Section 2–5.

$$\frac{6a^2 + 2a}{a} = \frac{6a^2}{a} + \frac{2a}{a} = 6a + 2$$

Example H

$$\frac{12b^2 - 6a^2}{3ab} = \frac{12b^2}{3ab} - \frac{6a^2}{3ab} = \frac{4b}{a} - \frac{2a}{b}$$

$$\frac{6c^2 - 8a + 6}{2c} = \frac{6c^2}{2c} - \frac{8a}{2c} + \frac{6}{2c} = 3c - \frac{4a}{c} + \frac{3}{c}$$

The denominator is divided into each term of the numerator; problems are generally worked as follows:

$$\frac{16c^2 + 20bc^2 - 8ac}{4ac\ \text{divide}} = \frac{4c}{a} + \frac{5bc}{a} - 2$$

$$\frac{8abc + a^2b^2 - 4c}{4c} = 2ab + \frac{a^2b^2}{4c} - 1$$

EXERCISE 2-6 Perform the indicated operations.

1. $\dfrac{12ab - c}{2a} = \ $ $6b - \dfrac{c}{2a}$

2. $\dfrac{V^2R + 7}{R} = \ $ $V^2 + \dfrac{7}{R}$

3. $\dfrac{4IR - R}{2} = \ $ $2IR - \dfrac{R}{2}$

4. $\dfrac{16i^2 + 7I}{7I} = \ $ $\dfrac{16i^2}{7I} + 1$

5. $\dfrac{3R^2 - 7ab^2 + 8}{aR} = \ $ $\dfrac{3R}{a} - \dfrac{7b^2}{R} + \dfrac{8}{aR}$

6. $\dfrac{14E^2 + IR - R}{R} = \dfrac{14E^2}{R} + I - 1$

7. $\dfrac{144V^3 + 84IV^2 - 60I^3}{12IV} = \dfrac{12V^2}{I} + 7V - \dfrac{5I^2}{V}$

8. $\dfrac{HV - vR + 16}{4V} = \dfrac{H}{4} - \dfrac{vR}{4V} + \dfrac{4}{V}$

9. $\dfrac{v^3 - 4v^2 + 6}{v^2} = v - 4 + \dfrac{6}{v^2}$

10. $\dfrac{IR - ir + I^2R - i^2r^2}{IRir} = \dfrac{1}{ir} - \dfrac{1}{IR} + \dfrac{I}{ir} - \dfrac{ir}{IR}$

11. $\dfrac{9F^3 - 3f^4 + 27Ff - 36f}{3Ff} = \dfrac{3F^2}{f} - \dfrac{f^3}{F} + 9 - \dfrac{12}{F}$

12. $\dfrac{81a^2b^2c^2 - 45a^3b^2c + 90b^4}{9abc} = 9abc - 5a^2b + \dfrac{10b^3}{ac}$

2–8 SIMPLIFYING ALGEBRAIC EXPRESSIONS. Expressions are sometimes a combination of collected terms, multiplication, and division. The order of the operations in simplifying expressions is first to multiply and divide, then to collect like terms (add or subtract). If this order is to be varied, then signs for grouping (parentheses, brackets, and braces) are used.

Signs of grouping: $\{[(\quad)]\}$

parentheses	(	)
brackets	[	]
braces	{	}

These signs of grouping indicate an operation of multiplication. The parentheses are the innermost sign of grouping and usually are removed first, followed by the bracket and then the brace. Like terms within the group can be collected before multiplication.

Example I

$$4b(3b + 6a - 4) + 6b^2$$

$$12b^2 + 24ab - 16b + 6b^2 \quad \text{(multiply)}$$

$$18b^2 + 24ab - 16b \quad \text{(collect like terms)}$$

If a sign of grouping is preceded by a minus sign ($-$), the quantity inside the grouping is multiplied by -1:

$$-(5a^2 - 3a + 4) = -5a^2 + 3a - 4$$

The result is that the sign in front of the parentheses changes the sign of each term within the grouping. Rules for this procedure are established:

Rule 1: If the sign of grouping is preceded by a minus sign, change the sign of each term and remove the sign of grouping.

Rule 2: If the sign of grouping is preceded by a plus sign, simply remove the sign of grouping.

Example J

$$3R - (4R + 6B - 3) - 4B$$

Remove parentheses and change the sign of each term:

$$3R - 4R - 6B + 3 - 4B$$

Collect like terms:

$$-R - 10B + 3$$

$$2a - [-3a + 4b + (6a - b + 8)]$$

Remove parentheses:

$$2a - [-3a + 4b + 6a - b + 8]$$

Remove brackets and change sign of each term:

$$2a + 3a - 4b - 6a + b - 8$$

Collect like terms:

$$-a - 3b - 8$$

$$3c - \{5c + 6 + [c - 4 - (8c + 7)]\}$$

Remove parentheses first and change sign:

$$3c - \{5c + 6 + [c - 4 - 8c - 7]\}$$

Remove brackets:

$$3c - \{5c + 6 + c - 4 - 8c - 7\}$$

Remove braces and change sign:

$$3c - 5c - 6 - c + 4 + 8c + 7$$

Collect like terms:

$$5c + 5$$

EXERCISE 2-7 Simplify the expressions by removing the signs of groupings and collecting like terms.

1. $3V - (2V + 6) + (5V - 4) =$ _6V−10_

2. $6I^2 - [3I - (2I^2 - I + 5)] =$ _8I²−4I+5_

3. $8 - [3E + (2E - 3) - (E + 14)] =$ _25−4E_

4. $4i^2 - \{2i^2 - [6i^2 + i - 7 - (i^2 + i) + 20]\} =$ _7i²+13_

5. $a + \{6a^2 - 3a - [8 - (a^2 + a - 2)]\} =$ $7a^2 - a - 10$

6. $IR - 3I(R - 2I - 6) =$ $6I^2 - 2IR + 18I$

7. $6v[2v + v(4v - 8)] =$ $24v^3 - 36v^2$

8. $8IV^2 - \{6IV^2 - 3V[IV - 4 + 2I(V^2 - V + 3)]\} =$ $-IV^2 - 12V + 6IV^3 + 18IV$

EVALUATION EXERCISE Perform the operations, simplifying all answers.

1. Identify the number of terms in each expression.

   $3a + 4 - 6c$ 3

   $6 + 2b$ 2

   $3a^2b + 7a - 6c^2 + 4$ 4

   $8V - V^2 + a^4 - 7 + a$ 5

   $6 + \dfrac{4T}{X} - 7 + 8v^2 - \dfrac{6T}{T^2}$ 5

2. Identify the number of literal factors in each expression. Also identify the factor generally considered the coefficient.

	Factors	Coefficient
$15a^2b^3$	a^2b^3	15
$6VI^2R$	VI^2R	6
I^2R	I^2R	1
$\dfrac{1}{2}T^3A^4B$	T^3A^4B	1/2
$\dfrac{R^2ir}{4}$	R^2ir	1/4

Chapter 2
Addition and
Subtraction of
Algebraic Expressions

Part 1
Review of Algebra

3. Identify the following. Are they mono-, bi-, tri-, or polynomial?

$a^3b - 7 + 4c$ TRI

$R - 6b + 7 - 8c$ POLY

$12b^2c - 6d + 8 - 7T + 6I$ POLY

$\dfrac{17a}{3b} + \dfrac{12D}{8c} - A + \dfrac{b}{D}$ POLY

$V^3 - \dfrac{6c}{V^3} + \dfrac{12}{2}$ TRI

4. Collect like terms and simplify.

$3d - 8b + 24d - 6b + 4 =$ $27d - 14b + 4$

$8a^2 - ab + a^2 - 18ab + 3a^2 =$ $12a^2 - 19ab$

$16 - V + I^2R + V - I^2R =$ 16

$18T^2R + 14T^2 - T^2R + T^2 =$ $17T^2R + 15T^2$

$iv - 3i^2v + 4iv - 2i^2v - 5iv + 5i^2v =$ 0

5. Perform the indicated multiplication.

$6a(7b^2a)(3a) =$ $126\,a^3b^2$

$R^2(IR)(I)(R)(I^2R) =$ R^5I^4

$4V^2v(vV)(4IV)(4vI) =$ $64V^4v^3I^2$

$5c^2b(4c^2b)^2(3cb)(b)(2) =$ $480\,c^7b^5$

$I(R)(3R^2)^2(I^{-3})(4I^{10})(6R^{-4})^3 =$ $\dfrac{7776\,I^8}{R^7}$

6. Perform the indicated division; leave your answers with positive exponents.

$\dfrac{12a^3b^4}{ab} = $ _$12a^2b^3$_

$\dfrac{E^2R}{R} = $ _E^2_

$\dfrac{18V^3R^2}{3V^{-2}R} = $ _$6V^5R$_

$\dfrac{20I^{-2}CR^2}{2I^2C^{-3}R} = $ _$\dfrac{10C^4R}{I^4}$_

$\dfrac{AB^{-3}C^4}{A^{-2}} = $ _$\dfrac{A^3C^4}{B^3}$_

7. Perform the indicated multiplication; simplify the answers by collecting like terms.

$4a(6a^2 + a - 7) = $ _$24a^3 + 4a^2 - 28a$_

$3vV(V^2 + 4 - 2v - 2v^2V) = $ _$3V^3v + 12Vv - 6Vv^2 - 6v^2V^3$_

$-3b(6b^2c + 4 - 12b - 4b^2c - 2bc) = $ _$-18b^3c - 12b + 36b^2 + 12b^3c + 6b^2c$_

$V(VIR - I + R - IR + V) = $ _$V^2IR - VI + VR - VIR + V^2$_

$-8c^2\left(\dfrac{3c}{2} + \dfrac{5a}{4} + \dfrac{c}{8}\right) = $ _$-13c^3 - 10ac^2$_

**Part 1
Review of Algebra**

8. Perform the indicated division.

$$\frac{16b^2c + c}{bc} = 16b + \frac{1}{b}$$

$$\frac{a^2b - 3b + c}{b} = a^2 - 3 + \frac{c}{b}$$

$$\frac{9IR - 3I}{3} = 3IR - I$$

$$\frac{V^3 - 4V^2 + 5}{V^2} = V - 4 + \frac{5}{V^2}$$

$$\frac{HR - hr + i^2r - i^2r^2}{HRir} = \frac{1}{ir} - \frac{h}{HRi} + \frac{i}{HR} - \frac{ir}{HR}$$

9. Simplify by removing the signs of grouping and collecting like terms.

$$2a - (2a + 4) + (3a - 3) = 3a - 7$$

$$3 - [3b - (3b^2 - 3b + 3)] = 3b^2 - 6b + 6$$

$$3R - \{2R - [R + (4V - 6)]\} = 2R + 4V - 6$$

$$A - \{16 + [2A - (3B - 10) + (-3B - 5)]\} = 6B - A - 21$$

CHAPTER 3
Multiplication and Division of Binomials and Polynomials

Objectives

After completing this chapter you will be able to:
- Multiply binomial expressions
- Square binomial expressions
- Multiply binomial expressions by polynomial
- Divide polynomial expressions by binomial expressions

Multiplication of binomials and polynomials, though similar to multiplying monomials, generally requires greater understanding and mental effort.

3-1 BINOMIAL TIMES BINOMIAL. There are several approaches to multiplying a quantity such as $(3a + 4)(2a + 3)$. A student should be able to multiply a problem like this mentally, but a method should first be developed.

Example A

$$(3a + 4)(2a + 3)$$

$$3a(2a + 3) = 6a^2 + 9a$$
$$4(2a + 3) = 8a + 12$$
$$\overline{6a^2 + 17a + 12}$$

Example B

$$\begin{array}{r} 3a + 4 \\ 2a + 3 \\ \hline 6a^2 + 8a \\ 9a + 12 \\ \hline 6a^2 + 17a + 12 \end{array}$$

Part 1
Review of Algebra

Example C

The multiplication in Example B can be accomplished mentally by inspection. The following procedure works well for most students. It follows the procedure of first term, outer term, inner term.

$$(3a + 4)(2a + 3)$$
$$6a^2$$

The product or the first two terms of the binomial gives the first term in the answer.

$$(3a + 4)(2a + 3)$$
$$12$$

The product of the second terms gives the last term in the answer.

$$(3a + 4)(2a + 3)$$
$$9a + 8a = 17a$$

The sum of the other two products gives the inner term. The mental process should then be:

$$(3a + 4)(2a + 3) = 6a^2 + 17a + 12$$

This skill should be practiced so problems of this nature can be solved by inspection.

EXERCISE 3-1 Multiply by inspection.

1. $(a + 8)(a + 2) =$ _$a^2 + 10a + 16$_

2. $(R - 6)(R - 3) =$ _$R^2 - 9R + 18$_

3. $(6E + 3)(3E + 5) =$ _$18E^2 + 39E + 15$_

4. $(T - 4)(3T + 8)$ = $3T^2 - 4T - 32$

5. $(5a - 7)(4a - 3)$ = $20a^2 - 43a + 21$

6. $(3c + 5)(4c + 3)$ = $12c^2 + 29c + 15$

7. $(3V + 7)(2V - 5)$ = $6V^2 - V - 35$

8. $(5a - 7)(4a + 3)$ = $20a^2 - 13a - 21$

9. $(7E - 3)(3E - 2)$ = $21E^2 - 23E + 6$

10. $(5a^2 + 3)(7a^2 - 2)$ = $35a^4 + 11a^2 - 6$

11. $(4b - 5c)(2b + 7c)$ = $8b^2 + 18bc - 35c^2$

12. $(T + 4b)(T + 3b)$ = $T^2 + 7Tb + 12b^2$

13. $(3V + 6)(5V - 6)$ = $15V^2 + 12V - 36$

14. $(8C - 2)(3C - 2)$ = $24C^2 - 22C + 4$

15. $(3ac + 5b)(4ac + 3b)$ = $12a^2c^2 + 29abc + 15b^2$

16. $(6T^2 + t)(2T^2 - t)$ = $12T^4 - 4tT^2 - t^2$

17. $(10D + 7)(D + 6)$ = $10D^2 + 67D + 42$

18. $(4bc - 7)(3bc + 3)$ = $12b^2c^2 - 9bc - 21$

27

Chapter 3
Multiplication and
Division of
Binomials and Polynomials

19. $(9 + 3a)(4 - 4a) =$ $36 - 24a - 12a^2$

20. $(6 - 5b)(4 - 3b) =$ $24 - 38b + 15b^2$

3–2 SPECIAL CASE OF BINOMIAL MULTIPLICATION. A special case of binomial multiplication is multiplication of the sum and difference of like quantities.

$(R + 4)$ and $(R - 4)$ In this special case we see the sum of R and 4 and the difference of R and 4. Here is what happens when they are multiplied using the conventional method:

$$
\begin{array}{r}
R + 4 \\
R - 4 \\
\hline
R^2 + 4R \\
- 4R - 16 \\
\hline
R^2 \quad 0 \; - 16 \\
R^2 - 16
\end{array}
$$

The center terms, $+ 4R$ and $- 4R$, always cancel.

Notice that the middle terms are equal and opposite in sign. In this type of multiplication, the middle terms always drop out. The remaining terms are the square of the original terms, as R^2 is the square of R and 16 is the square of 4. The sign between terms is always negative.

Example D

$$(a - 6)(a + 6) = a^2 - 36$$

$$(4b + 5)(4b - 5) = 16b^2 - 25$$

$$(8c - 3a^2b)(8c + 3a^2b) = 64c^2 - 9a^4b^2$$

EXERCISE 3–2 Multiply the following by inspection.

1. $(V - 6)(V + 6) =$ $V^2 - 36$

2. $(2b + 8)(2b - 8) =$ $4b^2 - 64$

3. $(12E - 12)(12E + 12) =$ $144E^2 - 144$

4. $(7T - 4)(7T + 4) =$ $49T^2 - 16$

5. $(9ab + 3c)(9ab - 3c) = $ __$81a^2b^2 - 9c^2$__

6. $(15I^2 + 3)(15I^2 - 3) = $ __$225I^4 - 9$__

7. $(VR + A)(VR - A) = $ __$V^2R^2 - A^2$__

8. $\left(\frac{1}{2}a + b\right)\left(\frac{1}{2}a - b\right) = $ __$\frac{1}{4}a^2 - b^2$__

9. $(V^3 - 3b^2)(V^3 + 3b^2) = $ __$V^6 - 9b^4$__

10. $\left(\frac{a}{6} - 5\right)\left(\frac{a}{6} + 5\right) = $ __$\frac{a^2}{36} - 25$__

3–3 SQUARING BINOMIALS. Squaring binomials is also considered a special case of binomial multiplication. Conventional multiplication gives the following results:

$$\begin{array}{r} (2R + 4)^2 = 2R + 4 \\ 2R + 4 \\ \hline 4R^2 + 8R \\ + 8R + 16 \\ \hline 4R^2 + 16R + 16 \end{array}$$

- The answer results in a trinomial: $(4R^2 + 16R + 16)$

- The first term is the square of the first term in the binomial: $(2R)(2R) = 4R^2$

- The third term is the square of the second term in the binomial: $(4)(4) = 16$

- The middle term is twice the product of the two terms in the binomial: $2[(2R)(4)] - 2[8R] - 16R$

Example E

$$(5a + 7)^2 = 25a^2 + 70a + 49$$

$$(a - 8)^2 = a^2 - 16a + 64$$

$$(2B - 3a)^2 = 4B^2 - 12aB + 9a^2$$

EXERCISE 3–3 Mentally square the following.

1. $(B + 6) = $ __$B^2 + 12B + 36$__

2. $(3c + 5) = $ __$9c^2 + 30c + 25$__

29

Chapter 3
Multiplication and
Division of
Binomials and Polynomials

3. $(4a - 2) = 16a^2 - 16a + 4$

4. $(6D - 4) = 36D^2 - 48D + 16$

5. $(P + 7) = P^2 + 14P + 49$

6. $(I - 5) = I^2 - 10I + 25$

7. $(3E + 4I) = 9E^2 + 24EI + 16I^2$

8. $(7 - 6B) = 49 - 84B + 36B^2$

9. $\left(\dfrac{a}{2} + 6\right) = \dfrac{a^2}{4} + 6a + 36$

10. $(2V - 11) = 4V^2 - 44V + 121$

11. $(12C - 6) = 144C^2 - 144C + 36$

12. $(11D^2 + 4b) = 121D^4 + 88D^2 b + 16b^2$

13. $(4ab + 3C) = 16a^2b^2 + 24abC + 9C^2$

14. $(5R^2b - 4I) = 25R^4b^2 - 40R^2bI + 16I^2$

15. $\left(I - \dfrac{R}{4}\right) = I^2 - \dfrac{IR}{2} + \dfrac{R^2}{16}$

3–4 BINOMIAL TIMES POLYNOMIAL. Skill in multiplying the binomial times the polynomial can be gained but is usually not required at this level. The long method is used. Multiply each term in the binomial and collect like terms.

$$(3B + 4)(2B + 3a + 2) = 3B(2B + 3a + 2) \text{ and } 4(2B + 3a + 2)$$

$$= 6B^2 + 9aB + 6B + 8B + 12a + 8$$

$$= 6B^2 + 9ab + 14B + 12a + 8$$

Example F

$$(2a + 3)(2a^2 + 2a - 3) = 4a^3 + 4a^2 - 6a + 6a^2 + 6a - 9$$

$$= 4a^3 + 10a^2 - 9$$

EXERCISE 3–4 Multiply the following and simplify by collecting like terms.

1. $(3B + 2)(2B^2 + 3B - 6) = \underline{6B^3 + 13B^2 - 12B - 12}$
$$6B^3 + 9B^2 - 18B$$
$$4B^2 + 6B - 12$$

2. $(a - 4)(a + 2 - b) = \underline{a^2 - 2a - ab + 4b - 8}$
$$a^2 + 2a - ab$$
$$-4a + 4b - 8$$

3. $(R + 3)(3R^2 - R - 4) = \underline{3R^3 + 8R^2 - 7R - 12}$
$$3R^3 - R^2 - 4R$$
$$9R^2 - 3R - 12$$

4. $(6V - 3)(V^2 - V + 5) = \underline{6V^3 - 9V^2 + 33V - 15}$
$$6V^3 - 6V^2 + 30V$$
$$-3V^2 + 3V - 15$$

5. $(2T^2 + T)(T^2 - 3T + 5) = \underline{2T^4 - 5T^3 + 7T^2 + 5T}$
$\underline{2T^4 - 5T^3 + 17T^2 + 5T}$
$$2T^4 - 6T^3 + 10T^2$$
$$T^3 - 3T^2 + 5T$$

6. $(I - 2)(I^3 - I^2 + 2I + 4) = \underline{I^4 - I^3 + 2I^2 + 4I}$
$$-2I^3 + 2I^2 - 4I - 8$$
$$\boxed{I^4 - 3I^3 + 4I^2 - 8}$$

7. $(2A + 3)(2A^2 + A - 4) = \underline{4A^3 + 8A^2 - 5A - 12}$
$$4A^3 + 2A^2 - 8A$$
$$6A^2 + 3A - 12$$

8. $(4R^2 + R)(2R^2 - R - 6) = \underline{8R^4 - 2R^3 - 25R^2 - 6R}$
$$8R^4 - 4R^3 - 24R^2$$
$$2R^3 - R^2 - 6R$$

3–5 DIVISION OF A POLYNOMIAL BY A BINOMIAL. Division of a polynomial by a binomial is similar to long division done in arithmetic.

Arithmetic division:
$$15 \overline{)347} \quad = 23\frac{2}{15} \text{(answer)}$$
$$\underline{30}$$
$$47$$
$$\underline{45}$$
$$2 \text{ (placed over divisor)}$$

Chapter 3
Multiplication and
Division of
Binomials and Polynomials

32

Part 1
Review of Algebra

Division of a polynomial by a binomial:

$$B + 2)\overline{B^2 + 4B + 7} B + 2 + \frac{3}{B+2}$$
$$\underline{B^2 + 2B}$$
$$2B + 7$$
$$\underline{2B + 4}$$
$$3 \text{ (remainder)}$$

1. arrange polynomial in descending order of power
2. $B^2 \div B = B$
3. $B(B + 2) = B^2 + 2B$
4. subtract by changing signs of subtrahend, then add
5. $2B \div B = 2$
6. $2(B + 2) = 2B + 4$
7. subtract (change sign)
8. place remainder over divisor $\dfrac{3}{B + 2}$

Example G

$$a - 3)\overline{a^2 - 5a - 9} a - 2 - \frac{15}{a-3}$$
$$\underline{a^2 - 3a}$$
$$-2a - 9$$
$$\underline{-2a + 6}$$
$$-15$$

EXERCISE 3–5 Perform the indicated division.

1. $I + 3)\overline{I^2 + 5I + 6}$ quotient: $I + 2$
   $\underline{I^2 + 3I}$
   $2I + 6$
   $\underline{2I + 6}$

   Answer: $I + 2$

2. $a + 3)\overline{a^2 + 8a + 15}$ quotient: $a + 5$
   $\underline{a^2 + 3a}$
   $5a + 15$
   $\underline{5a + 15}$

   Answer: $a + 5$

3. $R - 9)\overline{R^2 - R - 72}$ quotient: $R + 8$
   $\underline{R^2 - 9R}$
   $8R - 72$
   $\underline{8R - 72}$

   Answer: $R + 8$

4. $A - 2 \overline{\smash{)}3A^2 - A - 15}$ quotient: $3A+5-\dfrac{5}{A-2}$

$\underline{3A^2 + 6A}$
$\quad 5A - 15$
$\quad \underline{5A + 10}$
$\quad\quad -5$

$3A+5-\dfrac{5}{A-2}$

5. $B - 5 \overline{\smash{)}B^2 - 11B + 33}$ quotient: $B-6+\dfrac{3}{B-5}$

$\underline{B^2 - 5B}$
$\quad -6B + 33$
$\quad \underline{-6B + 30}$
$\quad\quad 3$

$B-6+\dfrac{3}{B-5}$

6. $3V + 2 \overline{\smash{)}12V^2 - 7V - 4}$ quotient: $4V-5+\dfrac{6}{3V+2}$

$\underline{12V^2 + 8V}$
$\quad -15V - 4$
$\quad \underline{-15V + 10}$
$\quad\quad 6$

$4V-5+\dfrac{6}{3V+2}$

7. $4c - 5 \overline{\smash{)}12c^2 + c - 9}$ quotient: $3c+4+\dfrac{11}{4c-5}$

$\underline{12c^2 - 15c}$
$\quad 16c - 9$
$\quad \underline{16c - 20}$
$\quad\quad +11$

$3c+4+\dfrac{11}{4c-5}$

8. $2R + 3 \overline{\smash{)}6R^2 - 5R - 17}$ quotient: $3R-7+\dfrac{4}{2R+3}$

$\underline{6R^2 + 9R}$
$\quad -14R - 17$
$\quad \underline{+14R + 21}$
$\quad\quad 4$

$3R-7+\dfrac{4}{2R+3}$

33

Chapter 3
Multiplication and
Division of
Binomials and Polynomials

EVALUATION EXERCISE Perform the operations.

1. $(B + 7)(B + 4)$ $B^2+11B+28$

2. $(3a - b)(4a - 3)$ $12a^2-4ab-9a+3b$

Part 1
Review of Algebra

3. $(a + b)(a - b)$
$a^2 - b^2$

$a^2 - b^2$

4. $(4V + 2)(4V + 2)$
$16V^2 + 16V + 4$

$16V^2 + 16V + 4$

5. $(7C - 2)(3C + 2)$
$-6C$
$+14C$

$21C^2 + 8C - 4$

6. $(B + 3)(2B - 2)$
$+6B$
$-2B$

$2B^2 + 4B - 6$

7. $(6I - 3)(6I + 3)$

$36I^2 - 9$

8. $(3ab + 2)(ab - 6)$
$+2ab$
$-18ab$

$3a^2b^2 - 16ab - 12$

9. $(5a^2 - 4)(4a^2 - 5)$
$-16a^2$
$-25a^2$

$20a^4 - 41a^2 + 20$

10. $(8IR + 3)(IR - 4)$
$+3IR$
$-32IR$

$8I^2R^2 - 29IR - 12$

11. $(7 - 6B)(7 + 6B)$

$49 - 36B^2$

12. $(10 - 5I)(10 - 5I)$
-50
-50

$100 - 100I + 25I^2$

13. $(9P + 4)(9P + 4)$ $\underline{81P^2+72P+16}$

14. $(3ac + 5)(2ac - 5)$ $\underline{6a^2c^2-5ac-25}$

15. $(R + V)^2$ $\underline{R^2+2RV+V^2}$

16. $(I + R)(I - R)$ $\underline{I^2-R^2}$

17. $\left(\dfrac{3D}{2} + 7\right)\left(\dfrac{3D}{2} - 7\right)$ $\underline{\dfrac{9D^2}{4}-49}$

18. $\left(\dfrac{a}{4} + 3\right)^2$ $\underline{\dfrac{a^2}{16}+1.5a+9}$ or $\underline{\dfrac{a^2}{16}+\dfrac{3a}{2}+9}$

19. $\left(\dfrac{R}{5} + \dfrac{I}{2}\right)\left(\dfrac{R}{5} - \dfrac{I}{2}\right) =$ $\underline{\dfrac{R^2}{25}-\dfrac{I^2}{4}}$

20. $\left(\dfrac{2a}{3} - 4\right)\left(\dfrac{2a}{3} - 4\right) =$ $\underline{.44a^2-5.33a+16}$ or $\underline{\dfrac{4a^2}{9}-\dfrac{16a}{3}+16}$

21. $(I + 4)(I^2 + I - 4) =$ $\underline{I^3+5I^2-16}$

22. $(2a - 2)(2a - a + 3) =$ $\underline{2a^2+4a-6}$

35

Chapter 3
Multiplication and
Division of
Binomials and Polynomials

36

Part 1
Review of Algebra

23. $(R + 3)(4I + R - 5) = R^2+4RI-2R+12I-15$
    $4RI+R^2-5R+12I+3R-15$

24. $(T + 3)(3T^2 - T + 6) = 3T^3+8T^2+3T+18$
    $3T^3-T^2+6T$
    $9T^2+3T+18$

25. $I + 3 \overline{)I^2 + 8I + 15}$ = $I+5$
    $\underline{I^2+3I}$
    $5I+15$

    (quotient shown above: $I+5$)

26. $R - 2 \overline{)3R^2 - R - 14}$ = $3R+5-\frac{4}{R-2}$
    $\underline{3R^2-6R}$
    $5R-14$
    $\underline{5R-10}$
    -4

    (quotient shown above: $3R+5-\frac{4}{R-2}$)

27. $3B + 2 \overline{)12B^2 - 7B - 4}$ = $4B-5+\frac{6}{3B+2}$
    $\underline{12B^2+8B}$
    $-15B-4$
    $\underline{-15B-10}$
    6

    (quotient shown above: $4B-5+\frac{6}{3B+2}$)

28. $2a + 3 \overline{)6a^2 - 5a - 17}$ = $3a-7+\frac{4}{2a+3}$
    $\underline{6a^2+9a}$
    $-14a-17$
    $\underline{-14a-21}$
    $+4$

    (quotient shown above: $3a-7+\frac{4}{2a+3}$)

CHAPTER 4

Factoring

Objectives

After completing this chapter you will be able to:
- Identify prime factors
- Determine the highest common factor
- Factor trinomial squares
- Factor the difference between perfect squares
- Factor trinomial expressions

The ability to factor usually indicates an understanding of basic algebra operations. Without the ability, it is difficult to progress. Factoring, which is the reverse of multiplication, relies on understanding the operations of algebraic multiplication.

4–1 PRIME FACTORS. By definition, a prime factor is a number that can be divided evenly only by itself and the numeral 1.

Examples of prime factors: 2, 3, 5, 7, 11, 17, 23, 29, etc.; a, b, c, d, etc.

What are the prime factors of $16a^4b^3$?

$$2\,(2)\,(2)\,(2)\,a\,a\,a\,a\,b\,b\,b$$

This is usually expressed in exponents, but it should be recognized as prime factors. The following is a prime-factor expression:

$$2^4 a^4 b^3$$

Literal numbers are expressed in a prime-factor notation such as a^4, b^3, c^5. The prime factors of numerals are easily recognized in smaller numbers like 9, 16, 25, 32, etc. When they are not recognizable, they can be determined by using the following procedure, such as, to determine the prime factors of 88.

```
 2 | 88
 2 | 44
 2 | 22
11 | 11
     1
```

First, divide by the smallest prime factor that will divide evenly into 88. Repeat step 1, each time using the smallest prime factor that will divide into the answer.

2^3 and 11 are primes of 88.

38

Part 1
Review of Algebra

Example A   Determine the prime factors of 154:

```
2 | 154
    7 | 77
        11 | 11
              1
```

2, 7, and 11 are the prime factors.

Example B   Determine the prime factors of 360:

```
2 | 360
  2 | 180
    2 | 90
      3 | 45
        3 | 15
          5 | 5
            1
```

2^3, 3^2, and 5 are the prime factors.

EXERCISE 4–1   Identify the prime factors.

1. 15   5(3)

2. 75 $3^1, 5^2$

3. 180 $2^2, 3^2, 5$

4. 156 $2^2, 3, 13$

5. 144 $2^4, 9$

6. $27a^2b^3$ $3^3, a^2b^3$

7. $40b^4c^5$ $2^3, 5\ b^4c^5$

39

Chapter 4
Factoring

8. $210R^3V^2$ $2,3,5,7, R^3, V^2$

4–2 FACTORING BY REMOVING THE COMMON EXPRESSION. First, remove the factor common to each expression.

Factor: $4a^2 + 12ab - 4a = 4aa + 4a(3) - 4a(1)$
$4a(a + 3b - 1)$

To check whether these are the factors, multiply the two expressions and return them to the original expression.

Example C

$$12R^3 + 6R^2 = 6R^2(2R + 1)$$

EXERCISE 4–2 Factor by determining the highest common factor. Mentally check your answer by multiplication.

1. $6a^3 - 18a^2$ $= 6a^2(a-3)$ $6a^2(a-3)$

2. $20b^2 - 15b$ $= 5b(4b-3)$ $5b(4b-3)$

3. $8I^2R^3 + 12I^3R^2 - 6IR$ $2IR(4IR^2+6I^2R-3)$
   $2IR(4IR^2+6I^2R-3)$

4. $12c^3b - 18c^3b + 3a^2b$ $3b(4c^3-6c^3+a^2)=3b(-2c^3+a^2)$
   $-6c^3b+3a^2b$

5. $3p^3 + 15p^2q - 6pq$ $3p(p^2+5pq-2q)$

40

Part 1
Review of Algebra

6. $10V^3R - 15V^2R^2 + 20VR$

   $5VR(2V^2-3VR+4)$

   $\underline{5VR(2V^2-3VR+4)}$

7. $24a^4b^2 - 36a^3b^3 + 12ab$

   $\underline{12ab(2a^3b-3a^2b^2+1)}$

   $12ab[a^2b(2-3b)+1]$

8. $a^4 + a^3 + a^2 + a$

   $\underline{a(a^3+a^2+a+1)}$
   $a[a(a^2+a+1)+1]$
   $a[a[a(a+1)+1]+1]$

9. $3c^3d - 12c^5d^4$

   $\underline{3c^3d(1-4c^2d^3)}$

10. $20M + 4 - M^2$

    $\underline{20m+4-m^2}$
    $m(20-m)+4$

11. $9I^2R + 4I^2R - I^2R$

    $\underline{I^2R(9+4-1)=}$
    $12I^2R$

12. $25R^2p^3 - 15R^2p^2 - 5Rp$

    $\underline{5Rp(5Rp^2-3Rp-1)}$

13. $5c^3 - 5c + 5c^2$

    $\underline{5c(c^2-1+c)}$

14. $2P^2I - 6PI^2 + 8P^3$

    $\underline{2P(PI-3I^2+4P^2)}$

15. $10a^5 - 8a^4 - 6a^3 + 2a^2 - a$

    $\underline{a(10a^4-8a^3-6a^2+2a-1)}$

4–3 FACTORING TRINOMIAL SQUARES. Factors common to each term should be removed first if there are any. If a trinomial is factorable, it will consist of the product of two binomials.

- Remove common factors.
- Arrange the trinomial in the order of descending power.

$$ax^2 + bx + c$$

- Check the first and third terms; if either one is a perfect square, it may be a perfect trinomial square. Remove the square root of the first and third terms and try them as factors.
- Multiply the determined factors; this should give the original trinomial. If not, then it is not a perfect trinomial square.

Example D

$\boxed{a^2}$ + 12a + $\boxed{36}$ = (a + 6)(a + 6) = (a + 6)^2

perfect square   perfect square   (a + 6)(a + 6) = a^2 + 12a + 36

remultiplying gives original trinomial

Other examples: $25a^2 - 40ab + 16b^2 = (5a - 4b)^2$

$B^2 + 6AB + 9 = $ N.P.S.

(*not a perfect trinomial square*)

$2c^2 - 24c + 72 = 2(c^2 - 12c + 36) = 2(c - 6)^2$

EXERCISE 4–3 If the following are perfect trinomial squares, remove the factors. If they are not perfect squares, indicate by writing N.P.S.

1. $I^2 + 10IR + 25R^2$ $(I + 5R)^2$

2. $B^2 - 12AB + 36A^2$ $(B - 6A)^2$

3. $V^2 - 2V + 1$ $(V - 1)^2$

Part 1
Review of Algebra

4. $25E^2 - 10ER + R^2$ $(5E-R)^2$
   $5E - R$

5. $49A^2 + 7AB + 9B^2$ NPS

6. $I^2 + 2C + C^2$ NPS

7. $2R^2 + 5R + 3R^2$ NPS
   $5R^2 + 5R$
   $5R(R+1)$

8. $M^2 - 2Mn + n^2$ $(M-n)^2$
   $M - n$

9. $16D^2 + 80BD + 100B^2$ $4(2D+5B)^2$
   $4(4D^2 + 20BD + 25B^2)$
   $4(2D+5B)^2$

10. $9A^2 - 30AE + 25E^2$ $(3A-5E)^2$
    $(3A - 5E)$
    $9A^2 - 15AE$

11. $I^2 - IR - R^2$ NPS
    $(I-R)(I$

12. $e^2 - 8e + 16$ $(e-4)^2$
    $(e-4)(e-4)$
    $e^2 - 4e$
    $ -4e + 16$

13. $4a^2 + 32a + 64$ $(2a+8)^2$
    $(2a+8)(2a+8)$
    $4a^2 + 16a$
    $ + 16a + 64$

14. $12I^2 - 12I + 27$ _NPS_

$3(4I^2 - 4I + 9)$
$2I - 3$

15. $2B^2 - 32B + 128$ $2(B-8)^2$

$2(B^2 - 16B + 64)$
$2(B-8)^2$

16. $12d^2 - 36d + 27$ $3(2d-3)^2$

$3(4d^2 - 12d + 9)$
$3(2d-3)(2d-3)$
$4d^2 - 6d$
$ -6d + 9$

17. $5A^2 + 125A + 5$ _NPS_

18. $3a^2 + 36a - 48$ _NPS_

$3(a^2 + 12a - 16)$

4-4 FACTORING TRINOMIAL EXPRESSIONS. The factors common to each term are separated first. If the expression can be factored further, it will factor into two binomials.

- Remove common factors.
- Arrange in order of descending power, such as $a^2 + b + c$
- If factorable, it will factor into two binomial expressions.
- Determine the factors of the first and third terms.
- The resulting binomials will consist of prime factors.

This is better illustrated by example.

Factor $\boxed{a^2}$ + 5a + $\boxed{6}$

(a · a)   (3) (2)   or   (6) (1)
Factors   Factors

The first term has only one set of factors, whereas the third term has two possibilities:

$(a + 3)(a + 2) = a^2 + 5a + 6$ (right)

$(a + 6)(a + 1) = a^2 + 7a + 6$ (wrong)

The second set of factors does not give the correct middle term, but the first set does.

Example E

$$6a^2 + 42a + 60 =$$
$$6(a^2 + 7a + 10) =$$
$$6(a + 5)(a + 2)$$

EXERCISE 4-4 Factor by first removing common factors; then look for the binomial expressions that make up the trinomial. If they are not factorable, indicate "not factorable" (N.F.) or prime number.

1. $I^2 - 9I + 20 =$ $(I-5)(I-4)$

2. $R^2 + 8R + 15 =$ $(R+5)(R+3)$

3. $a^2 - 11a + 28 =$ $(a-7)(a-4)$

4. $16c^2 - 16c + 16 =$ $16(c^2 - c + 1)$
$16(c^2 - c + 1)$
$c - \frac{k}{c}$
$-2c$

5. $E^2 + 7E - 30 =$ $(E+10)(E-3)$

6. $3B^2 + 18B - 48 =$ $3(B+8)(B-2)$
$3(B^2 + 6B - 16)$

7. $I^2 - 7I + 12 =$ $(I-4)(I-3)$

8. $2R^2 + 26R + 72 =$ $2(R+4)(R+9)$
$2(R^2 + 13R + 36)$
$(R+4)(R+9)$

9. $a^2 - 8a + 20 =$ NF
-2
-10

10. $5T^2 - 11T + 6 =$ $(5T-6)(T-1)$

$(5T-6)(T-1)$
$5T^2 - 5T$
$-6T +6$

11. $8I^2 + 47I - 6 =$ $(8I-1)(I+6)$

$(8I-1)(I+6)$

12. $2F^2 - 9F + 10 =$ $(2F-5)(F-2)$

$(2F-5)(F-2)$

13. $3B^2 + 17B + 20 =$ $(3B+5)(B+4)$

$(3B+5)(B+4)$

14. $6c^2 + 11c - 10 =$ $(3c-2)(2c+5)$

$(3c-2)(2c+5)$ $6c^2 - 4c$
$+15c - 10$

15. $3a^2 - 10a + 7 =$ $(3a-7)(a-1)$

$(3a-7)(a-1)$

16. $8R^3 + 36R^2 + 36R =$ $4R(2R+3)(R+3)$

$4R(2R^2+9R+9)$ $(2R+3)(R+3)$
$8R(R^2+$

17. $6V^2 - 4V - 5 =$ NF

$6V+5$
$V-1$

18. $3e^2 - 18e + 27 =$ $3(e-3)^2$

$3(e^2-6e+9)$ $(e-3)(e-3)$

19. $16P^2 - 66P - 8 =$ $2(8p^2-33p-4)$

$2(8p^2-33p-4)$ $(8p+1)(p-4)$

20. $32a^2 + 56a - 16 =$ $8(4a-1)(a+2)$

$4(8a^2+14a-4)$ $(4a-1)(2a+4)$

21. $27i^2 + 15i - 2 =$ $(9i-1)(3i+2)$

$(9i-1)(3i+2)$
$+18$
-3

45

Chapter 4
Factoring

22. $20B^2 - 22B + 6 = \underline{2(5B-3)(2B-1)}$
    $2(10B^2 - 11B + 3)\ (5B-3)(2B-1)$

23. $4c^2 + 35c + 49 = \underline{(4c+7)(c+7)}$
    $(4c+7)(c+7)$

24. $12p^2 - 40p + 32 = \underline{(12p-16)(p-2) \text{ or } 4(3p-4)(p-2)}$

4–5 FACTORING THE DIFFERENCE OF TWO SQUARES INTO THE SUM AND DIFFERENCE OF LIKE QUANTITIES. Such expressions as $(R + 4)$ and $(R - 4)$ are the sum or difference of like quantities. Here it is the sum of R and 4 and the difference of R and 4. Remember that the product of the sum and difference of like quantities results in the difference of their squares.

$$(R + 4)(R - 4) = R^2 - 16$$

In a binomial expression, if the first term is a perfect square and the second term is a perfect square, and they are separated by a minus sign, the factors will be the sum and difference of the squares of the terms.

Example F

$$(a^2 - 25) = (a + 5)(a - 5)$$

$$(9R^2 - 36) = (3B + 6)(3B - 6)$$

$$(25a^2 - 36B^2) = (5a + 6B)(5a - 6B)$$

$$(18c^2 - 200) = 2(9c^2 - 100) = 2(3c + 10)(3c - 10)$$

EXERCISE 4–5 Factor by first separating common factors and then finding the factors of the remaining expressions, if any.

1. $25I^2 - 9 = \underline{(5I-3)(5I+3)}$

2. $36V^2 - 25 = \underline{(6V-5)(6V+5)}$

3. $(9i^2 - 1) = \underline{(3I-1)(3I+1)}$

4. $(4E^2 - R^2)$ = $(2E-R)(2E+R)$

47

Chapter 4
Factoring

5. $(a^2 - 1)$ = $(a-1)(a+1)$

6. $(6B^2 - 6)$ = $6(B+1)(B-1)$

7. $(25A^2B^2 - D^2)$ = $(5AB-D)(5AB+D)$

8. $(27I^4 - 3I^2)$ = $3(3I^2-I)(3I^2+I)$
$3I^2(9I^2-1)$ $\boxed{3I^2(3I+1)(3I-1)}$

9. $(36p^2 - 400)$ = $(6p-20)(6p+20)$ or $4(3p-10)(3p+10)$

10. $(6A^2F^2 - 6)$ = $6(AF-1)(AF+1)$

11. $(3e^3 - \overset{25}{75}e)$ = $3e(e-5)(e+5)$

12. $(72I^2 + 18)$ = $18(4I^2+1)$
$18(4I^2+1)$
$18(2I+1)(2I+1)$

13. $(A^4 - 81)$ = $(A^2-9)(A^2+9) = (A^2+9)(A+3)(A-3)$

14. $(16B^4 - C^4)$ = $(4B^2-C^2)(4B^2+C^2) = (4B^2+C^2)(2B+C)(2B-C)$

48

Part 1
Review of Algebra

EVALUATION EXERCISE Factor the expressions to their prime factors. It is not necessary to express the primes of each numerical coefficient.

1. $B^2 + 2B + 1 =$ $(B+1)(B+1) = (B+1)^2$

2. $c^2 + 10c + 12 =$ NF

3. $I^2 - 1 =$ $(I-1)(I+1)$

★ 4. $16a^2 - 14a - a =$ $a(16a^2 - 14a - 1) = a(16a - 15)$ [−15a written above]
   $a(16a^2 - 14a - 1)$ $(4a+2)(\;)$

5. $3T^3 + 27T =$ $3T(T^2 + 9)$
   $3T(T^2+9)$

6. $12d^2 - 2d - 4 =$ $2(3d-2)(2d+1)$
   $2(6d^2 - d - 2)$
   $2(3d+2)(d-1)$

7. $6I^2 + 11I + 3 =$ $(3I+1)(2I+3)$
   $(3I+1)(2I+3)$

8. $a^2 - 4aB + 4B^2 =$ $(a-2B)^2$

9. $25c^2d - 36d =$ $d(5c-6)(5c+6)$
   $d(25c^2 - 36)$

10. $e^2 + 34e - 72 =$ $(e-2)(e+36)$
    $(e-2)(e+36)$

11. $2m^2 + 2m - 60 = \underline{2(m-5)(m+6)}$
$2(m^2 + m - 30)$
$(m-5)(m+6)$

12. $5t^2 + 50t + 60 = \underline{5(t^2 + 10t + 12)}$
$5(T^2 + 10t + 12)$

13. $75i^3 - 27i = \underline{3i(5i-3)(5i+3)}$
$3i(25i^2 - 9)$

14. $3a^3 + 42a^2 + 147a = \underline{3a(a+7)^2}$
$3a(a^2 + 14a + 49)$

15. $64e^3 - 16e = \underline{16e(2e+1)(2e-1)}$
$16e(4e^2 - 1)$

16. $15R^2 + R - 40 = \underline{(5R-8)(3R+5)}$

17. $16i^2 - 40i + 25 = \underline{(4i-5)^2}$

18. $18E^2 - 105E - 18 = \underline{3(6E+1)(E-6)}$
$3(6E^2 - 35E - 6)$

19. $16i^2 - 2ir - 5r^2 = \underline{(8i - 5r)(2i + r)}$

20. $24e^3 + 141x^2 - 18 = \underline{3(8e^3 + 47x^2 - 6)} = 3(8e^2-1)(e+6)$
$3(8e^3 + 47x^2 - 6)$

49

Chapter 4
Factoring

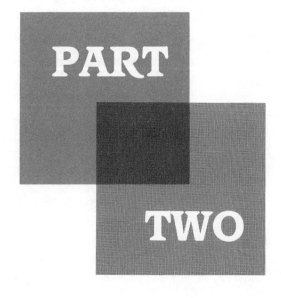

PART TWO

Mathematics for DC and AC Circuits

CHAPTER 5
Powers of Ten

Objectives

After completing this chapter you will be able to:
- Place numbers in power-of-10 notation
- Identify scientific notation
- Multiply using powers of 10
- Divide using powers of 10
- Add and subtract using powers of 10
- Square power-of-10 notation
- Extract square roots of power-of-10 notations
- Identify significant figures
- Round numbers using significant figures

Electronics is often confronted by large quantative values such as the number of electrons in a coulomb—6,280,000,000,000,000,000—or the number of amperes in 5.0 microamperes: 0.000005. This data can be simplified by a notation referred to as "power of 10."

5–1 MULTIPLES OF 10 can be expressed:

$$100 = 10 \times 10 \quad \text{or } 10^2$$
$$1{,}000 = 10 \times 10 \times 10 \quad \text{or } 10^3$$
$$10{,}000 = 10 \times 10 \times 10 \times 10 \text{ or } 10^4$$
etc.

5–2 OTHER QUANTITIES can be expressed:

$$628 = 6.28 \times 10 \times 10 \quad \text{or } 6.28 \times 10^2$$
$$6{,}280 = 6.28 \times 10 \times 10 \times 10 \quad \text{or } 6.28 \times 10^3$$
$$62{,}800 = 6.28 \times 10 \times 10 \times 10 \times 10 \text{ or } 6.28 \times 10^4$$

5-3 SCIENTIFIC NOTATION.

Numbers can be expressed in an almost unlimited variety of ways and still be in a correct power of 10 notation. The number 6280 can be expressed:

$$6.28 \times 10^3$$
$$62.8 \times 10^2$$
$$628 \times 10^1$$
$$0.628 \times 10^4$$

All are numerically correct.

To avoid confusion in recording scientific data or the final answer to problems, it is standard practice to use "scientific notation."

Scientific notation is defined as a number between 1 and 10 multiplied by the proper power of 10. Scientific notation can also be thought of as moving the decimal point in a number so there will be one nonzero digit to the left of the decimal point, and the number would also be expressed to the correct power of 10.

Example A

The notation 55,000 volts can be expressed 5.5×10^4 V. It was changed by moving the decimal point to the left and counting the number of moves required to move the decimal point.

$$5\,5\,0\,0\,0 = 5.5 \times 10^{④}$$
$$④\,3\,2\,1$$

$$6\,2\,5\,0 = 6.25 \times 10^{③}$$
$$③\,2\,1$$

If the number is less than 1, as in 0.0026 A, move the decimal point to the right and affix the correct power of 10. This power of 10 is then given a negative value.

Example B

$$0.0\,2\,6 = 2.6 \times 10^{-③}$$
$$1\,2\,③$$

$$0.0\,0\,0\,0\,6\,7\,9 = 6.79 \times 10^{-⑥}$$
$$1\,2\,3\,4\,5\,⑥$$

When the decimal point is moved from right to left, collect the power of 10 as a positive number; when it is moved from left to right, collect the power of 10 as a negative number.

EXERCISE 5–1 Place the numbers in scientific notation.

1. $8675 = \underline{8.675 \times 10^3}$
2. $1126 = \underline{1.126 \times 10^3}$
3. $0.0178 = \underline{1.78 \times 10^{-2}}$
4. $0.000178 = \underline{1.78 \times 10^{-4}}$
5. $86.75 = \underline{8.675 \times 10^1}$
6. $0.8675 = \underline{8.675 \times 10^{-1}}$
7. $0.0000100 = \underline{1 \times 10^{-5}}$
8. $1.500 = \underline{1.5 \times 10^0}$
9. $6{,}280{,}000 = \underline{6.28 \times 10^6}$
10. $135.67 = \underline{1.3567 \times 10^2}$

5–4 CHANGING TO SCIENTIFIC NOTATION. A number already in a power of 10 notation can be changed to a scientific notation by moving the decimal point to the place where there is one whole number to the left of the decimal point. Count the number of places moved with the correct sign, positive or negative. Algebraicly add this number to the number already indicated as the power of 10 notation.

Example C

$$6.28 \times 10^3 = 6.28 \times 10^{3+2} = 6.28 \times 10^5$$

$$0.025 \times 10^{-4} = 2.5 \times 10^{-4-2} = 2.5 \times 10^{-6}$$

$$0.0028 \times 10^6 = 2.8 \times 10^{6-3} = 2.8 \times 10^3$$

$$7.26 \times 10^{-5} = 7.26 \times 10^{-5+2} = 7.26 \times 10^{-3}$$

EXERCISE 5–2 Place the numbers in scientific notation.

1. $865 \times 10^5 = \underline{8.65 \times 10^7}$
2. $97.3 \times 10^1 = \underline{9.73 \times 10^2}$

3. $0.0016 \times 10^{-3} = 1.6 \times 10^{-6}$

4. $829{,}000 \times 10^6 = 8.29 \times 10^{11}$

5. $0.092 \times 10^2 = 9.2 \times 10^0$

6. $0.0004 \times 10^{-12} = 4 \times 10^{-16}$

7. $1600 \times 10^{-8} = 1.6 \times 10^{-5}$

8. $47{,}000{,}000 \times 10^{-6} = 4.7 \times 10^1$

9. $50{,}000 \times 10^{-3} = 5 \times 10^1$

10. $10{,}000 \times 10^{-4} = 1 \times 10^0$

11. $118 \times 10^{-3} = 1.18 \times 10^{-1}$

12. $777.7 \times 10^{-16} = 7.777 \times 10^{-14}$

13. $628 \times 10^{-6} = 6.28 \times 10^{-4}$

14. $0.0076 \times 10^3 = 7.6 \times 10^0$

15. $0.0156 \times 10^{12} = 1.56 \times 10^{10}$

16. $0.000006 \times 10^{-12} = 6 \times 10^{-18}$

5–5 SIGNIFICANT FIGURES. Meter readings and other measurements are usually accurate to a certain degree and are generally determined by application and type of equipment. One method of referring to accuracy is, how many significant figures a measurement contains. Only the figures or numbers that can be determined on the measuring instrument are significant. When reading 0.002 amp on a 0 to 5 scale, the observer sees only the number 2. The 0.00 is added as a placeholder because the observer knows how the instrument operates.

Four Rules for Determining Significant Figures

Rule 1: All nonzero numbers are significant.
Rule 2: All zeros between nonzeroed numbers are significant.

Rule 3: Zeros to the right of a decimal point are significant, but zeros to the right of nonzeroed numbers and to the left of the decimal point may or may not be significant. If they can be seen on the measuring instrument, they are significant; if they are used only as placeholders, they are not significant. This is determined in application and will not be considered in this text.

Rule 4: Zeros to the left of decimal points are used to place a decimal point and are not significant.

Example D

Number	Significant Figure	Rule
326	3	1
82.657	5	1
2006	4	2
1984.072	7	2
868.80	5	3
100.00	5	3
0.055	2	4
0.0060	2	4 and 3
4.42×10^5	3	3

EXERCISE 5-3 Determine the number of significant figures.

1. 96 = _____
2. 175.6 = _____
3. 0.001 = _____
4. 68.6 = _____
5. 821754 = _____
6. 3060.7 = _____
7. 4.007 = _____
8. 5.050 = _____
9. 0.000726 = _____
10. 1.15×10^3 = _____
11. 3.65×10^{-4} = _____
12. 0.05×10^6 = _____

5-6 ROUNDING WITH SIGNIFICANT FIGURES. In most applications of electronics accuracy to three significant figures is sufficient.

Example E
Round to three significant figures

$$37.6\boxed{4} = 37.6$$

dropped because it is less than 5

$$37.6\cancel{6} = 37.7$$

increases the third significant figure because it is greater than 5

When the determiner is exactly 5, such as 37.65, and if the number to be rounded is an even number, leave it even and drop the five.

$$37.6\cancel{5} = 37.6$$
drop

When the determiner is 5 and the number to be rounded is an odd number, raise the number by 1 and make it an even number.

$$37.7\cancel{5} = 37.8$$
determiner ↗   ↖ left as an even number

EXERCISE 5-4 Round the numbers to three significant figures.

1. 178.6 = _179_
2. 8.743 = _8.74_
3. 0.06321 = _.063_
4. 0.8429 = _.843_
5. 199.8 = _200_
6. 300.62 = _301_
7. 290.526 = _290_
8. 0.04003 = _.040_
9. 99.95 = _100_
10. 2.045 = _2.04_

5-7 MULTIPLICATION OF POWERS OF 10. To multiply quantities in power-of-10 notation, first multiply the significant figures, then algebraicly add the exponents. Numbers can be multiplied in any form, but putting them in scientific notation may be easier.

Example F

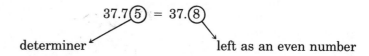

$$(7.20 \times 10^{3})(3.60 \times 10^{2})$$

multiply, add

$$25.92 \times 10^{5}$$

2.59×10^{6} (rounded to three significant figures and left in scientific notation)

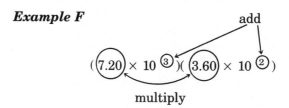

$$(2.52 \times 10^{-4})(5.40 \times 10^{2})$$

multiply, add

$$13.608 \times 10^{-2} = 1.36 \times 10^{-1}$$

EXERCISE 5-5 Rewrite the following numbers in scientific notation and round to three significant figures before multiplying. All numbers do not contain three significant figures; however, round all answers to three significant figures. Use a calculator for multiplying significant figures, but do not use it for collecting the powers of 10.

Chapter 5
Powers of Ten

1. $1000 \times 0.0100 \times 100 =$ $(1 \times 10^3)(1 \times 10^{-2})(1 \times 10^2) = 1 \times 10^3$
2. $10^6 \times 10^{-4} \times 10^{12} =$ $(1 \times 10^6)(1 \times 10^{-4})(1 \times 10^{12}) = 1 \times 10^{14}$
3. $3.70 \times 10^4 \times 868 \times 0.764 =$ $(3.7 \times 10^0)(1 \times 10^4)(8.68 \times 10^2)(7.64 \times 10^{-1}) = 2.45 \times 10^7$ 245.36
4. $0.000760 \times 10^3 \times 46.5 \times 10^{-2} =$ $(7.6 \times 10^{-4})(1 \times 10^3)(4.65 \times 10^1)(1 \times 10^{-2}) = 35.34 = 3.53 \times 10^1$
5. $6.28 \times 10^3 \times 3.1415 \times 0.002 =$ $(6.28 \times 10^0)(1 \times 10^3)(3.14 \times 10^0)(2 \times 10^{-3}) = 39.43 \times 10^0 = 3.94 \times 10^1$
6. $0.000950 \times 0.00300 \times 10^4 \times 0.6784 =$ $(9.5 \times 10^{-4})(3 \times 10^{-3})(1 \times 10^4)(6.78 \times 10^{-1}) = 193 \times 10^{-4} = 1.93 \times 10^{-2}$
7. $10^{-4} \times 6.00 \times 10^3 \times 1.50 \times 10^{-12} =$ $(1 \times 10^{-4})(6 \times 10^0)(1 \times 10^3)(1.5 \times 10^0)(1 \times 10^{-12}) = 9 \times 10^{-13}$
8. $82{,}657{,}000 \times 4860 \times 32 =$ $(8.26 \times 10^7)(4.86 \times 10^3)(3.2 \times 10^1) = 1.28 \times 10^{13}$ 128×10^{11}
9. $0.00500 \times 10^3 \times 500 \times 10^{-3} \times 16 =$ $(5 \times 10^{-3})(1 \times 10^3)(5 \times 10^2)(1 \times 10^{-3})(1.6 \times 10^1) = 4 \times 10^1$ [check] 40×10^0
10. $0.100 \times 10^{-12} \times 0.150 \times 10^{-8} \times 100 =$ $(1 \times 10^{-1})(1 \times 10^{-12})(1.5 \times 10^{-1})(1 \times 10^{-8})(1 \times 10^2) = 1.5 \times 10^{-20}$

5-8 DIVISION USING POWER-OF-10 NOTATION. In division the significant figure can be divided by using a calculator; power-of-10 notations are subtracted as outlined in Example G.

Example G

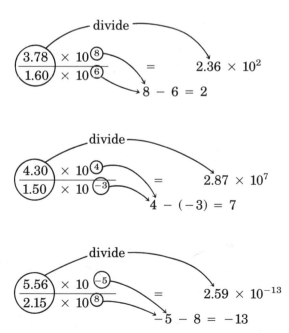

EXERCISE 5–6 Round each value to three significant figures and place it in scientific notation. Then perform the indicated operation, leaving the final answer in scientific notation.

1. $\dfrac{8670}{3.20 \times 10^2} = \dfrac{8.67 \times 10^3}{3.2 \times 10^2} = 2.71 \times 10^1$

2. $\dfrac{0.00075}{6.42 \times 10^4} = \dfrac{7.5 \times 10^{-4}}{6.42 \times 10^4} = 1.17 \times 10^{-8}$

3. $\dfrac{10^2 \times 10^4}{5.00 \times 10^3} = \dfrac{(1 \times 10^2)(1 \times 10^4)}{5 \times 10^3} = \dfrac{1 \times 10^6}{5 \times 10^3} = .2 \times 10^3 = 2 \times 10^2$

4. $\dfrac{3.70 \times 10^{-2} \times 12.5 \times 10^3}{86.7 \times 0.000730} = \dfrac{4.62 \times 10^2}{6.33 \times 10^{-2}} = 7.3 \times 10^3$

   $(8.67 \times 10^1)(7.3 \times 10^{-4})$
   63.29×10^{-3}
   $.7298 \times 10^{+3}$

5. $\dfrac{1}{6.28 \times 0.00150 \times 10^6} = \dfrac{1}{(6.28 \times 10^6)(1.5 \times 10^{-3})} = 1.06 \times 10^{-4}$

   9.42×10^3

6. $\dfrac{6.28 \times 50.0 \times 10^{-6}}{6.28 \times 10^8} = \dfrac{(6.28 \times 10^{-6})(5 \times 10^1)}{6.28 \times 10^8} = \dfrac{3.14 \times 10^{-4}}{6.28 \times 10^8} = 5 \times 10^{-13}$

   $.5$

7. $\dfrac{165 \times 10^3 \times 0.0100}{300 \times 10^{-4} \times 10^7} = \dfrac{(1.65 \times 10^5)(1 \times 10^{-2})}{(3 \times 10^9)(1 \times 10^{-4})} = 5.5 \times 10^{-3}$

   1.65×10^3
   3×10^5
   $.55 \times 10^{-2}$

8. $\dfrac{999.9 \times 10^4 \times 10^8}{10^6 \times 10^{-4} \times 1000} = \dfrac{(1 \times 10^7)(10^8)}{(1 \times 10^{-1})(10^6)} = 1 \times 10^{10}$

   1.000
   1×10^5
   1×10^5

5-9 SQUARES AND SQUARE ROOTS WITH POWERS OF 10. To square a number in power-of-10 notation, first square the significant figures, then double the exponent of 10.

Example H

$$(3 \times 10^{4})^{2} = 9 \times 10^{8}$$

(square the 3; double the 4)

$$(5 \times 10^{-3})^{2} = 25 \times 10^{-6}$$

(square the 5; double the -3)

To extract the square root, the number must first be expressed as a power-of-10 notation with an *even* exponent. Then extract the square root of the significant figures and divide the exponent of 10 by 2.

Example I

$$\sqrt{2.5 \times 10^{5}} = \sqrt{25 \times 10^{4}}$$

(extract square root of 25; divide 4 by 2)

$$= 5 \times 10^{2}$$

$$\sqrt{0.036 \times 10^{-5}} = \sqrt{0.36 \times 10^{-6}} = 0.6 \times 10^{-3}$$

or

$$\sqrt{0.036 \times 10^{-5}} = \sqrt{36 \times 10^{-8}} = 6 \times 10^{-4}$$

EXERCISE 5-7 Place all numbers in scientific notation; then square the number.

1. $786 = (7.86 \times 10^{2})^{2} = 61.8 \times 10^{4} = 6.18 \times 10^{5}$
2. $0.0187 = (1.87 \times 10^{-2})^{2} = 3.5 \times 10^{-4}$
3. $13.4 \times 10^{-5} = (1.34 \times 10^{-4})^{2} = 1.8 \times 10^{-8}$
4. $0.88 \times 10^{-3} = (8.8 \times 10^{-4})^{2} = 77.4 \times 10^{-8} = 7.74 \times 10^{-7}$

5. $1096 \times 10^4 = (1.1 \times 10^7)^2 = 1.21 \times 10^{14}$
6. $987 \times 10^{-3} = (9.87 \times 10^{-1})^2 = 97.4 \times 10^{-2} = 9.74 \times 10^{-1}$
7. $0.000425 = (4.25 \times 10^{-4})^2 = 1.81 \times 10^{-7}$
8. $10^5 = (1 \times 10^5)^2 = 1 \times 10^{10}$

Place the numbers in 9 through 16 in a convenient power-of-10 notation with an even-numbered exponent and extract the square root. Leave the answers in scientific notation and round to three significant figures.

9. $\sqrt{168 \times 10^5} = \sqrt{16.8 \times 10^6} = 4.1 \times 10^3$

10. $\sqrt{0.014 \times 10^{-3}} = \sqrt{14 \times 10^{-6}} = 3.74 \times 10^{-3}$

11. $\sqrt{1142 \times 10^7} = \sqrt{1.142 \times 10^{10}} = 1.07 \times 10^5$

12. $\sqrt{928,647} = \sqrt{92.8647 \times 10^4} = 9.64 \times 10^2$

13. $\sqrt{10^{-9}} = \sqrt{1 \times 10^{-9}} = \sqrt{.1 \times 10^{-8}} = .316 \times 10^{-4} = 3.16 \times 10^{-5}$

14. $\sqrt{1 \times 10^{13}} = \sqrt{.1 \times 10^{14}} = .316 \times 10^7 = 3.16 \times 10^6$

15. $\sqrt{0.000068} = \sqrt{6.8 \times 10^{-6}} = 8.25 \times 10^{-3}$

16. $\sqrt{36 \times 10^7} = \sqrt{3.6 \times 10^8} = 1.9 \times 10^4$

5-10 ADDITION AND SUBTRACTION OF NUMBERS IN POWERS-OF-10 NOTATION.
A basic rule of addition and subtraction is: only like things can be added or subtracted. Power-of-10 notation is no exception. What makes power-of-10 notations similar is, they have the same exponent.

Example J

Add
$$\begin{aligned} 230 \times 10^4 &= 230 \times 10^4 \\ 180 \times 10^3 &= 18 \times 10^4 \\ \hline &248 \times 10^4 \end{aligned}$$

Subtract
$$\begin{aligned} 0.015 \times 10^{-2} &= 150 \times 10^{-6} \\ (-)125 \times 10^{-6} &= 125 \times 10^{-6} \\ \hline &25 \times 10^{-6} \end{aligned}$$

EXERCISE 5–8 Add or subtract as indicated, leaving the answers in scientific notation. Do not round off.

1. $(11.5 \times 10^7) + (830 \times 10^5) =$ $\dfrac{11.5 \times 10^7}{+8.3 \times 10^7} = 19.8 \times 10^7 = 1.98 \times 10^8$

2. $(656 \times 10^3) + (42 \times 10^1) + 950 =$ $6560 \times 10^2 + 4.2 \times 10^2 + 9.5 \times 10^2 = 6573.7 \times 10^2 = 6.5737 \times 10^5$

3. $3.3 \times 10^{-6} + 0.00055 + 0.006 \times 10^{-3} =$ $\begin{array}{c}3.3 \times 10^{-6}\\550.0 \times 10^{-6}\\6.0 \times 10^{-6}\end{array} = 559.3 \times 10^{-6} = 5.593 \times 10^{-4}$

4. $150 \times 10^{10} - 300 \times 10^9 =$ $\begin{array}{c}150\\-30\end{array} \; 10^{10} = 120 \times 10^{10} = 1.2 \times 10^{12}$

5. $0.0075 - 2 \times 10^{-3} =$ $\begin{array}{c}7.5\\-2\end{array} \; \begin{array}{c}10^{-3}\\10^{-3}\end{array} = 5.5 \times 10^{-3}$

6. $10^3 + 10^5 + 10^2 =$ $.01 + 1 + .001 = 1.011 \times 10^5$
$.01 \times 10^3 \quad 1 \times 10^5 \; .001 \times 10^2$

7. $10^6 - 10^4 =$ $\dfrac{1 \times 10^6}{-.01 \times 10^6} = .99 \times 10^6 = 9.9 \times 10^5$
$.01 \times 10^4$

8. $10^{-10} - 10^{-11} =$ $\dfrac{1 \times 10^{-10}}{-.1 \times 10^{-10}} = .9 \times 10^{-10} = 9 \times 10^{-11}$

EVALUATION EXERCISE Place all numbers in scientific notation rounded to three significant figures and perform the indicated operations. Leave the final answers in scientific notation rounded to three significant figures.

1. $8.7 \times 10^3 \times 3.5 \times 10^{-2} =$ $30.45 \times 10^1 = 3.04 \times 10^2$

2. $\dfrac{1686}{7.24 \times 10^{-6}} =$ $\dfrac{1.69 \times 10^3}{7.24 \times 10^{-6}} = .2334 \times 10^9 = 2.33 \times 10^8$

3. $\dfrac{14.5 \times 10^6 \times 0.008}{3.2 \times 78 \times 10^2} =$ $\dfrac{(1.45 \times 10^7)(8 \times 10^{-3})}{(3.2 \times 10^2)(7.8 \times 10^1)} = \dfrac{11.6 \times 10^4}{24.96 \times 10^3} = \dfrac{1.16 \times 10^5}{2.5 \times 10^4} = .464 \times 10^1 = 4.64 \times 10^0$

4. $\dfrac{1}{6.28(60)0.015 \times 10^{-6}} =$ $\dfrac{1 \times 10^0}{(6.28 \times 10^{-6})(6 \times 10^1)(1.5 \times 10^{-2})} = .1769 \times 10^6 = 1.77 \times 10^5$
$\dfrac{}{56.52 \times 10^{-7}}$
5.65×10^{-6}

62
Part 2
Mathematics for DC
and AC Circuits

5. $\dfrac{6.28 \times 10^6 \times 50}{10^6} =$ $\dfrac{(6.28\times 10^6)(5\times 10^1)}{1\times 10^6}$ $\overset{31.4\times 10^7}{\dfrac{3.14\times 10^8}{1\times 10^6}} = 3.14\times 10^2$

6. $(456 \times 10^{-7})^2 =$ $\dfrac{(4.56\times 10^{-5})^2 = 2.08\times 10^{-9}}{20.79^{-10}}$

7. $\sqrt{0.000726 \times 10^4} =$ $\sqrt{7.26\times 10^0} = 2.7\times 10^0$

8. $(17.5 \times 10^2 \times 3.14 \times 10^6)^2 =$ $\dfrac{(5.5\times 10^9)^2 = 3.02\times 10^{19}}{30.25\times 10^{18}}$
   $1.75\times 10^3 \times 3.14\times 10^6$

9. $\sqrt{32 \times 10^4 \times 4 \times 10^3} =$ $\sqrt{12.8\times 10^8} = 3.58\times 10^4$
   $3.2\times 10^5 \times 4\times 10^3$

10. $\left(\dfrac{1}{6.28\sqrt{5 \times 10^3 \times 0.05 \times 10^{-6}}}\right)^2 =$ 1.02×10^2
    $6.28\sqrt{5\times 10^3 \quad .5\times 10^{-7}}$
    $6.28\sqrt{2.5\times 10^{-4}} = 6.28(1.58\times 10^{-2}) = \dfrac{1\times 10^{0}}{9.92\times 10^{-2}} = .1\times 10^2$
    $\dfrac{1\times 10^{11}}{1\times 10^1}$

11. $\left(\dfrac{6.28 \times 10^9}{10^6}\right)^2 =$ $(6.28\times 10^3)^2 = 39.4\times 10^6 = 3.94\times 10^7$

12. $(\sqrt{10^6 \times 10^5})^2 =$ $10\times 10^{10} = 1\times 10^{11}$

$(1\times 10^6)(1\times 10^5)$
$(10\times 10^5)(1\times 10^5)$
$\sqrt{10\times 10^{10}} = (3.16\times 10^5)^2$

13. $0.0016 + 3 \times 10^{-3} + 10^{-3} =$ __5.6×10^{-3}__
    $1.6 + 3 \quad + 1$

14. $8200 + 4 \times 10^2 + 10^2 =$ __$87 \times 10^2 = 8.7 \times 10^3$__
    $82 + 4 \quad + 1$

15. $5.5 \times 10^3 - 3000 =$ __2.5×10^3__
    $5.5 - 3$

16. $10 \times 10^{-4} - 0.00006 =$ __9.4×10^{-4}__
    $10 - .6$

CHAPTER 6

Metric Units with Applications

Objectives

After completing this chapter you will be able to:
- Identify metric units used in electronics applications
- Convert metric units to convenient metric prefixes
- Convert metric prefixes to basic units by using powers of 10
- Identify engineering notation
- Apply metric units in applications of Ohm's Law
- Apply metric units to simple-series dc circuits

Science has long used metric units to accomplish measurement, and the field of electronics is no exception. Such prefixes as micro, pico, kilo, and centi are used every day in electronics. They must be thoroughly learned and understood by the student. Because metric units are a new topic for most students, they are sometimes viewed with anxiety. This chapter should give the basic concepts of the system as it applies to the field of the electronics technician. Learning and understanding will become thorough as this concept is applied in other applications.

6–1 METRIC SYSTEM. The metric system has been a legal system of measurement in the United States for more than a century. Through world agreement the SI (Systeme International) metric system was established. This system establishes the meter as a unit of length, the kilogram as a unit of mass, and the liter as a unit of volume. These are the units required for everyday living, but they need to be expanded to units commonly used in electronics. Table 6–1 illustrates these units and is expanded to include the ohm, volt, ampere, henry, farad, and second. A complete table of prefixes and units can be found on pages ix and x, following the Preface.

A great deal of information is given to a new student in electronics technology. All metric units and prefixes are not needed by the beginning student. This chapter concentrates on the units and prefixes needed in the first few weeks of study. Once they are mastered, the additional units and prefixes can be easily handled.

6–2 CONVERTING UNITS WITHIN THE METRIC SYSTEM. Units can be converted by using a multiplication factor indicated in Table 6–1. Use of a multiplication factor results in simply moving the decimal point left or right. Table 6–1 is conveniently set up to aid in moving the decimal point. After programmed use of this table, movement of the decimal point will become easy.

Table 6–1 can be used in the following way: Convert 10 m to cm. Basic units are indicated in the center of the table. Since cm is two moves to the right of the basic unit, the decimal point moves two places to the right.

$$10.0 \text{ m} = 1\,0\,0\,0\,0 \text{ cm} = 1000 \text{ cm}$$

$$\phantom{10.0 \text{ m} = 10}1\,2$$

TABLE 6–1   Metric Prefixes

Basic unit:
meter (m), liter (L), gram (g), ohm (Ω), volt (V), ampere (A), henry (H), farad (F), second (s), watts (W)

10^{12}	10^{9}	10^{6}	10^{3}	10^{2}	10^{1}		10^{-1}	10^{-2}	10^{-3}	10^{-6}	10^{-9}	10^{-12}
tera	giga	mega	kilo	hecto	deka		deci	centi	milli	micro	nano	pico
T	G	M	k	h	da		d	c	m	μ	n	p

Convert 1,500,000 m to km (use Table 6–1).

Because km is three places to the left of m, move the decimal point three places to the left.

$$1\,5\,0\,0\,0\,0\,0\,\text{m} = 1500 \text{ km}$$

$$3\,2\,1$$

Convert 50 mA to A (use Table 6–1).

Because the basic unit is three places to the left of mA, move the decimal point three places to the left.

$$0\,5\,0\,\text{mA} = 0.05 \text{ A}$$

$$3\,2\,1$$

Convert 15,000,000,000 μV to kV (use Table 6–1).

Because kV is nine places to the left of μV, move the decimal point nine places to the left.

$$1\,5\,0\,0\,0\,0\,0\,0\,0\,0\,0\,\mu\text{V} = 15 \text{ kV}$$

$$9\,8\,7\,6\,5\,4\,3\,2\,1$$

EXERCISE 6–1   Convert the units to the new metric prefix, as indicated.

1. 36 mm to ____.036____ m   2. 850 μV to ____.00085____ V

3. 1050 kV to __1,050,000__ V   4. 0.0005 μA to __.0000000005__ A

5. 0.002 ms to __.000002__ s   6. 1.75 L to __1750__ mL (175000)

7. 1600 V to __1.6__ kV   8. 136.8 cm to __1.368__ m

9. 13000 mA to __.013__ kA   10. 8.8 kV to __8,800,000__ mV

11. 6.5 μS to __.0000065__ S   12. 8400 g to __8.4__ kg

13. 55,000,000 Ω to __55__ MΩ

14. 120,000,000 Ω to __120,000__ kΩ

6–3 CONVERSION BY USING POWERS OF 10. Converting units from one prefix to another is more easily accomplished by adding a power-of-10 notation than by moving decimal points.

Convert 50 μA to A

The decimal point should move six places to the left in accordance with Table 6–1. Rather than physically moving the decimal point, the same result can be accomplished by affixing a power-of-10 notation.

$$50 \text{ μA} = 50 \times 10^{-6} \text{ A}$$

accomplishes the same result as moving the decimal point six places to the left

Convert 15,000 V to mV

$$15,000 \times 10^{3}$$

move the decimal point three places to the right

Example A
Convert 50,000,000 mA to kA
$$50,000,000 \times 10^{-6} \text{ kA}$$

Convert 5×10^3 A to mA

change A to mA

$$5 \times 10^{3+3} = 5 \times 10^6 \text{ mA}$$

Convert 1.5×10^2 μV to V:

convert μV to V
$1.5 \times 10^{2+(-6)} = 1.5 \times 10^{-4}$ V

67

Chapter 6
Metric Units with Applications

EXERCISE 6-2 Convert the units to the indicated prefix by affixing the correct power-of-10 notation. Then place the final answer in scientific notation.

1. 150 mV to V $\underline{150 \times 10^{-3}}$ V $\underline{1.5 \times 10^{-1}}$ V

2. 75 μA to A $\underline{75 \times 10^{-6}}$ A $\underline{7.5 \times 10^{-5}}$ A

3. 300 mA to A $\underline{300 \times 10^{-3}}$ A $\underline{3 \times 10^{-1}}$ A

4. 165 A to mA $\underline{165 \times 10^{3}\text{ A}}$ mA $\underline{1.65 \times 10^{5}}$ mA

5. 36 cm to m $\underline{36 \times 10^{-2}}$ m $\underline{3.6 \times 10^{-1}}$ m

6. 820 kΩ to Ω $\underline{820 \times 10^{6}}$ Ω $\underline{8.2 \times 10^{8}}$ Ω ✗

7. 1000 kV to mV $\underline{1000 \times 10^{6}}$ mV $\underline{1 \times 10^{9}}$ mV

8. 10 MΩ to Ω $\underline{10 \times 10^{6}}$ Ω $\underline{1 \times 10^{7}}$ Ω

9. 150,000 Ω to kΩ $\underline{150,000 \times 10^{+5-3}}$ kΩ $\underline{1.5 \times 10^{+2}}$ kΩ ✗

10. 26 mm to m $\underline{26 \times 10^{-3}}$ m $\underline{2.6 \times 10^{-2}}$ m

11. 0.004 A to mA $\underline{.004 \times 10^{3}}$ mA $\underline{4 \times 10^{0}}$ mA

12. 0.000125 A to μA $\underline{.000125 \times 10^{-6}}$ μA $\underline{125 \times 10^{0}}$ μA 1.2×10^{-4}

13. 0.250 mV to kV $\underline{.250 \times 10^{-6}}$ kV $\underline{2.5 \times 10^{-7}}$ kV

14. 1.75 L to mL $\underline{1.75 \times 10^{3}}$ mL $\underline{1.75 \times 10^{3}}$ mL

15. 15,000 kV to V $\underline{15,000 \times 10^{3}}$ V $\underline{1.5 \times 10^{7}}$ V

V → mV × 10^{3}
mV → kV × 10^{-6}
mV → V × 10^{-6}
kV → V × 10^{+3}

16. 5.6×10^{-2} Ω to kΩ 5.6×10^{1} kΩ 5.6×10^{-5} kΩ

17. 10^9 μV to V 10^3 V 10^3 V

18. 1.2×10^3 MΩ to kΩ 1.2×10^6 kΩ 1.2×10^6 kΩ

19. 6.3×10^{-3} V to mV 6.3×10^0 mV 6.3 mV

20. 150×10^2 cm to m 1.50×10^2 m $1.50 \times 10^{+2}$ m

6-4 ENGINEERING NOTATION. The exercises in Exercise 6–2 illustrate that the metric prefixes are generally grouped in units of three (such as kilo, 10^3; milli, 10^{-3}; micro, 10^{-6}). It is convenient to express power-of-10 notation with the power in groups of three; then use the appropriate prefix, usually in convenient numerical numbers. The power of 10 denotes the prefix.

$$0.0000165 \text{ A} = 16.5 \times 10^{-6} \text{ A} = 16.5 \text{ μA}$$

(the -6 denotes the prefix micro)

$$320{,}000 \text{ Ω} = 320 \times 10^3 \text{ Ω} = 320 \text{ kΩ}$$

$$60{,}000{,}000 \text{ V} = 60 \times 10^6 \text{ V} = 60 \text{ MV}$$

EXERCISE 6-3 Place the quantities in convenient numbers expressed as an engineering notation. Then express them in a convenient prefix notation, eliminating the power of 10.

		Engineering notation (in basic unit)	Prefix notation
1.	0.00005 A	50×10^{-6} A	50 μA
2.	12,000 V	12×10^3 V	12 kV
3.	12×10^{-3} mA	12×10^{-6}	12 μA
4.	10^8 Ω	$1 \times 10^8 = 100 \times 10^6$	100 MΩ
5.	7.5×10^{-8} V	75×10^{-9}	75 nA

6. 330×10^5 Ω 33×10^6 33 MΩ

7. 0.00175 A 1.75×10^{-3} 1.75 mA

8. 470,000 kΩ 470×10^6 470 MΩ

9. 1250×10^2 mA 125×10^3 125 A

10. 0.000955 kV 955×10^{-6} 955 mV

6-5 USING OHM'S LAW WITH PREFIX NOTATION. The basic concepts of Ohm's Law and series circuits can be understood without an in-depth study of electronics. The basic equation has three variations, which are generally used to solve circuit problems.

$$V = IR \qquad I = \frac{V}{R} \qquad R = \frac{V}{I}$$

where V is measured in volts
I is measured in amperes
R is measured in ohms

These equations and the use of prefix notation are applied in Section 6-5. Units used in Ohm's Law are in the basic unit; V is in volts (V), I in amperes (A), and R in ohms (Ω). Units should be converted to basic units before manipulating an equation.

Determine the voltage applied to a circuit if the current is 12 mA and the resistance is 270 kΩ.

$V = IR$ (Select the correct form of Ohm's Law.)

$V = (12 \times 10^{-3})(270 \times 10^3)$ (Substitute in the equation, changing the given prefix units to basic units by affixing the correct power of 10; do not move the decimal point.)

$V = 3240$ V (Either leave the answer in a basic unit or change to a convenient prefix unit; if left in power-of-10 notation, leave in scientific notation.)

Example B
Determine the current in a circuit with an applied voltage of 9 V and a resistance of 150,000 Ω.

$$I = \frac{V}{R}$$

$$I = \frac{9}{150 \times 10^3}$$

$I = 0.06 \times 10^{-3}$ A $= 60 \times 10^{-6}$ A $= 60$ μA (preferred answer)

Chapter 6
Metric Units with Applications

EXERCISE 6–4 All of the following are applications of Ohm's Law. Find the missing quantity. Leave answers in a convenient prefix notation, such as milli, micro, kilo, or basic unit if it is between common prefixes. Show equations, substitutions, and final answers.

	V	I	R
1.	_10.8 mV_	30 µA	360 Ω
2.	12 V	_80 µA_	150 kΩ
3.	15 V	75 mA	_200 Ω_
4.	_2.13 V_	26 µA	82 kΩ
5.	6.3 V	_42 nA_ / _42×10⁻⁹_	150 MΩ

6.   13.2 V   18 mA   733 Ω

71
*Chapter 6
Metric Units with
Applications*

7.   81.6 mV   120 μA   680 Ω

8.   1.5 V   55.5 μA   27 kΩ

9.   6 V   5 μA   1.2 MΩ

10. 3 V   13.6 μA   220 kΩ

11. 4.5 V   32 μA   140625
                               141 KΩ

6-6 SERIES RESISTANCE CIRCUITS.

A series circuit is a circuit in which the components are connected so they share the same current. Figure 6-1 illustrates a series resistive circuit connection showing the physical connection and the schematic drawing. Ohm's Law applies to each resistor and to the total characteristics of the circuit. The series resistive circuit can be described by the following equations.

Total resistance

$$R_T = R_1 + R_2 + R_3 + \text{etc.}$$

Total current

$$I_T = I_1 = I_2 = I_3 = \text{etc.}$$

Total voltage

$$V_T = V_1 + V_2 + V_3 + \text{etc.}$$

These equations and Ohm's Law can be used to solve series circuit problems. Determine the total current in the circuit given in Figure 6-1.

Determine the total resistance:

$$R_T = R_1 + R_2 + R_3$$
$$R_T = 100 + 390 + 220$$
$$R_T = 710 \ \Omega$$

By Ohm's Law:

$$I_T = \frac{V_T}{R_T}$$
$$I_T = \frac{9}{710}$$
$$I_T = 12.7 \text{ mA}$$

Determine the voltage across each resistor in the circuit in Figure 6-1.

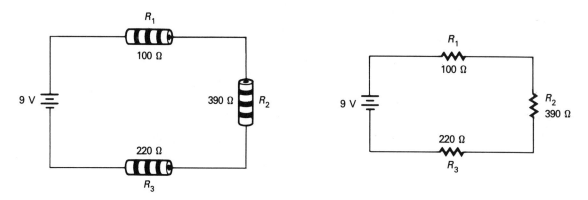

FIGURE 6-1   Physical connections and schematic

By Ohm's Law:

$$V_1 = I_T R_1$$
$$V_1 = (12.7 \times 10^{-3})(100)$$
$$V_1 = 1270 \times 10^{-3}$$
$$V_1 = 1270 \text{ mV or } 1.27 \text{ V (either one is a good prefix)}$$

$$V_2 = I_T R_2$$
$$V_2 = (12.7 \times 10^{-3})(390)$$
$$V_2 = 4953 \times 10^{-3}$$
$$V_2 = 4953 \text{ mV or } 4.91 \text{ V}$$

$$V_3 = I_T R_3$$
$$V_3 = (12.7 \times 10^{-3})(220)$$
$$V_3 = 2794 \times 10^{-3}$$
$$V_3 = 2794 \text{ mV or } 2.79 \text{ V}$$

Examples C and D illustrate additional applications of Ohm's Law and the series circuit.

Example C

A series circuit contains three resistors, $R_1 = 22$ kΩ, $R_2 = 82$ kΩ, and $R_3 = 56$ kΩ. The voltage across $R_2 = 2$ V. What is the applied voltage?

Draw the schematic and label the known values as shown in Figure 6–2. If any two Ohm's Law quantities are known, the third can be determined. In the case of R_2, V and R are known and I can be determined.

Determine the total current by Ohm's Law.

$$I_2 = \frac{V_2}{R_2}$$

$$I_2 = \frac{2}{(82 \times 10^3)}$$
$$I_2 = 0.0244 \times 10^{-3}$$
$$I_2 = 24.4 \text{ μA}$$
$$I_T = 24.4 \text{ μA}$$

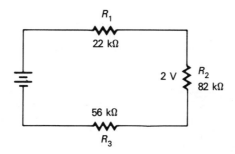

FIGURE 6–2   Circuit for Example C

Total resistance:

$$R_T = R_1 + R_2 + R_3$$
$$R_T = 22 + 82 + 56$$
$$R_T = 160 \text{ k}\Omega$$

Total voltage by Ohm's Law:

$$V_T = I_T R_T$$
$$V_T = (24.4 \times 10^{-6})(160 \times 10^3)$$
$$V_T = 3904 \times 10^{-3}$$
$$V_T = 3.90 \text{ V}$$

Example D
Determine the resistance value of R_2 in Figure 6–3.

Total current:

$$I_T = \frac{V_1}{R_1}$$
$$I_T = \frac{5}{(33 \times 10^3)}$$
$$I_T = 0.1515 \times 10^{-3}$$
$$I_T = 152 \text{ μA}$$

Voltage across R_3:

$$V_3 = I_T R_3$$
$$V_3 = (152 \times 10^{-6})(15 \times 10^3)$$
$$V_3 = 2280 \times 10^{-3}$$
$$V_3 = 2.28 \text{ V}$$

Voltage across R_2:

$$V_T = V_1 + V_2 + V_3$$
$$22 = 5 + V_2 + 2.28$$
$$V_2 = 22 - 7.28$$
$$V_2 = 14.7 \text{ V}$$

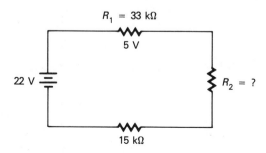

FIGURE 6–3   Circuit for Example D

Value of R_2:

$$R_2 = \frac{V_2}{I_T}$$

$$R_2 = \frac{14.7}{(152 \times 10^{-6})}$$

$$R_T = 0.967 \times 10^6$$

$$R_2 = 967 \text{ k}\Omega$$

6-7 POWER DISSIPATION.
Power is dissipated only in the form of heat and is measured in watts. Three equations can be used to determine the wattage dissipated by a resistive component.

$$P = IV$$

where P is power dissipated, in watts (W)
I is current through, in amperes (A)
V is voltage across the resistor, volts (V)

$$P = I^2 R$$

where R is the resistance of the resistor, in ohms (Ω)

$$P = \frac{V^2}{R}$$

The equations can be applied either to each resistor in the circuit or to the total values of the circuit. The total power is also the sum of the power dissipated by each component.

$$P_T = P_1 + P_2 + P_3 + \text{etc.}$$

EXERCISE 6-5
1. Three resistors are in series across a supply of 18 V: $R_1 = 39$ kΩ, $R_2 = 47$ kΩ, and $R_3 = 12$ kΩ. What is the voltage across each resistor?

    [Handwritten work: 39K 47K 12K, 18v, $R_T = 98K$, $I_T = 183.67\mu A = 184 \mu A$, $V_1 = 39K \times 183.67 \mu A$ / $184 \mu A$]

    $V_1 = $ __7.14 V__
    $V_2 = $ __8.658 V__   8.648
    $V_3 = $ __2.21 V__

2. In the circuit in Exercise 1, if the voltage across R_1 is 750 mV, what is the applied voltage?

    $V_T = $ __1.88 V__

[Handwritten: $I_T = \frac{750 mV}{39K} = 19.2 \mu A$]

Chapter 6
Metric Units with Applications

3. Two resistors are connected in series across a 6 V supply: $R_1 = 3.3$ kΩ, and $R_2 = 5.6$ kΩ. What is the power dissipated by each resistor? What is the total power dissipated by the circuit?

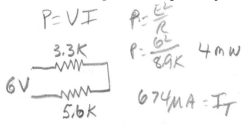

$P = VI$ $P = \dfrac{E^2}{R}$

$P = \dfrac{6^2}{8.9K}$ 4 mW

674 μA = I_T

$P_1 = $ __1.5 mW__

$P_2 = $ __2.5 mW__

$P_T = $ __4 mW__

4. Using the components in Exercise 3, if the power dissipated by R_1 is 30 mW, what is the power dissipated by R_2 and the voltage applied to the circuit?

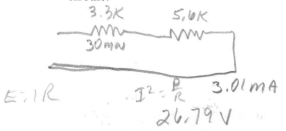

$E = IR$ $I^2 = \dfrac{P}{R}$ 3.01 mA

26.79 V

$P_2 = $ __50.7 mW__

$V_T = $ __26.79 V__

5. Three resistors are connected in series across a 250 mV source. The voltage across R_1 is 50 mV, and its resistive value is 1.5 kΩ. The value of R_2 is 680 Ω. What is the value of R_3?

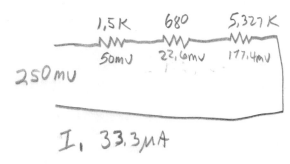

250 mV

I_1 33.3 μA

$R_3 = $ __5.32 KΩ__

6. Two resistors are connected in series across a 3 V source. The circuit dissipates 120 mW. If R_1 has a resistance value of 15 Ω, what is the value of R_2?

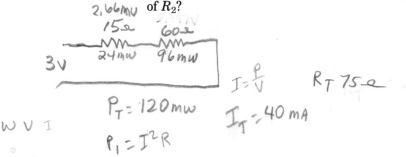

W V I

$P_T = 120$ mW $I = \dfrac{P}{V}$ R_T 75 Ω

$P_1 = I^2 R$ $I_T = 40$ mA

$R_2 = $ __60 Ω__

EVALUATION EXERCISE Convert the units to the indicated prefixes by affixing the correct power-of-10 notation; then place the final answer in scientific notation.

77

Chapter 6
Metric Units with Applications

1. 1250 μA to mA $1250 \times 10^{-6+3}$ mA 1.25×10^{-3} mA

2. 280 cm to mm 280×10^{-2} mm 2.8×10^{0} mm

3. 0.0035 mV to μV $.0035 \times 10^{-3}$ μV 3.5×10^{-6} μV

4. 680,000 Ω to kΩ $680,000 \times 10^{0}$ kΩ 680×10^{3} kΩ

5. 8620 mm to m 8620×10^{-3} m 8.62×10^{0} m

6. 2.2 MΩ to Ω 2.2×10^{6} Ω $2,200,000 \times 10^{0}$ Ω
   2.2×10^{6}

7. Three resistors are connected in series across a 12 V source: R_1 = 470 kΩ, R_2 = 150 kΩ, R_3 = 330 kΩ. What is the total current in the circuit? (Leave the answer in preferred prefix.)

   $12.6 mA = I_T$

   470K 150K 330K
   12V

   $I_T = \dfrac{12}{R_T} = \dfrac{12}{470K + 150K + 330K}$

   $\dfrac{12}{950K}$

   $I_T = 12.6 mA$

8. Three resistances in series are connected across a source: R_1 = 180 kΩ, R_2 = 1.2 MΩ, R_3 = 390 kΩ. What is the source voltage if the total current is 25 μA? (Leave the answer in preferred prefix.)

   180K 1.2m 390K

   $I_T = 25 \mu A$ $R_T = 1.77 M\Omega$

   $V = IR$
   $V = (25\mu A)(1.77 M\Omega)$

   $V = 44.25 V$

9. Referring to the circuit in Exercise 8, what would be the voltage across R_1 if the voltage across R_2 were 150 mV? (Leave the answer in preferred prefix.)

180K 1.2M 390K

150mV

$I_T = .125 mA$

$V_{R_1} = IR$

V_{R_1} ___22.5 mV___

10. Referring to the circuit in Exercise 7, what would be the power dissipated by R_3 if the current were 85 mA? (Leave the answer in preferred prefix.)

$P = VA$

330K
470K
150K
950K = R_T

$R_3 = 330K\Omega$
$I = 85 mA$
$P = I^2 R$

___2384 W___

11. In Exercise 8, what would be the power dissipated by R_1 if the power dissipated by R_2 were 750 μW? (Leave the answer in preferred prefix.)

$R_2 = 1.2M\Omega$ $R_1 = 180K$

750μW $I_T = 25mA$ $P = I^2 R$

$P = I^2 R$ $P = 25mA^2 \cdot 180K$

$I^2 = \frac{P}{R} = \frac{750\mu W}{1.2M}$

___1125 μW___

12. In Exercise 7, what would be the power dissipated by R_2 if the total power dissipated were 9.5 mW? (Leave the answer in preferred prefix.)

$P = \frac{E^2}{R}$ $P_T = 9.5 mW$

$E^2 = PR$ $P_{R_2} = 1.5 mW$

$E^2 = 9.5 mW \times 950K$

(1.71 mA) $E = 95 V$

$I_T = 71 mA$

___1.5 mW___

$P = VA$ $R_2 = 150K$

$E_{R_2} = 15V$

P_{R_2} 1.5 mW

CHAPTER 7

Equations

Objectives

After completing this chapter you will be able to:
- Balance equations using basic axioms
- Solve equations by transposing terms
- Solve equations that contain signs of grouping
- Solve equations by transposing factors
- Solve literal equations used in electronics applications

Algebra is used to solve equations. The study of electronics requires the manipulation and understanding of a variety of equations. Equations such as $V = IR$ and $X_c = \dfrac{1}{2\pi fC}$ are common. An equation is not just a tool used to find a mathematical solution to a problem; it also states the theory of an operation. The equation $V = IR$ can be used to calculate voltage, but it also states that the voltage is directly proportional to the current and resistance. One should be able to develop theory from an equation and to develop an equation from theory.

7–1 THE EQUATION. An equation is the equality of two parts. The left side of the equality is called the "left member"; the right side is called the "right member."

$$\text{left member} = \text{right member}$$

The important thing to remember is, always keep the equation equal or balanced. A balance scale is often used to illustrate this principle, as shown in Figure 7–1. Illustrations can be made for subtraction, multiplication, and division. Four basic rules, called axioms, can be stated about equations. An axiom is a truth that is self-evident and needs no further explanation.

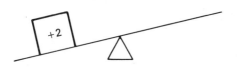

Adding something on the left side unbalances the scale in favor of the left side.

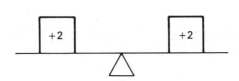

Adding the same amount to the right side keeps the scale balanced.

FIGURE 7–1 Balancing equations

1. Equal quantities can be added to both members of an equation and the equation will remain balanced.

2. Equal quantities can be subtracted from both members of an equation and the equation will remain balanced.

3. Both members of an equation can be multiplied by equal quantities and the equation will remain balanced.

4. Both members of an equation can be divided by equal quantities and the equation will remain balanced.

7-2 SOLVING AN EQUATION OF THE TYPE $I + 4 = 6$. The numbers I, 4, and 6 are all terms of the expression. Terms are handled by the operations of addition and subtraction.

To solve $I + 4 = 6$, add -4 to both sides of the equation.

$$\begin{aligned} I + 4 &= 6 \\ -4 & -4 \\ \hline I + 0 &= 2 \\ I &= 2 \end{aligned}$$

7-3 TO SOLVE TYPE $3I = 6$. Three (3) is a factor of $3I$. Factors are handled by division or multiplication. In this type, divide both sides of the equation by 3 and cancel 3 on the left side.

$$\frac{\cancel{3}I}{\cancel{3}} = \frac{6}{3}$$

$$I = 2$$

7-4 TO SOLVE TYPE $\dfrac{I}{3} = 12$. To solve for I, remove the factor 3. Multiply each side of the equation by 3 and cancel the 3 on the left side.

$$(\cancel{3}) \frac{I}{\cancel{3}} = 12(3)$$

$$I = 36$$

7-5 SOLVING EQUATIONS BY TRANSPOSING TERMS. Sections 7-2 through 7-4 describe how equations are solved using the basic axioms.

Transposing terms is usually a more convenient method of solving equations. The four axioms are still being applied, just in a different manner.

$$B \;\boxed{+\; 6} \;=\; 10$$
$$B \;=\; 10 \;-\; 6$$

When a term is moved from one member of an equation to the other, change the sign.

$$2a + 6 = 4a + 10$$

$$2a - 4a = 10 - 6$$

Move the unknowns containing a to the left side of the equation and the constants to the right side, thus changing the algebraic signs.

$$-2a = 4$$
$$a = -2$$

Collect like terms and solve for a.

Example A

$$14 = I - 6 \qquad\qquad 2R - 8 = 3R + 12$$
$$14 + 6 = I \qquad\qquad 2R - 3R = 12 + 8$$
$$20 = I \qquad\qquad\qquad -R = 20$$
$$I = 20 \qquad\qquad\qquad R = -20$$

EXERCISE 7-1 Solve for the unknown.

1. $b + 7 = 3$ \qquad\qquad $b = \underline{\;-4\;}$

    $b = 3 - 7$
    $b = -4$

2. $4.3 - I = 6$ \qquad\qquad $I = \underline{\;-1.7\;}$

    $-I = 6 - 4.3$
    $-I = 1.7$

3. $0.5 - R = 6$ \qquad\qquad $R = \underline{\;-5.5\;}$

    $-R = 6 - .5$
    $-R = 5.5$
    $R = -5.5$

4. $a + 0.5 = 3$ \qquad\qquad $a = \underline{\;2.5\;}$

    $a = 3 - .5$
    $a = 2.5$

Chapter 7
Equations

5. $4I = 16$ $I = \underline{4}$

   $I = 4$

6. $-E + 4 = 6.5$ $E = \underline{}$

7. $3R - 8 = 14$ $R = \underline{7.33}$

   $3R = 22$
   $R = 7\frac{1}{3}$

8. $25 - 6B = 3$ $B = \underline{3\frac{2}{3}}$

   $6B = 25 - 3$
   $6B = 22$
   $B = 3\frac{2}{3}$

9. $0.54 - 0.3V = -0.75$ $V = \underline{4.3}$

   $.3V = .75 + .54$
   $.3V = 1.29$
   $V = 4.3$

10. $0.84 = 3R - 1$ $R = \underline{}$

11. $9I = 3I + 18$ $I = \underline{3}$

    $6I = 18$
    $I = 3$

12. $7a + 4 = 4a - 13$ $a = \underline{-\frac{17}{3}}$

    $7a - 4a = -13 - 4$
    $3a = -17$
    $a = -5.66$

13. $5R - 9 = 7R - 9$ $R = \underline{0}$

    $-9 + 9 = 7R - 5R$
    $0 = R$

14. $2P + 14 = 6P + 7$ $P = \underline{}$

15. $4c - 6 + 5c = (-13 - 2)$ $c = \underline{-1}$

$$9c = -15 + 6$$
$$9c = -9$$
$$c = -1$$

16. $9 + 5I = I + 8$ $I = \underline{}$

17. $9i + 25 - i = 5 + 6i + 3$ $i = \underline{-8.5}$

$$9i - i - 6i = 5 + 3 - 25$$
$$2i = -17$$
$$i =$$

18. $n - 5 + 2n + 7 = 2 - 2n - 5n$ $n = \underline{\tfrac{0}{10} = 0}$

$$n + 2n + 2n + 5n = 2 - 7 + 5$$
$$10n = 0$$
$$3n + 2 = 2 - 7n$$

19. $3e - 5 = 5e + 13$ $e = \underline{-4.5}$

$$-5 - 13 = 5e - 3e$$
$$-18 = 4e$$
$$-4\tfrac{1}{2} = e$$

20. $5E - 12 = 16 + 3E$ $E = \underline{}$

21. $4 - 3a = 13 - a$ $a = \underline{-4.5}$

$$4 - 13 = -a + 3a$$
$$-9 = 2a$$

22. $4v + 7 = 39 - 3v$ $v = \underline{}$

23. $2I - 5 + I = 4 - 6I + 2$ $I = \underline{1.22}$

$$2I + I + 6I = 4 + 2 + 5$$
$$9I = 11$$
$$I = 1.22$$

24. $V - 28 = 6V - 5 - 5V - 3 - V$ $V = \underline{}$

83

Chapter 7
Equations

84

Part 2
Mathematics for DC and AC Circuits

7–6 EQUATIONS WITH REGIME OF GROUPING. Sometimes equations contain signs of grouping, such as parentheses, brackets, or braces. First, the grouping signs are removed; then like terms are collected. Finally, the equations are solved using the conventional method.

Example B

$$3B - 2(6B + 3) = 4(B - 5)$$
$$3B - 12B - 6 = 4B - 20$$
$$-9B - 6 = 4B - 20$$
$$-9B - 4B = -20 + 6$$
$$-13B = -14$$
$$B = 1.08$$

Example C

$$x - (x + 2)(x + 3) = 14 - x^2$$
$$x - (x^2 + 5x + 6) = 14 - x^2$$
$$x - x^2 - 5x - 6 = 14 - x^2$$
$$-4x = 14 + 6$$
$$-4x = 20$$
$$x = -5$$

Example D

$$2a - [(a + 2)(a - 3)] = a(5 - a)$$
$$2a - [a^2 - a - 6] = 5a - a^2$$
$$2a - a^2 + a + 6 = 5a - a^2$$
$$3a - 5a = -6$$
$$-2a = -6$$
$$a = 3$$

EXERCISE 7–2 Solve the following for the unknown.

1. $4(I - 5) - 3(I - 2) = 2(I - 1)$

    $4I - 20 - 3I + 6 = 2I - 2$
    $I - 14 = 2I - 2$
    $-14 + 2 = 2I - I$
    $-12 = I$

    $I = \underline{-12}$

2. $3E - E(2E - 5) + 5 = 13 + 2E(5 - E)$

Chapter 7
Equations

$E = $ _____

3. $a(a - 5) + 19 = 4(2a - 3) + a^2$

$a^2 - 5a + 19 = 8a - 12 + a^2$
$\cancel{a^2} - \cancel{a^2} - 5a - 8a = -12 - 19$
$-13a = -31$
$a =$

$a = \underline{2.38}$

4. $2 + 3(r - 7) = 18 + 2(5r + 1)$

$r = \underline{1.5}$

5. $3i - (4i - 5) = 7 - (5i - 4)$

$3i - 4i + 5 = 7 - 5i + 4$
$3i - 4i + 5i = 7 + 4 - 5$
$4i = 6$
$i =$

$i = \underline{1.5}$

6. $3e - 4[-(-2e - 3)] - 5 = 7 - 3(4 - 2e) - 5e$

$3e - 4[+2e + 3] - 5 = 7 - 12 + 6e - 5e$
$3e - 8e - 12 - 5 = 7 - 12 + 6e - 5e$
$-5e - 17 = -5 + e$
$-17 + 5 = e + 5e \qquad e = \frac{-12}{6}$
$-12 = 6e$

$e = \underline{-2}$

7. $(I + 2)(I - 5) - I(3I + 2) = 2 - 2I(I + 3) + 5I$

$I^2 - 3I - 10 - 3I^2 - 2I = 2 - 2I^2 - 6I + 5I$
$-2I^2 - 5I - 10 = 2 - 2I^2 - I$
$-\cancel{2I^2} + \cancel{2I^2} - 5I + I = 10 + 2$
$-4I = 12$

$I = \underline{-3}$

8. $(R - 4)(R - 1) = 6 + (R - 5)(R + 2) - 4R$

$R = $ _____

9. $3v(5 - v) - 2v(3 - v) = -[v(v + 3)]$

$15v - 3v^2 - 6v + 2v^2 = -v^2 - 3v$
$15v - 3v^2 - 6v + 2v^2 + v^2 + 3v = 0$
$18 - 6 \quad 12v = 0$

$v = \underline{\quad 0 \quad}$

10. $6p - [(p - 2)(p - 3)] = 9 - [(p - 4)(p - 5)]$

$p = \underline{\qquad}$

11. $a - (2a - 1)(3a + 2) = 5 - 2(a - 3)(3a + 1)$

$a - (6a^2 + a - 2) = 5 - 2(3a^2 - 8a - 3)$
$a - 6a^2 - a + 2 = 5 - 6a^2 + 16a + 6$
$-16a = 5 + 6 - 2$
$-16a = 9$

$a = \underline{\; -.5625 \;}$

12. $4i^2 + (i - 4)(i + 5) = 32 + 2(i - 4) + (5i + 3)(i - 4)$

$i = \underline{\qquad}$

7–7 SOLVING EQUATIONS BY TRANSPOSING FACTORS.

Solve

$2N = 6$
$N = \dfrac{6}{2}$

$\dfrac{N}{D} \times \dfrac{N}{D}$

Factors can be transposed from the numerator to the denominator of opposite sides of the equation. The sign of the factor is not changed.

$8 = 4I \qquad 16 = \dfrac{a}{2} \qquad \dfrac{4}{a} = 12$

$\dfrac{8}{4} = I \qquad 16(2) = a \qquad \dfrac{4}{12} = a$

$I = 2 \qquad a = 32 \qquad a = \dfrac{1}{3}$

7–8 SOLVING EQUATIONS WITH LITERAL NUMBERS.
In electronics a variety of equations are used. It becomes necessary to solve literal equations. In Ohm's Law, $V = IR$ and is often given as $I = \dfrac{V}{R}$ or $R = \dfrac{V}{I}$. These are examples of solving literal equations. The methods are the same as solving the equations in previous sections, but the results are also literal equations. They are solved by transposing terms and factors or by applying the axioms (as covered in Sections 7–1 and 7–7).

Solve for n: transpose term b and change the sign: $-b$. (See Section 7–4.)

$$n + b = c$$
$$n = c - b$$

Solve for n: transpose factor b, but do not change the sign. (See Section 7–7.)

$$\frac{n}{b} = c$$
$$n = cb$$

Solve the equation for R.

$$T = \frac{R^2}{A} \qquad \text{(transpose factor } A\text{)}$$

$$\sqrt{TA} = \sqrt{R^2} \qquad \text{(extract the square root of both sides of the equation)}$$

$$R = \sqrt{TA}$$

Example E
Solve the equation for R_1.

$$R_T = R_1 + R_2$$
$$R_T - R_2 = R_1$$
$$R_1 = R_T - R_2$$

Solve the equation for A.

$$R = \frac{L}{\mu A}$$

$$A = \frac{L}{\mu R}$$

Solve the equation for I.

$$P = I^2 R$$
$$I^2 = \frac{P}{R}$$
$$\sqrt{I^2} = \sqrt{\frac{P}{R}}$$
$$I = \sqrt{\frac{P}{R}}$$

Solve the equation for V.

$$H = \frac{0.5V}{DE}$$
$$HDE = 0.5V$$
$$\frac{HDE}{0.5} = V$$
$$V = \frac{HDE}{0.5}$$

EXERCISE 7-3 Solve the literal equations for the unknown indicated.

all

1. $Z = \dfrac{V}{I}$ $V = ZI$ $V = \underline{ZI}$

2. $Z = \dfrac{V}{I}$ $IZ = V$ $I = \underline{V/Z}$
   $I = \dfrac{V}{Z}$

3. $Q = \dfrac{X_L}{R}$ $RQ = X_L$ $X_L = \underline{QR}$

4. $Q = \dfrac{X_L}{R}$ $\dfrac{QR}{Q} = \dfrac{X_L}{Q}$ $R = \underline{\dfrac{X_L}{Q}}$

5. $X_L = 2\pi f L$ $\dfrac{X_L}{2\pi L} = f$ $f = \underline{\dfrac{X_L}{2\pi L}}$

6. $X_L = 2\pi f L$ $L = \underline{\dfrac{X_L}{2\pi f}}$

7. $P = \dfrac{V^2}{R}$ $RP = V^2$ $R = \underline{\dfrac{V^2}{P}}$
   $R = \dfrac{V^2}{P}$

8. $P = \dfrac{V^2}{R}$ $V^2 = RP$ $V = \underline{\sqrt{RP}}$
   $V = \sqrt{RP}$

9. $\mu = g_m r_p$ $g_m = \dfrac{\mu}{r_p}$ $g_m = \underline{\dfrac{\mu}{r_p}}$

10. $\mu = g_m r_p$ $r_p = \underline{\dfrac{\mu}{g_m}}$

11. $\beta = \dfrac{I_c}{I_b}$ $I_c = \beta I_b$ $I_c = \underline{\beta I_b}$

12. $\beta = \dfrac{I_c}{I_b}$ $I_b = \dfrac{I_c}{\beta}$ $I_b = \underline{\dfrac{I_c}{\beta}}$

13. $A_v = \dfrac{A_o}{A_i}$ $A_o = \underline{A_v A_i}$

14. $A_v = \dfrac{A_o}{A_i}$ $A_i = \underline{\dfrac{A_o}{A_v}}$

Chapter 7 Equations

15. $X_c = \dfrac{1}{2\pi fC}$ $\qquad f = \dfrac{1}{2\pi C X_c}$

16. $X_c = \dfrac{1}{2\pi fC}$ $\qquad C = \dfrac{1}{2\pi f X_c}$

17. $V_c = V_{cc} - i_c R_L$ $\qquad V_{cc} = V_c + i_c R_L$

18. $V_c = V_{cc} - i_c R_L$ $\qquad i_c = \dfrac{V_{cc} - V_c}{R_L}$

19. $V_{\text{rms}} = 0.707\, V_m$ $\qquad V_m = \dfrac{V_{\text{rms}}}{.707}$

20. $V_m = \dfrac{V_{p-p}}{2}$ $\qquad V_{p-p} = 2 V_m$

21. $T = \dfrac{1}{f}$ $\qquad f = \dfrac{1}{T}$

22. $T_c = RC$ $\qquad C = \dfrac{T_c}{R}$

23. $V_{\text{ave}} = 0.636\, V_m$ $\qquad V_m = \dfrac{V_{\text{ave}}}{.636}$

24. $V_{\text{ave}} = \dfrac{0.636\, V_{p-p}}{2}$ $\qquad V_{p-p} = \dfrac{2 V_{\text{ave}}}{.636}$

$.636\, V_{p-p} = 2 V_{\text{ave}}$

25. $N_s = \dfrac{E_s N_p}{E_p}$ $\quad E_s N_p = N_s E_p \qquad E_s = \dfrac{N_s E_p}{N_p}$

26. $N_s = \dfrac{E_s N_p}{E_p}$ $\quad E_s N_p = N_s E_p \qquad N_p = \dfrac{N_s E_p}{E_s}$

27. $R = \dfrac{E - e}{I}$ $\quad E - e = RI \qquad E = RI + e$
$\qquad\qquad\qquad E = RI + e$

28. $R = \dfrac{E - e}{I}$ $\qquad e = E - RI$
$\qquad -e = RI - E$
$\qquad e = E - RI$

EVALUATION EXERCISE all

1. $2a + 14 = 8a + 7$ $a = \underline{1.166}$

    $14 - 7 = 8a - 2a$
    $6a = 7$
    $a =$

2. $e - 5 + 2e + 7 = 4 - 4e + 5$ $e = \underline{1}$

    $e + 2e + 4e = 4 + 5 + 5 - 7$
    $7e = 7$
    $e = 1$

3. $3R - 10 + R = 8 - 12R + 4$ $R = \underline{1.375}$

    $3R + R + 12R = 8 + 4 + 10$
    $16R = 22$
    $R =$

4. $E(E - 5) + 19 = 4(3E - 3) + E^2$ $E = \underline{1.823}$

    $E^2 - 5E + 19 = 12E - 12 + E^2$
    $E^2 - 5E - 12E - E^2 = -12 - 19$
    $-17E = -31$
    $E =$

5. $4 + 6(r - 3) = 14 + 2(5r + 1)$ $r = \underline{-6.5}$

    $4 + 6r - 18 = 14 + 10r + 2$
    $6r - 10r = 14 + 2 + 18 - 4$
    $-4r = 26$
    $r =$

6. $(a - 4)(a - 1) = 12 + (a - 5)(a + 2) - 4a$ $a = \underline{-1}$

    $a^2 - 5a + 4 = 12 + a^2 - 3a - 10 - 4a$
    $a^2 - 5a - a^2 + 3a + 4a = 12 - 10 - 4$
    $2a = -2$

7. $e - (2e - 1)(3e + 2) = 10 - 2(e - 3)(3e + 1)$ $e = \underline{-.875}$

    $e - [6e^2 + e - 2] = 10 - 2[3e^2 - 8e - 3]$
    $e - 6e^2 - e + 2 = 10 - 6e^2 + 16e + 6$
    $e - 6e^2 - e + 6e^2 - 16e = 10 + 6 - 2$
    $-16e = 14$

8. $F = 1.8C + 32$ $C = \underline{\dfrac{F-32}{1.8}}$

    $1.8C = F - 32$
    $C = \dfrac{F - 32}{1.8}$

    $1.8 = \dfrac{9}{5}$

9. $i_p = \dfrac{\mu e_g}{r_p + r_b}$ $\qquad r_p = \dfrac{\mu e_g}{i_p} - r_b$

10. $a = \dfrac{R_t + R_o}{R_a}$ $\qquad R_o = a R_a - R_t$

$$i_p = \dfrac{\mu e_g}{r_p + r_b}$$

$$i_p(r_p + r_b) = \mu e_g$$
$$i_p r_p + i_p r_b = \mu e_g$$
$$i_p r_p = \mu e_g - i_p r_b$$
$$r_p = \dfrac{\mu e_g - i_p r_b}{i_p}$$
$$r_p = \dfrac{\mu e_g}{i_p} - r_b$$

$$a = \dfrac{R_t + R_o}{R_a}$$
$$a R_a = R_t + R_o$$
$$a R_a - R_t = R_o$$

CHAPTER 8
Fractions

Objectives

After completing this chapter you will be able to:
- Identify the properties of the numerator and denominator in algebraic fractions
- Reduce algebraic fractions to their lowest terms
- Identify and change the algebraic signs of fractions
- Add and subtract algebraic fractions
- Multiply algebraic fractions
- Divide algebraic fractions

Algebraic fractions are like numerical fractions and can be treated the same as they are in arithmetic. The fraction contains a numerator and a denominator:

$$\frac{\text{numerator}}{\text{denominator}}$$

8–1 RULES FOR FRACTIONS. The numerator can be multiplied or divided by a quantity as long as the denominator is multiplied or divided by the same quantity; then the value of the fraction will not change.

$$\frac{2(4)}{3(4)} = \frac{8}{12}$$

$$\frac{a(a)}{b(a)} = \frac{a^2}{ab}$$

$$\frac{5 \div 5}{15 \div 5} = \frac{1}{3}$$

$$\frac{B^2 \div B}{Bc \div B} = \frac{B}{c}$$

EXERCISE 8–1 Find the missing term.

1. $\dfrac{3}{4} = \dfrac{18}{24}$

2. $\dfrac{5}{12} = \dfrac{30}{72}$

3. $\dfrac{a}{b} = \dfrac{ab^2}{b^3}$

4. $\dfrac{6I^2}{2I^2 V} = \dfrac{12I^4}{2I^2 V}$

5. $\dfrac{2i}{i-4} = \dfrac{2i^2+8i}{i^2-16}$ *i+4*

6. $\dfrac{I^2}{R} = \dfrac{I^2R}{R^2}$

7. $\dfrac{R-1}{R+2} = \dfrac{R^2-R}{R^2+2R}$ *R*

8. $\dfrac{v+2}{v+3} = \dfrac{v^2+4v+4}{v^2+5v+6}$ *v+2*

9. $\dfrac{e-3}{e-5} = \dfrac{e^2-8e+15}{e^2-10e+25}$ *e−5*

10. $\dfrac{L^3}{L+7} = \dfrac{L^4-7}{L^2-49}$ *L−7*

8–2 REDUCING FRACTIONS TO THEIR LOWEST TERMS. To reduce a fraction to its lowest terms divide the numerator and denominator by the highest quantity that divides into both evenly. In some fractions this quantity can be identified by inspection. If it can't, then factor both numerator and denominator into prime factors and cancel out like factors.

Example A

$$\dfrac{9}{12} = \dfrac{(3)(3)}{(4)(3)} = \dfrac{3}{4}$$

$$\dfrac{a^2}{ab} = \dfrac{(a)(a)}{(a)(b)} = \dfrac{a}{b}$$

$$\dfrac{B^2-1}{B-1} = \dfrac{(B+1)(B-1)}{(B-1)} = B+1$$

$$\dfrac{a^2+4a+4}{a+2} = \dfrac{(a+2)(a+2)}{(a+2)} = a+2$$

EXERCISE 8–2 Reduce the fraction to its lowest terms by dividing the numerator and denominator by the same quantity or canceling like factors.

1. $\dfrac{5}{25} =$ ___ $1/5$ ___

2. $\dfrac{64}{128} =$ ___ $8/16 = 1/2$ ___

3. $\dfrac{R^2}{R^3} =$ ___ $1/R$ ___

4. $\dfrac{6i}{15i^2} =$ ___ $2/5i$ ___

5. $\dfrac{E^2I}{I^2E} =$ ___ E/I ___

6. $\dfrac{v+1}{v^2+v} =$ ___ $1/v$ ___   *v(v+1)*

7. $\dfrac{(a+b)^2}{a^2-b^2} =$ ___ $\dfrac{a+b}{a-b}$ ___   *a+b a−b*

8. $\dfrac{(p+s)^3}{(p+s)^5} =$ ___ $\dfrac{1}{(p+s)^2}$ ___

9. $\dfrac{I^2(I+6)}{I^2+I} = \dfrac{I(I+6)}{I+1}$ 10. $\dfrac{a^2+2ab+b^2}{2a^2+6ab+4b^2} = \dfrac{(a+b)(a+b)}{2(a+2b)(a+b)} = \dfrac{a+b}{2(a+2b)}$

11. $\dfrac{i^2-r^2}{i^2-2ir+r^2} = \dfrac{i+r}{i-r}$ $(i-r)(i-r)$

12. $\dfrac{v^2-5v-6}{5v^2-v-6} = \dfrac{(v-6)(v+1)}{(5v-6)(v+1)}$

$4(E^2-16)$
$4(E-4)(E+4)$

13. $\dfrac{4E^2-64}{4E^2-32E+64} = \dfrac{E+4}{E-4}$
$4(E^2-8E+16)$
$4(E-4)(E-4)$

14. $\dfrac{3(i^2-1)(i-1)}{3i^2-6i+3} = \dfrac{3(i-1)(i+1)(i-1)}{3(i-1)(i-1)} = \dfrac{3(i+1)}{1}$

8–3 SIGNS OF FRACTIONS. A fraction indicates a problem in division. The fraction $\dfrac{3}{4}$ means divide 3 by 4. The following illustrates how the signs of a fraction can be changed without changing the result of the division. Three signs are associated with a fraction: the sign of the numerator, the sign of the denominator, and the sign preceding the fraction.

$$+\dfrac{+3}{+4}$$

The fraction $-\dfrac{3}{4}$ should result in -0.75 after division. Keeping that in mind, note how the signs of the fractions can be changed without changing the result, -0.75.

1. Changing the sign in front of the fraction and the numerator.

$$+\dfrac{-3}{4} = -0.75$$

2. Changing the sign in front of the fraction and the denominator.

$$+\dfrac{3}{-4} = -0.75$$

3. Changing the sign of the numerator and the denominator.

$$-\dfrac{-3}{-4} = -0.75$$

These illustrate the rule that any two signs of a fraction can be changed without changing the result of the division. This could also be stated: the signs of any two factors associated with a fraction can be changed.

Example B

$$\dfrac{a}{(a-b)} = -\dfrac{-a}{(a-b)} = -\dfrac{a}{(b-a)} = \dfrac{-a}{(b-a)}$$

$$\dfrac{ab}{(a-b)} = -\dfrac{-ab}{(a-b)} = -\dfrac{ab}{(b-a)} = \dfrac{(-a)(-b)}{(a-b)}$$

EXERCISE 8–3 Express each fraction in exercises 1 through 6 as a positive fraction by changing the sign of the numerator.

1. $-\dfrac{-4}{5} = $ _____ $4/5$

2. $-\dfrac{-I^2R}{I+R} = $ _____ $\dfrac{-I^2R}{I+R}$

3. $-\dfrac{E-R}{E^2} = $ _____ $\dfrac{-(E-R)}{E^2} = \dfrac{R-E}{E^2}$

4. $-\dfrac{-(a+b)}{a-b} = $ _____ $\dfrac{a+b}{a-b}$

5. $-\dfrac{(-a^2)(-b)}{a-b} = $ _____ $\dfrac{(-a^2b)}{a-b}$; $-a^2b$

6. $-\dfrac{ir+r+i}{i+r} = $ _____ $\dfrac{-(ir+r+i)}{i+r}$

Express each fraction in exercises 7 through 12 as a negative fraction by changing the sign of the denominator.

7. $\dfrac{-3}{5} = $ _____ $-3/5$

8. $\dfrac{ab}{a-b} = $ _____ $-\dfrac{ab}{b-a}$

9. $\dfrac{i+r}{i-r} = $ _____ $-\dfrac{i+r}{r-i}$

10. $\dfrac{(-E)(-R)}{-(E+R)} = $ _____ $-\dfrac{(-E)(-R)}{E+R}$; $-E-R$

11. $\dfrac{a}{a-b+c} = $ _____ $-\dfrac{a}{(b-c-a)}$

12. $\dfrac{I^2R-R}{I^2R+R} = $ _____

Reduce the fractions in exercises 13 through 18 to their lowest terms.

13. $\dfrac{a-b}{b-a} = $ _____ $-\dfrac{a-b}{a-b} = -1$

14. $\dfrac{E^2-1}{1-E} = $ _____

15. $-\dfrac{-I^2R}{I^2} = $ _____ $\dfrac{I^2R}{I^2} = R$

16. $\dfrac{e^2-2e+1}{1-e^2} = $ _____

17. $\dfrac{(a-b)^2}{(b-a)^2} = $ _____ 1

$\dfrac{(a-b)(a-b)}{(b-a)(b-a)}$

$(a-b)(a-b)$

18. $\dfrac{i^2-2ir+r^2}{r^2-2ir+i^2} = $ _____

8-4 ADDING AND SUBTRACTING FRACTIONS. The procedure for adding and subtracting algebraic fractions is the same as that for numerical fractions. Only like things can be added or subtracted. What makes one fraction like another is that they must have the same denominator. Determine a common denominator and add or subtract:

$$\frac{3}{4} + \frac{5}{6} = \frac{9}{12} + \frac{10}{12} = \frac{19}{12}$$

$$\frac{a}{b} + \frac{b}{d} = \frac{ad}{bd} + \frac{b^2}{bd} = \frac{ad + b^2}{bd}$$

$$\frac{3R}{i} - \frac{4I}{r} = \frac{3Rr}{ir} - \frac{4Ii}{ir} = \frac{3Rr - 4Ii}{ir}$$

When the denominators are not easily identifiable as they are in the previous examples, then:

1. Factor the denominators into their prime factors.

2. The common denominator is determined by the greatest number of times the prime appears in any one denominator.

3. The product of these primes is the least common denominator.

$$\frac{a}{a^2 - b^2} + \frac{c}{a^2 + ab}$$

$$\frac{a}{(a - b)(a + b)} + \frac{c}{a(a + b)} \quad \text{(factor denominators)}$$

$a(a - b)(a + b)$ is the least common denominator

4. Multiply the numerator of the fraction by the prime missing in its denominator.

In the fraction $\frac{a}{(a - b)(a + b)}$ the prime number a is missing in the common denominator. The numerator should then be multiplied by a and placed over the common denominator:

$$\frac{a}{(a - b)(a + b)} \frac{(a)}{(a)} = \frac{a^2}{a(a - b)(a + b)}$$

In the fraction $\frac{c}{a(a + b)}$ the missing prime is $a - b$. Multiply the numerator by $a - b$ and put it over the common denominator.

$$\frac{c}{a(a + b)} \frac{(a - b)}{(a - b)} = \frac{ac - bc}{a(a + b)(a - b)}$$

5. Combine the numerators over the common denominator:

$$\frac{a^2 + ac - bc}{a(a + b)(a - b)}$$

Example C

$$\frac{I+3}{I^2-4} + \frac{I-3}{I^2-4I+4}$$
$$(I-2)(I+2) \quad (I-2)(I-2) \qquad \text{(factor denominator)}$$

The common denominator is $(I-2)(I-2)(I+2)$

$$\frac{(I+3)(I-2) + (I-3)(I+2)}{(I-2)^2(I+2)} =$$

$$\frac{I^2+I-6+I^2-I-6}{(I-2)^2(I+2)} =$$

$$\frac{2I^2-12}{(I-2)^2(I+2)}$$

Example D

$$\frac{I^2}{I^2-1} - \frac{I+1}{I-1}$$
$$(I-1)(I+1) \quad (I-1) \qquad \text{(factor)}$$

$$\frac{I^2 \ominus [(I+1)(I+1)]}{(I-1)(I+1)} = \quad \text{← caution*}$$

$$\frac{I^2 - [I^2+2I+1]}{(I-1)(I+1)} =$$

$$\frac{I^2 - I^2 - 2I - 1}{(I-1)(I+1)} =$$

$$\frac{-2I-1}{(I-1)(I+1)}$$

*Caution must be taken when the sign of operation is to subtract. It will change the sign of terms in the numerator. (Brackets used in this example will help you remember this operation.)

Chapter 8
Fractions

EXERCISE 8-4 Perform the indicated operations, leaving the answers in the lowest form. The denominator can be left in prime factors.

1. $\dfrac{3}{16} + \dfrac{5}{8} - \dfrac{3}{4} = \dfrac{3}{16} + \dfrac{10}{16} - \dfrac{12}{16} = \dfrac{1}{16}$

2. $\dfrac{5}{12} - \dfrac{3}{8} + \dfrac{5}{9} = \dfrac{30}{72} - \dfrac{27}{72} + \dfrac{40}{72} = \dfrac{43}{72}$ (70)

3. $\dfrac{a}{b} + \dfrac{b}{c} - \dfrac{c}{D} = \dfrac{acD}{bcD} + \dfrac{b^2 D}{bcD} - \dfrac{c^2 b}{bcD} = \dfrac{acD + b^2 D - c^2 b}{bcD}$

4. $\dfrac{3a}{a^2 b} - \dfrac{2a}{ab^2} = \dfrac{3ab - 2a^2}{a^2 b^2}$

5. $\dfrac{6}{I} + \dfrac{4}{R} + \dfrac{3}{IR} = \dfrac{6R + 4I + 3}{IR}$

6. $\dfrac{8}{i-r} - \dfrac{6}{i^2 - r^2} = \dfrac{8i + 8r - 6}{(i-r)(i+r) = i^2 - r^2}$

7. $\dfrac{e-r}{e+r} - \dfrac{e+r}{e-r} = \dfrac{(e-r)(e-r) - (e+r)(e+r)}{(e+r)(e-r)} = \dfrac{-4er}{(e+r)(e-r)}$

$\dfrac{e^2 - 2er + r^2 - e^2 - 2er - r^2}{e^2 - r^2} \quad \dfrac{(e^2 - 2er + r^2) - (e^2 + 2er + r^2)}{e^2 - r^2}$

$\dfrac{-4er}{e^2 - r^2}$

8. $\dfrac{10}{a-3} + \dfrac{6}{a-2} = \dfrac{16a - 38}{(a-3)(a-2)}$

$\dfrac{10a - 20 + 6a - 18}{(a-3)(a-2)}$

9. $\dfrac{1}{R_1} + \dfrac{1}{R_2} + \dfrac{1}{R_3} = \dfrac{R_2 R_3 + R_1 R_3 + R_1 R_2}{R_1 R_2 R_3}$

★ 10. $\dfrac{R_1 R_2}{R_1 + R_2} + \dfrac{1}{R_2} = \dfrac{R_1 R_2^2 + R_1 + R_2}{R_2(R_1 + R_2)}$

$\dfrac{R_2(R_1 R_2) + R_1 + R_2}{R_2(R_1 + R_2)} = \dfrac{R_1 R_2^2 + R_1 + R_2}{R_1 R_2 + R_2^2}$

11. $\dfrac{3er}{(e-r)(e-2r)} - \dfrac{3}{r-e} + \dfrac{2}{2r-e} = \dfrac{\frac{3er-4re+e}{(e-r)(e-2r)}}{}$

$\dfrac{-3er}{(r-e)(2r-e)} - \dfrac{3}{r-e} + \dfrac{2}{2r-e} = \dfrac{-3er - 3(2r-e) + 2(r-e)}{(r-e)(2r-e)} = \dfrac{-3er - 6r + 3e + 2r - 2e}{}$

$+ \dfrac{3er - 4r + e}{(r-e)(2r-e)} = \dfrac{3er - 4r + e}{(e-r)(e-2r)}$

12. $\dfrac{6}{a^2 - b^2} + \dfrac{4a}{b-a} - \dfrac{3}{b+a} = \dfrac{6 - 4a^2 - 4ab - 3a + 3b}{(a-b)(a+b)}$

$\dfrac{-6}{(a-b)(a+b)} - \dfrac{4a}{a-b} - \dfrac{3}{a+b} = 6 - 4a(a+b) - 3(a-b) = 6 - 4a^2 - 4ab - 3a + 3b$

$\dfrac{-6}{(b-a)(b+a)} + \dfrac{4a}{b-a} - \dfrac{3}{b+a}$

13. $\dfrac{2}{I^2 - 4} + \dfrac{I}{I^2 + 4I + 4} - \dfrac{5}{I+2} = \dfrac{24 - 4I^2}{(I+2)^2(I-2)}$

$\dfrac{2}{(I-2)(I+2)} + \dfrac{I}{(I+2)^2} - \dfrac{5}{I+2} =$

$2(I+2) + I(I-2) - 5(I+2)(I-2) = 2I + 4 + I^2 - 2I - 5I^2 + 20 =$
$\qquad\qquad\qquad\qquad\qquad -5(I^2 - 4) \qquad\qquad\qquad 24 - 4I^2$

★ 14. $\dfrac{3a - 3b}{2a - 2b} - \dfrac{2a^2 - 2ab}{a^2 - 2ab + b^2} = \dfrac{\frac{a+3b}{2(b-a)}}{}$

$\dfrac{3(a-b)}{2(a-b)} - \dfrac{2a(a-b)}{(a-b)(a-b)} = \dfrac{3}{2} - \dfrac{2a}{a-b} = \dfrac{3a - 3b - 4a}{2(a-b)} = \dfrac{-a - 3b}{2(a-b)} = \dfrac{a + 3b}{2(b-a)}$

8–5 MULTIPLICATION. When multiplying fractions, multiply the numerator, then the denominator.

$$\dfrac{2}{1} \times \dfrac{2}{3} \times \dfrac{2}{5} = \dfrac{8}{15}$$

If possible, the process of cancellation should be applied first. That will help simplify the multiplication and result in the fraction being reduced to its lowest terms. Only factors can be cancelled.

$$\frac{3}{5} \times \frac{2}{3} \times \frac{5}{12}$$

$$\frac{\cancel{3}^1}{\cancel{5}_1} \times \frac{\cancel{2}^1}{\cancel{3}_1} \times \frac{\cancel{5}^1}{\cancel{12}_6} = \frac{1}{6}$$
3s cancel
5s cancel
2 and 12 divide by 2

This process, when applied to arithmetic, is also applicable to algebra.

Example E

$$\frac{a}{b} \times \frac{a}{c} \times \frac{d}{d} = \frac{a^2 d}{bcd}$$

Example F

$$\frac{\cancel{2a^2}^{1a}}{\cancel{3c}_{1(1)}} \times \frac{\cancel{c}^1}{\cancel{d}_1} \times \frac{\cancel{9ab}^{3ab}}{\cancel{10d}_5} = \frac{3a^2 b}{5d}$$

Example G

$$\frac{a^2 - b^2}{3a} \times \frac{ab^2}{a^2 - 2ab + b^2} \times \frac{a+b}{b^2} = \quad \text{(factor)}$$

$$\frac{\cancel{(a-b)}^1 (a+b)}{3a} \times \frac{a\cancel{b^2}^{(1)1}}{\cancel{(a-b)}_1 (a-b)} \times \frac{(a+b)}{\cancel{b^2}_1} = \quad \text{(cancel and multiply)}$$

$$\frac{(a+b)(a+b)}{3(a-b)} = \frac{a^2 + 2ab + b^2}{3a - 3b}$$

EXERCISE 8-5 Multiply, leaving all answers in reduced form.

1. $\dfrac{3a^2}{b} \times \dfrac{b^2}{a} = \dfrac{3a\cancel{(a)} \times \cancel{b} \cdot b}{\cancel{b} \times \cancel{a}} = 3ab$

2. $\dfrac{I^2}{IR} \times \dfrac{R^2}{I} = \underline{\quad R \quad}$

   $\dfrac{I \cdot I}{I \cdot R} \times \dfrac{R \cdot R}{I}$

3. $\dfrac{e+r}{r^2} \times \dfrac{r - r^2}{1 - r} = \dfrac{e+r}{r \cdot r} \times \dfrac{r(1-r)}{1-r} = \dfrac{e+r}{r}$

101

**Chapter 8
Fractions**

4. $\dfrac{i+I}{3I^2} \times \dfrac{I}{I^2 - i^2} = \dfrac{1}{3I(I-i)}$

$\dfrac{i+I}{3I \cdot I} \times \dfrac{I}{(I-i)(I+i)} = \dfrac{1}{3I(I-i)}$

5. $\dfrac{4a}{a+b} \times \dfrac{b^2}{a^2} \times \dfrac{(a+b)^2}{b} = \dfrac{4a}{a+b} \times \dfrac{b \cdot b}{a \cdot a} \times \dfrac{(a+b)(a+b)}{b} = \dfrac{4b(a+b)}{a}$

6. $\dfrac{2v-6}{v^2} \times \dfrac{v-2}{v^2 - 5v + 6} = \dfrac{2}{v^2}$

$\dfrac{2(v-3)}{v^2} \times \dfrac{v-2}{(v-2)(v-3)}$

$(I-i)(I-1)$

7. $\dfrac{I^2 - 2I + 1}{I} \times \dfrac{3I + 3}{I^2 - 1} = 3$

$\dfrac{I(I-2)+1}{I} \times \dfrac{3(I+1)}{(I-1)(I+1)} = \dfrac{I-2+1}{1} \times \dfrac{3}{I-1} = \dfrac{(I-1)(3)}{I-1} = 3$

8. $\dfrac{P^2 r}{16} \times \dfrac{4r}{P^2 - P} \times \dfrac{P^2 - 1}{r^3} = \dfrac{P(P^2-1)}{4r(P-1)}$

9. $\dfrac{E^2 - 12IE + 12I^2}{E - 2I} \times \dfrac{E^2 - 4I^2}{E + 2I} = E^2 - 12IE + 12I^2$

$\dfrac{E^2 - 12IE + 12I^2}{E-2I} \times \dfrac{(E-2I)(E+2I)}{E+2I}$

10. $\dfrac{(a+b)^2}{(a-b)^2} \times \dfrac{(a-b)^2}{a^2 + 2ab + b^2} = 1$

$\dfrac{(a+b)(a+b)}{(a-b)(a-b)} \times \dfrac{(a-b)(a-b)}{(a+b)(a+b)} =$

8–6 DIVISION. In the operation of division the divisor is inverted, and the problem becomes one of multiplication. The caution at this point is: *do not cancel until the divisor is inverted.*

Example H

$$\frac{a^2b}{c} \div \frac{ab^2}{c^2} =$$

$$\frac{\overset{a}{\cancel{a^2b}}\,\overset{1}{}}{\cancel{c}} \times \frac{\overset{c}{\cancel{c^2}}}{\cancel{ab^2}} = \frac{ac}{b}$$

EXERCISE 8–6 Perform the operations (answers may be left in factored form).

1. $\dfrac{I^2R}{R^2} \div \dfrac{R}{I} = \dfrac{I^2R}{R^2} \times \dfrac{I}{R} = \dfrac{I^3}{R^2}$

2. $\dfrac{I+R}{I-R} \div \dfrac{I+R}{I-R} = 1$

3. $\dfrac{E^2-1}{IR} \div \dfrac{E-1}{R^2} = \dfrac{E^2-1}{IR} \times \dfrac{R^2}{E-1} = \dfrac{R(E+1)}{I}$

$\dfrac{(E-1)(E+1)}{I \cdot R} \times \dfrac{R \cdot R}{E-1}$

4. $\dfrac{i^2-i}{i-1} \div \dfrac{i}{i^2-1} = i^2-i \text{ or } i(i-1)$

5. $P \div \dfrac{P-1}{P+1} = P \times \dfrac{P+1}{P-1} = \dfrac{P^2+P}{P-1}$

6. $\dfrac{e^2 + 2ie + i^2}{(e-1)} \div \dfrac{e^3 - e}{e-1} = \dfrac{(e+i)(e+i)}{e(e^2-1)}$

103

Chapter 8
Fractions

7. $\dfrac{e^2 + 2e - 15}{3i^2 - 2i - 8} \div \dfrac{e^2 + 10e + 25}{2i^2 - i - 6} = \dfrac{\cancel{(2i+3)(e-3)}}{\cancel{(3i+4)(e+5)}}$

$\dfrac{e^2+2e-15}{3i^2-2i-8} \times \dfrac{2i^2-i-6}{e^2+10e+25} = \dfrac{\cancel{(e+5)}(e-3)}{(3i+4)\cancel{(i-2)}} \times \dfrac{(2i+3)\cancel{(i-2)}}{\cancel{(e+5)}(e+5)} = \dfrac{(2i+3)(e-3)}{(3i+4)(e+5)}$

8. $\dfrac{ab(b+1)^2}{(b-1)^2} \div \dfrac{a^2b^2(b+1)}{b^3(b^2-1)} = \dfrac{b(b+1)(b+1)}{a(b-1)}$

EVALUATION EXERCISE Reduce the fractions in exercises 1 through 8 to their lowest terms.

1. $\dfrac{a^2b}{aB} = \dfrac{a \cdot a \cdot b}{aB} = \dfrac{ab}{B}$

2. $\dfrac{(eH)^2}{e^2 - 1} = \dfrac{e^2H^2}{(e-1)(e+1)}$ ✗

3. $\dfrac{2e^2 + 4ei + 2i^2}{e^2 + 2ei + i^2} = \dfrac{2(e^2+2ei+i^2)}{e^2+2ei+i^2} = 2$

4. $\dfrac{R^3 - R}{R^2 + R} = \dfrac{R(R^2-1)}{R(R+1)} = \dfrac{R^2-1}{R+1} = \dfrac{(R-1)(R+1)}{R+1} = R-1$

Part 2 — Mathematics for DC and AC Circuits

5. $\dfrac{5(a^2 - 1)(a - 1)}{5a^2 - 10a + 5} = \dfrac{5(a^2-1)(a-1)}{5(a^2-2a+1)} = \dfrac{5(a-1)(a+1)(a-1)}{5(a-1)(a-1)} = a+1$

6. $\dfrac{(R - r)^2}{(r - R)^2} = \dfrac{(R-r)(R-r)}{(R-r)(R-r)} = -1$

7. $\dfrac{a^2 - 2a + 1}{1 - a^2} = \dfrac{(a-1)^2}{1-a^2} = \dfrac{(a-1)(a-1)}{(1-a)(1+a)} = \dfrac{-(1-a)(1-a)}{(1-a)(1+a)} = \dfrac{a-1}{1+a} = \dfrac{1-a}{1+a}$

8. $\dfrac{E - e}{e - E} = \dfrac{E-e}{E-e} = -1$

Perform the operation indicated in exercises 9 through 18, leaving the answers in their lowest form.

9. $\dfrac{2e}{e^2 i^2} - \dfrac{3e}{i^2 e^2} = \dfrac{-e}{e^2 i^2}$

10. $\dfrac{5}{a} + \dfrac{7}{b} + \dfrac{3}{ab} = \dfrac{5b + 7a + 3}{ab}$

11. $\dfrac{a - b}{a + b} - \dfrac{a + b}{a - b} = \dfrac{(a-b)^2 - (a+b)^2}{(a+b)(a-b)} = \dfrac{a^2-2ab+b^2 - a^2-2ab-b^2}{(a+b)(a-b)} = \dfrac{-4ab}{(a+b)(a-b)}$

12. $\dfrac{2}{R^2 - 4} + \dfrac{R}{R^2 + 4R + 4} - \dfrac{5}{R + 2} = \dfrac{24 - 4R^2}{(R+2)^2(R-2)}$

$\dfrac{2}{(R-2)(R+2)} + \dfrac{R}{(R+2)^2} - \dfrac{5}{R+2}$

13. $\dfrac{4}{E^2 - e^2} + \dfrac{6}{e - E} - \dfrac{3}{e + E} = \dfrac{4 - 9E - 3e}{(E-e)(E+e)}$ $\boxed{4 - 3E - 3e}$ ✗

$\underline{4}$

14. $\dfrac{5c^2}{d} \times \dfrac{d^2}{c} = \dfrac{5 \cdot \cancel{c} \cdot c}{\cancel{d}} \times \dfrac{\cancel{d} \cdot d}{\cancel{c}} = 5cd$

15. $\dfrac{a+b}{b^2} \times \dfrac{b - b^2}{1 - b} = \dfrac{a+b}{b \cdot b} \times \dfrac{b(1-b)}{1-b} = \dfrac{a+b}{b}$

16. $\dfrac{(e+i)^2}{(e-i)^2} \times \dfrac{(e+1)^2}{e^2 + 2ei + i^2} = \dfrac{(e+i)^2}{(e-i)^2} \times \dfrac{(e+1)^2}{(e+i)^2} = \dfrac{(e+1)^2}{(e-i)^2}$

17. $\dfrac{a^2 - 1}{ab} \div \dfrac{a-1}{b^2} = \dfrac{(a-1)(a+1)}{(a)(b)} \times \dfrac{(b)(b)}{a-1} = \dfrac{b(a+1)}{a}$

18. $\dfrac{c^2 + 2c - 15}{3d^2 - 2d - 8} \div \dfrac{c^2 + 10c + 25}{2d^2 - d - 6} = \dfrac{(2d+3)(c-3)}{(3d+4)(c+5)}$

$\dfrac{c^2 + 2c - 15}{3d^2 - 2d - 8} \times \dfrac{2d^2 - d - 6}{c^2 + 10c + 25} = \dfrac{(c+5)(c-3)}{(3d+4)(d-2)} \times \dfrac{(2d+3)(d-2)}{(c+5)(c+5)}$

CHAPTER 9

Fractional Equations

Objectives

After completing this chapter you will be able to:
- Solve fractional equations with numerical denominators
- Solve fractional equations with unknowns in the denominators
- Solve literal fractional equations
- Apply fractional equations to parallel dc circuits
- Apply fractional equations to series-parallel dc circuits

Fractional equations such as those used in parallel circuits

$$\frac{1}{R_T} = \frac{1}{R_1} + \frac{1}{R_2} + \frac{1}{R_3}$$

are often encountered in electronics.

9-1 FRACTIONAL EQUATIONS WITH NUMERICAL DENOMINATORS.

To solve fractional equations first clear the denominator, then solve the equation using conventional methods.

Solve $\quad \dfrac{a}{2} + \dfrac{5a}{4} = \dfrac{7}{12}$

1. First determine the least common denominator: in this problem the common denominator is 12.
2. Multiply each term by the least common denominator.
3. This cancels the denominator, simplifying the equation.

$$\frac{(12)a}{2} + \frac{(12)5a}{4} = \frac{(12)7}{12} \quad \text{(multiply by 12)}$$

$$\frac{(\cancel{12})a}{\cancel{2}}^{6} + \frac{(\cancel{12})5a}{\cancel{4}}^{3} = \frac{(\cancel{12})7}{\cancel{12}}^{1} \quad \text{(cancel)}$$

$$6a + 15a = 7 \quad \text{(simplify)}$$

$$21a = 7 \quad \text{(solve by conventional method)}$$

$$a = \frac{7}{21} = \frac{1}{3}$$

Chapter 9
Fractional Equations

Example A

$$\frac{a+4}{3} - \frac{a-3}{9} = 5$$

$$\frac{\overset{3}{\cancel{9}}(a+4)}{\cancel{3}} \ominus \frac{\overset{1}{\cancel{9}}(a-3)}{\cancel{9}} = (9)(5)$$

caution: the sign in the numerator changes

$$3(a+4) \ominus (a-3) = (9)(5)$$
$$3a + 12 - a + 3 = 45$$
$$2a = 45 - 15$$
$$2a = 30$$
$$a = 15$$

EXERCISE 9–1 Solve for the unknown in each exercise.

1. $\dfrac{I}{8} + \dfrac{2}{6} = \dfrac{10}{12} - \dfrac{I}{16}$ $\qquad I = \underline{2.67}$

$\dfrac{I}{8} + \dfrac{I}{16} = \dfrac{10}{12} - \dfrac{2}{6}$

48 CD $\qquad 6I + 3I = 40 - 16 \qquad 9I = 24 \quad I = 2.67$

2. $\dfrac{3R}{3} - \dfrac{5R}{7} = 16$ $\qquad R = \underline{56}$

21 CD
$21R - 15R = 336 \qquad R = ?$
$6R = 336$

3. $\dfrac{1}{5} - \dfrac{i}{3} = \dfrac{i-3}{15}$ $\qquad i = \underline{1}$

$3 - 5i = i - 3 \qquad 6 = 6i$
$3 + 3 = i + 5i \qquad 1 = i$

4. $\dfrac{a-6}{3} = \dfrac{a-4}{4}$ $\qquad a = \underline{12}$

$4a - 24 = 3a - 12$
$4a - 3a = 24 - 12$
$a = 12$

5. $\dfrac{P}{3} - \dfrac{P+1}{4} = 1$ $\qquad P = \underline{15}$

$4P - 3P - 3 = 12$
$P = 15$

6. $\dfrac{3e-5}{5} - \dfrac{3-3e}{7} = \dfrac{5e+2}{2}$ $e = -1.65$

70 CD

7. $6 - \dfrac{n+1}{2} = \dfrac{2n-6}{3}$ $n = 6.43$

6 CD

8. $\dfrac{2}{3}(p+1) - \dfrac{3p+2}{6} = \dfrac{1}{2}(2-p)$ $p = 1$

6 CD

9. $\dfrac{L-15}{8} = \dfrac{13}{2} + L$ $L = -9.57$

8 CD

10. $\dfrac{4a+3}{10} - \dfrac{a+5}{5} = \dfrac{a}{3}$ $a = -5.25$

30 CD

9-2 EQUATIONS WITH UNKNOWNS IN THE DENOMINATOR. This type of equation is also solved by determining the common denominator and multiplying each term by that denominator. If each side is a simple, single fraction, this can be done by cross-multiplication.

Example B

$$\dfrac{3}{a-5} = \dfrac{4}{a+5}$$

$$\dfrac{3(a+5)\cancel{(a-5)}^{1}}{\cancel{(a-5)}_{1}} = \dfrac{4(a-5)\cancel{(a+5)}^{1}}{\cancel{(a+5)}_{1}}$$

$$3(a+5) = 4(a-5)$$

$$3a + 15 = 4a - 20$$

$$3a - 4a = -20 - 15$$

$$-a = -35$$

$$a = 35$$

Example C

$$\frac{a+4}{a+3} = \frac{a^2-6}{a^2+5a+6} \quad \text{(factor denominator)}$$

$$\frac{a+4}{a+3} = \frac{a^2-6}{(a+2)(a+3)}$$

$$\frac{(a+4)(a+2)\cancel{(a+3)}}{\cancel{(a+3)}} = \frac{(a^2-6)\cancel{(a+2)}\cancel{(a+3)}}{\cancel{(a+2)}\cancel{(a+3)}}$$

$$(a+4)(a+2) = a^2 - 6$$

$$a^2 + 6a + 8 = a^2 - 6$$

$$a^2 - a^2 + 6a = -6 - 8$$

$$6a = -14$$

$$a = -2\frac{1}{3}$$

EXERCISE 9–2   Solve for the unknown.

1. $\dfrac{6}{I} + \dfrac{5}{I} = 33$ $I = \underline{\ .333\ }$

   $6 + 5 = 33I$
   $11 = 33I$

2. $\dfrac{a-5}{3a} - \dfrac{10+a}{2a} = 8$ $a = \underline{\ -.816\ }$

   $6a$ CD $2a - 10 - 30 - 3a = 48a + 3a - 2a$
   $-40 = 49a \quad a = \dfrac{-40}{49}$

3. $\dfrac{4}{V-1} + \dfrac{5}{V-1} - \dfrac{7}{1-V} = 15$ $V = \underline{\ 2.07\ }$

   $\dfrac{4}{V-1} + \dfrac{5}{V-1} + \dfrac{7}{V-1} = 15$
   $4 + 5 + 7 = 15V - 15$ $15V = 4+5+7+15 \quad 15V = 31$

4. $\dfrac{6}{2e+8} = \dfrac{4}{10-2e}$ $e = \underline{\ 1.4\ }$

   $\dfrac{6}{2(e+4)} = \dfrac{4}{2(5-e)}$ $\dfrac{3}{e+4} = \dfrac{2}{5-e}$ $15 - 3e = 2e + 8$
   $15 - 8 = 2e + 3e$
   $7 = 5e$

5. $v - \dfrac{6}{v-1} = 24 + v$ $v = .75$

$v(v-1) - 6 = 24(v-1) + v(v-1)$
$v^2 - v - 6 = 24v - 24 + v^2 - v$
$v^2 - v - 24v - v^2 + v = -24 + 6$
$-24v = -18$
$v = .75$

6. $\dfrac{i+5}{i-5} - \dfrac{6}{5-i} = 10$ $i = 6.78$

$\dfrac{i+5}{i-5} + \dfrac{6}{i-5} = 10$

$i + 5 + 6 = 10i - 50$
$50 + 6 + 5 = 10i - i$
$61 = 9i$

7. $\dfrac{e^2 - e}{e^2 + 5e + 6} - \dfrac{e-7}{e+2} = \dfrac{5}{e+3}$ $e = 5.5$

$\dfrac{e^2-e}{(e+2)(e+3)} - \dfrac{e-7}{e+2} = \dfrac{5}{e+3}$

$e^2 - e - \dfrac{(e-7)(e+3)}{e^2-4e-21} = 5(e+2)$
$e^2 - e - e^2 + 4e + 21 = 5e + 10$

$4e - e - 5e = 10 - 21$
$-2e = -11$
$e = 5.5$

8. $\dfrac{4V - 15}{V + 3} - \dfrac{4V - 10}{V - 9} = 0$ $V = 3.11$

$(4V-15)(V-9) - (4V-10)(V+3) = 0$
$4V^2 + 2V - 30$
$4V^2 - 51V + 135 - 4V^2 - 2V + 30 = 0$
$-53V = -165$
$V =$

9. $\dfrac{i+4}{4} - \dfrac{i-4}{4+i} = \dfrac{4+i}{i-4}$ $i = 6.93$

$\dfrac{i+4}{4} - \dfrac{4-i}{i-4} = \dfrac{4+i}{i-4}$

$(i+4)(i-4) - 4(4-i) = 4(4+i)$
$i^2 - 16 - 16 + 4i = 16 + 4i$
$i^2 + 4i - 4i = 16 + 32$

$i^2 = 48$
$i = 6.93$

10. $\dfrac{3r}{r^2 - 2r + 1} = \dfrac{r+1}{2(r-1)^2} - \dfrac{3}{5(r-1)}$ $r = .355$

$(r-1)(r-1) \quad 2(r-1)(r-1)$

$10(r-1)(r-1)$
CD

$10(3r) = 5(r+1) - 6(r-1)$
$30r = 5r + 5 - 6r + 6$
$31r = 11 \quad r = .3548$

9-3 FRACTIONAL LITERAL EQUATIONS.
Many equations used in electronics are of the fractional type. It is necessary to transpose these equations from one form to another. The method used is the same as that used in Section 9-2.

Solve for C

$$X_c = \frac{1}{2\pi f C} \qquad \text{(multiply each side by } C \text{ and cancel)}$$

$$(C)X_c = \frac{1}{2\pi f \cancel{C}} \cdot \frac{(\cancel{C})}{1}$$

$$CX_c = \frac{1}{2\pi f} \qquad \text{(divide each side by } X_c \text{ and cancel)}$$

$$\frac{C\cancel{X_c}}{\cancel{X_c}} = \frac{1}{2\pi f(X_c)}$$

$$C = \frac{1}{2\pi f X_c}$$

The preceding operation can also be done by transposing factors. X_c can be moved from the numerator on the left side of the equation to the denominator on the right side; C be transposed from the denominator on the right side to the numerator on the left side.

Solve for C_T:

$$\frac{1}{C_T} = \frac{1}{C_1} + \frac{1}{C_2}$$

Take the reciprocal of both sides of the equation

$$C_T = \frac{1}{\frac{1}{C_1}} + \frac{1}{\frac{1}{C_2}} \qquad \text{or} \qquad C_T = \frac{1}{\frac{1}{C_1} + \frac{1}{C_2}}$$

Solve for C_1:

$$\frac{1}{C_T} = \frac{1}{C_1} + \frac{1}{C_2}$$

Combine the fractions over a common denominator

$$\frac{1}{C_T} = \frac{C_1 + C_2}{C_1 C_2}$$

Part 2
Mathematics for DC and AC Circuits

Take the reciprocal of both sides

$$C_T = \frac{C_1 C_2}{C_1 + C_2}$$

Cross-multiply

$$C_T(C_1 + C_2) = C_1 C_2$$
$$C_T C_1 + C_T C_2 = C_1 C_2$$

Transpose and change the sign of each term

$$C_1 C_2 - C_T C_1 = C_2 C_T$$

Factor out C_1

$$C_1(C_2 - C_T) = C_2 C_T$$

Divide both sides by $(C_2 - C_T)$

$$C_1 = \frac{C_2 C_T}{C_2 - C_T}$$

Example D
Solve the equation for a

$$L = \frac{a}{(1 + a)}$$

$$L(1 + a) = a$$
$$L + La = a$$

$$L = a - La$$
$$L = a(1 - L)$$

$$a = \frac{L}{(1 - L)}$$

Solve the equation for I

$$V = \frac{A}{I} - P$$

$$V + P = \frac{A}{I}$$

$$I = \frac{A}{V + P}$$

EXERCISE 9–3 Solve for the unknown indicated.

1. $X_c = \dfrac{1}{2\pi f C}$ $\qquad f = \dfrac{1}{2\pi X_c C}$

**Chapter 9
Fractional Equations**

2. $X_L = 2\pi f L$ $L = \dfrac{X_L}{2\pi f}$

3. $X_L = 2\pi f L$ $f = \dfrac{X_L}{2\pi L}$

4. $\dfrac{1}{R_T} = \dfrac{1}{R_1} + \dfrac{1}{R_2}$ $R_T = \dfrac{1}{\frac{1}{R_1}+\frac{1}{R_2}} = \dfrac{R_1 R_2}{R_1 + R_2}$

5. $\dfrac{1}{R_T} = \dfrac{1}{R_1} + \dfrac{1}{R_2}$ $R_1 = \dfrac{R_2 R_T}{R_2 - R_T}$

6. $F = 1.8C + 32$ $C = \dfrac{F - 32}{1.8}$

7. $V_c = V_{bb} - I_s R_s$ $I_s = \dfrac{V_{bb} - V_c}{R_s}$

8. $B = \dfrac{a}{1-a}$ $a = \dfrac{B}{1+B}$

   $B - Ba = a \qquad B = a(1-B) \qquad \dfrac{B}{1-B} = a$
   $B = a + Ba$

9. $C_T = \dfrac{C_1 C_2}{C_1 + C_2}$ $C_1 = \dfrac{C_2 C_T}{C_2 - C_T}$

10. $v_1 = L \dfrac{d_i}{d_t}$ $L = \dfrac{v_1}{\frac{di}{dt}} = \dfrac{v_1 \, dt}{di}$

11. $f_2 - f_1 = \dfrac{f_r}{Q}$ $Q = \dfrac{f_r}{f_2 - f_1}$

    $Q(f_2 - f_1) = f_r$
    $Q = \dfrac{f_r}{f_2 - f_1}$

12. $A = \dfrac{\mu R_L}{R_L + r_p}$ $r_p = \dfrac{\mu R_L - R_L}{A}$

13. $A = \dfrac{\mu R_L}{R_L + r_p}$ $\mu = \dfrac{A(R_L + r_p)}{R_L}$

14. $A = \dfrac{\mu R_L}{R_L + r_p}$ $R_L = \dfrac{A r_p}{\mu - A}$

15. $R_1 = \dfrac{V_T}{I} - R_2$ $R_2 = \dfrac{V_T}{I} - R_1$

16. $R_1 = \dfrac{V_T}{I} - R_2$ $I = \dfrac{V_T}{R_1 + R_2}$

$R_1 + R_2 = \dfrac{V_T}{I}$

$I(R_1 + R_2) = V_T$

9–4 PARALLEL RESISTIVE CIRCUITS. A parallel resistive circuit is one in which the voltage is common to all components and the current is divided inversely between the branch components. The following equations describe the conditions of the parallel resistive circuit. Figure 9–1 illustrates a parallel resistive circuit showing the physical connection and a schematic drawing.

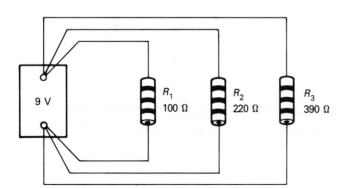
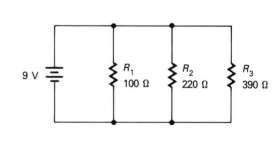

FIGURE 9–1   Parallel circuit

Total resistance:

$$R_T = \frac{1}{\frac{1}{R_1} + \frac{1}{R_2} + \frac{1}{R_3} + \text{etc.}}$$

In the case of only two resistances in parallel the equation can be transposed into the product over the sum:

$$R_T = \frac{R_1 R_2}{R_1 + R_2}$$

In the case of equal resistances, the equation can be simplified into the following:

$$R_T = \frac{R}{n} \quad \text{(where } n \text{ is any number of equal resistors)}$$

Total current:

$$I_T = I_1 + I_2 + I_3 + \text{etc.}$$

Total voltage:

$$V_T = V_1 = V_2 = V_3 = \text{etc.}$$

Parallel circuit equations and Ohm's Law can be used to solve problems related to parallel circuits. Equations used to determine power consumption are the same as those for a series circuit.

$$P = I^2 R \qquad P = \frac{V^2}{R} \qquad P = IV$$

$$P_T = P_1 + P_2 + P_3 + \text{etc.}$$

Determine the total resistance in the circuit in Figure 9–1.

$$R_T = \frac{1}{\frac{1}{R_1} + \frac{1}{R_2} + \frac{1}{R_3}}$$

$$R_T = \frac{1}{\frac{1}{100} + \frac{1}{220} + \frac{1}{390}}$$

Determine the reciprocal of each resistor. The "reciprocal" key ("1/x") on a calculator will help in this operation. Further use of the calculator memory simplifies the calculation. Consult the calculator instruction manual for this operation.

$$R_T = \frac{1}{0.01 + 0.00454 + 0.00256}$$

Determine the sum of the reciprocals:

$$R_T = \frac{1}{0.0171}$$

Determine the reciprocal of the sum:

$$R_T = 58.5 \ \Omega$$

Determine the branch currents:

$$I_1 = \frac{V}{R_1}$$
$$I_1 = \frac{9}{100}$$
$$I_1 = 90 \text{ mA}$$

$$I_2 = \frac{V}{R_2}$$
$$I_2 = \frac{9}{220}$$
$$I_2 = 40.9 \text{ mA}$$

$$I_3 = \frac{V}{R_3}$$
$$I_3 = \frac{9}{390}$$
$$I_3 = 23.1 \text{ mA}$$

Determine the total circuit current:

$$I_T = I_1 + I_2 + I_3$$
$$I_T = 90 + 40.9 + 23.1$$
$$I_T = 154 \text{ mA}$$

Total current can also be found by Ohm's Law:

$$I_T = \frac{V_T}{R_T}$$
$$I_T = \frac{9}{58.5}$$
$$I_T = 154 \text{ mA}$$

Determine the power dissipated by each component:

$$P_1 = I_1 V_1$$
$$P_1 = (0.09)(9)$$
$$P_1 = 810 \text{ mW}$$

$$P_2 = I_2 V_2$$
$$P_2 = (0.0409)(9)$$
$$P_2 = 368 \text{ mW}$$

$$P_3 = I_2 V_3$$
$$P_3 = (0.0231)(9)$$
$$P_3 = 208 \text{ mW}$$

Determine the total power dissipated by the circuit:

$$P_T = I_T V_T$$
$$P_T = (0.154)(9)$$
$$P_T = 1.39 \text{ W}$$

This can be checked by summing the individual branch power.

$$P_T = P_1 + P_2 + P_3$$
$$P_T = 810 + 368 + 208$$
$$P_T = 1386 \text{ mW} = 1.39 \text{ W}$$

Examples E and F illustrate parallel circuit problems.

Example E

Two resistors, $R_1 = 100 \ \Omega$ and $R_2 = 220 \ \Omega$, are connected in parallel, and the current through the 220 Ω resistor is 35 mA. Determine the total resistance of the circuit—the total voltage and the current through the 100 Ω resistor.

Draw the schematic and label the known values.

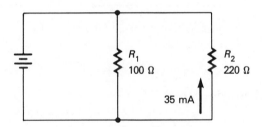

FIGURE 9-2 Circuit for Example E

Total resistance

$$R_T = \frac{R_1 R_2}{R_1 + R_2}$$

$$R_T = \frac{(100)(220)}{100 + 220}$$

$$R_T = 68.8 \ \Omega$$

Total voltage

$$V_2 = I_2 R_2$$
$$V_2 = (0.035)(220)$$
$$V_2 = 7.7 \text{ V}$$
$$V_2 = V_T = 7.7 \text{ V}$$

Current through R_1

$$I_1 = \frac{V_1}{R_1}$$
$$I_1 = \frac{7.7}{100}$$
$$I_1 = 77 \text{ mA}$$

Example F

Three 390 Ω resistors are connected in parallel; the total current in the circuit is 65 mA. How much power is dissipated by the circuit?

Draw the circuit diagram and label the known values.

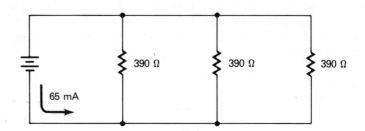

FIGURE 9–3 Circuit for Example F

Total resistance

$$R_T = \frac{R}{n}$$
$$R_T = \frac{390}{3}$$
$$R_T = 130 \text{ Ω}$$

Total power dissipated

$$P_T = I_T{}^2 R_T$$
$$P_T = (0.065)^2(130)$$
$$P_T = 549 \text{ mW}$$

EXERCISE 9–4

1. Three resistors are connected in parallel across a 300 mV source: $R_1 = 470$ Ω, $R_2 = 56$ Ω, and $R_3 = 82$ Ω. What is the total resistance of the circuit?

   31.07

   $R_T = \underline{31.1 \text{ Ω}}$

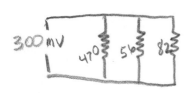

2. What are the individual branch currents in the circuit used in Exercise 1?

I_1 = __638 μA__

I_2 = __5.36 mA__

I_3 = __3.66 mA__

Chapter 9
Fractional Equations

3. What is the total parallel resistance of a 560 kΩ resistor and a 330 kΩ resistor?

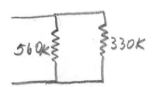

R_T = __207.6 KΩ__
        __208 KΩ__

4. In Exercise 3 the current through the 560 kΩ resistor is 850 μA. What is the power dissipated by the 330 kΩ resistor?

$E = IR$
$E = 850 μA \cdot 560K$
$E = 476 V$

$W = \dfrac{E^2}{R}$

P_{330} = __686.5 mW__

5. Two resistors are connected in parallel. The total current is 120 mA and the current through R_1 = 37 mA. What is the value of R_2 if R_1 = 1 kΩ?

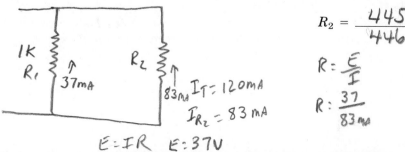

$I_T = 120 mA$
$I_{R_2} = 83 mA$
$E = IR \quad E = 37V$

R_2 = __445.7 Ω__
        __446 Ω__

$R = \dfrac{E}{I}$

$R = \dfrac{37}{83 mA}$

6. The total resistance of two resistors in parallel is 67 Ω. One resistor has a resistance of 120 Ω. What is the value of the other resistor?

$R_T = 67\,\Omega$

$R_2 = 120\,\Omega$

$R_1 = 152\,\Omega$
   151.69

R = _152 Ω_

Book Answer

7. A 3.3 MΩ, a 4.7 MΩ and a 5.6 MΩ resistor are connected in parallel. The total power dissipated by the circuit is 2.2 W. What is the total current and the applied voltage of the circuit?

$R_T = 1.44\,M\Omega$

$P_T = 2.2\,W$

$W = I^2 R$

$I^2 = \dfrac{W}{R}$

$I = \sqrt{\dfrac{W}{R}}$

$E = IR$

$I_T =$ _1.24 mA_

$V_T =$ _1786 V_
   1785.6

8. The total resistance of two resistors in parallel is 8.92 kΩ. If $R_1 = 22$ kΩ, what is the value of R_2?

$R_T = 8.92\,K$

$R_1 = 22\,K$

$R_2 = 15\,K$

$R_2 =$ _15K_

9–5 SERIES PARALLEL CIRCUITS. In the application of electronics the combination of series circuits and parallel circuits is common. The equations used to solve series circuits and those used to solve parallel circuits are applied to the series-parallel circuit. Examples G and H illustrate a simple series-parallel solution.

Example G

Determine the total resistance of the circuit given in Figure 9–4.

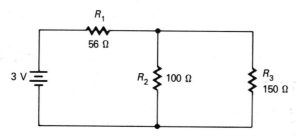

FIGURE 9-4 Circuit for Example G

$$R_T = R_1 + \frac{R_1 R_2}{R_1 + R_2}$$

$$R_T = 56 + \frac{(100)(150)}{100 + 150}$$

$$R_T = 56 + 60$$
$$R_T = 116 \ \Omega$$

Example H

Determine the voltage across the parallel branch in Figure 9-4.

$$I_T = \frac{V_T}{R_T}$$
$$I_T = \frac{3}{116}$$
$$I_T = 25.9 \text{ mA}$$

The total resistance of the 100 Ω and 150 Ω resistors is 60 Ω, as determined in Example G.

$$V_{PAR} = I_T R_{PAR}$$
$$V_{PAR} = (0.0259)(60)$$
$$V_{PAR} = 1.55 \text{ V}$$

EXERCISE 9-5

1. Determine the total resistance of the circuit shown in Figure 9-5.

   $R_T = $ _1431 Ω_

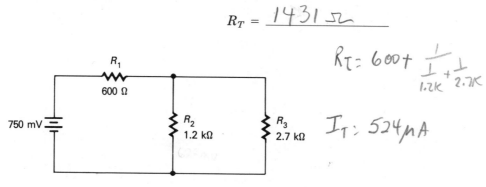

FIGURE 9-5 Circuit for Exercise 9-5 (1)

$R_T = 600 + \dfrac{1}{\frac{1}{1.2k} + \frac{1}{2.7k}}$

$I_T = 524 \mu A$

436 mV

2. Determine the voltage across the parallel branch in the circuit shown in Figure 9–5.

V_{PAR} = __435 mV__

3. Repeat Exercise 1 with $R_1 = 3.5\ \Omega$, $R_2 = 39\ \Omega$, and $R_3 = 62\ \Omega$.

R_T = __27.4 Ω__

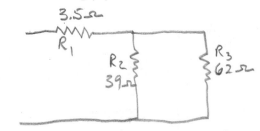

$R_T = 3.5 + \dfrac{1}{\frac{1}{39} + \frac{1}{62}}$

$3.5 + 23.9$

$I_T = 219\ mA$

4. Repeat Exercise 2 with an applied voltage of 6 V.

$R_{IV} = 766\ mV$

V_{PAR} = __5.23 V__

5. Find the total resistance of the circuit in Figure 9–6.

R_T = __21.2 Ω__

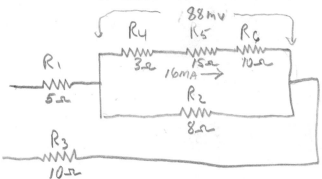

$R_T = 5 + 6.22 + 10$

$= 21.2\ \Omega$

$\left.\begin{array}{r} R_{4,5,6} = 28 \\ R_2 = 8 \end{array}\right\} R_{II} = 6.22\ \Omega$

123

Chapter 9
Fractional Equations

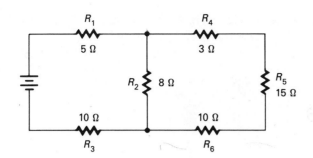

FIGURE 9-6   Circuit for Exercise 9-6 (5)

6. In the circuit shown in Figure 9-6, the current through R_5 is 16 mA. What is the current through R_2.

$$I_2 = \underline{11 \text{ mA}} \quad \boxed{56 \text{ mA}}$$

$E_{R_5} = 24 \text{ mv}$ $E_{PAR} = 88 \text{ mv}$

$E_{R_4} = 48 \text{ mv}$ $I = \dfrac{E}{R} = \dfrac{88 \text{ mv}}{8\Omega}$

$E_{R_6} = 16 \text{ mv}$

7. Determine the total resistance of the circuit in Figure 9-7.

$$R_T = \underline{28\,\Omega}$$

8. In the circuit shown in Figure 9-7, what power is dissipated by the load?

$$P_L = \underline{1.86 \text{ W}} \quad \boxed{8 \text{ W}}$$

$W = \dfrac{E^2}{R} = \dfrac{(9.64)^2}{50}$

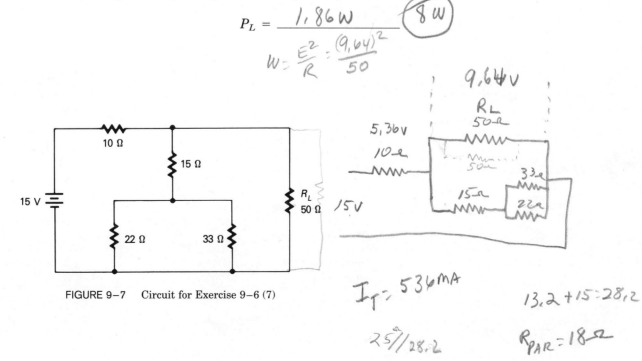

FIGURE 9-7   Circuit for Exercise 9-6 (7)

9.64 V

$I_T = 534 \text{ mA}$

$25/28.2$

$13.2 + 15 = 28.2$

$R_{PAR} = 18\,\Omega$

9. If the load in the circuit in Figure 9–7 were paralleled by a 50 Ω resistor, what would be the total circuit resistance?

$R_{PAR} = 13.3 \Omega$

$R_T = 13.3 + 10 = 23.3$

$R_T = \underline{23.3 \Omega}$

10. In Exercise 9 what is the voltage across the load?

$I = \frac{E}{R}$

$I_T = \frac{15}{23.3} = 644 \text{ mA}$

$E = .644 \times 10 = 6.44 v$

$V_L = 15 - 6.44$

$V_L = \underline{8.56 V}$

EVALUATION EXERCISE Solve for the unknown indicated.

1. $\dfrac{a}{5} - \dfrac{a}{3} = \dfrac{a-6}{15}$ $a = \underline{2}$

$3a - 5a = a - 6$
$3a - 5a - a = -6$
$-3a = -6$ $a = 2$

2. $\dfrac{3i-5}{5} - \dfrac{3-3i}{7} = \dfrac{5i+2}{5}$ $i = \underline{64}$

$7(3i-5) - 5(3-3i) = 7(5i+2)$
$21i - 35 - 15 + 15i = 35i + 14$
$21i + 15i - 35i = 14 + 15 + 35$
$i = 64$

3. $10 - \dfrac{R+1}{4} = \dfrac{2R-6}{3}$ $R = \underline{12.8}$

$120 - 3R - 3 = 8R - 24$
$120 - 3 + 24 = 8R + 3R$
$144 - 3 = 141 = 11R$

4. $\dfrac{B-10}{3B} - \dfrac{5+B}{2B} = 16$ $B = \underline{-.361}$

$2(B-10) - 3B(5+B) = 16(6B)$
$2B - 20 - 15 - 3B = 96B$... 9.9^{12}
$-35 = 96B + 3B - 2B$
$-35 = 97B$

5. $\dfrac{a+4}{3} - \dfrac{a-4}{4-a} = 18$ $a = \underline{50}$

$\dfrac{a+4}{3} + \dfrac{a-4}{a-4} = 18$

$a + 4 = 54$
$a = 50$

6. $\dfrac{3e}{e^2 - 2e + 1} = \dfrac{e+1}{2(e-1)^2}$ $e = \underline{\ .2\ }$

7. $\dfrac{1}{C_T} = \dfrac{1}{C_1} + \dfrac{1}{C_2}$ $C_1 = \underline{\ \dfrac{C_2 C_T}{C_2 - C_T}\ }$

8. $V_a = V_b - IR$ $I = \underline{\ \dfrac{V_b - V_a}{R}\ }$

9. $G = \dfrac{IR_L}{R_a - R_L}$ $R_L = \underline{\ \dfrac{G R_a}{I + G}\ }$

10. $R_b = \dfrac{V}{I} - R_a$ $I = \underline{\ \dfrac{V}{R_b + R_a}\ }$

11. Three resistances are connected in parallel: $R_1 = 4700\ \Omega$; $R_2 = 6.8\ k\Omega$; and $R_3 = 10\ k\Omega$. What is the total resistance of the circuit?

$R_T = \underline{\ 2175\ \Omega\ }$

12. The current through one branch of a three-parallel branch circuit is 12 mA and the resistance of that branch is 22 kΩ. The resistance of branch 2 is 15 kΩ and that of branch 3 is 680 Ω. What is the current through branches 2 and 3?

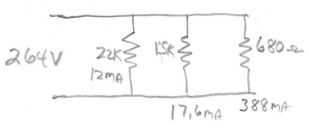

I_2 = __17.6 mA__

I_3 = __388 mA__

$R_T = 632 \Omega$
$I_T = 418 mA$

13. The total resistance of two resistors in parallel is 487 Ω. One of the resistances is 1.2 kΩ. What is the value of the other resistor?

$R_T = 487 \Omega$

$R_2 = 1.2K$

$R_1 = \dfrac{R_2 R_T}{R_2 - R_T} = \dfrac{584.4 K}{713} =$

R = __819.6__
    __820 Ω__

14. A series-parallel circuit consists of two parallel resistors, 150 Ω and 68 Ω, in series with a 22 Ω resistor across a 6.3 V source. What is the total resistance of the circuit?

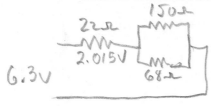

$R_{PAR} = 46.8 \Omega$
$I_T = 91.59 mA$

R_T = __68.78__
      __68.8 Ω__

15. In Exercise 14 what is the voltage across the parallel branch?

V_{PAR} = __4.28 V__

16. Two resistors are connected in parallel: $R_1 = 620$ kΩ and $R_2 = 470$ kΩ. The power dissipated by $R_1 = 25$ mW. What power is dissipated by R_2?

$P_2 = $ __32.7 mW__

[Circuit diagram: 620K resistor with 25mW in parallel with 470K resistor]

$W = I^2 R$ $W = \dfrac{E^2}{R}$ $W = \dfrac{E^2}{R}$

$I^2 = \dfrac{W}{R}$ $E^2 = WR$ $W = 32.7$ mW

$I = \sqrt{\dfrac{W}{R}}$ $E = \sqrt{WR}$

$I_{R_1} = 200$ μA $E_{R_1} = 124.49$ V $I_{R_2} = \sqrt{\dfrac{W}{R}}$
              124 V   = 263.7 μA

$R_T = 267339$

$R_T = 267.3$ kΩ

$E_T = 124$ V

$I_T = 465.6$ μA

CHAPTER 10
Right-Angle Trigonometry

Objectives

After completing this chapter you will be able to:
- Identify the standard right triangle and the impedance triangle
- Solve the impedance triangle using the Pythagorean Theorem
- Identify trigonometric ratios of sine, cosine, and tangent
- Determine the trigonometric ratios with a calculator
- Determine angles using trigonometric ratios
- Solve the impedance triangle with sine, cosine, and tangent

Right-angle trigonometry is the base used in the study of alternating current. This chapter presents the fundamentals of right-angle trigonometry; later chapters expand on the right triangle and the study of vectors and phasor algebra.

10–1 THE STANDARD RIGHT TRIANGLE. The standard right triangle used in geometry is defined as shown in Figure 10–1. The altitude (a) of the right triangle and the base (b) subtend the right angle (90°), and the hypotenuse (c) is the longest side of the triangle. In this triangle the angles are described by the same letter as the angle that is opposite the corresponding side. The side opposite the right angle is the hypotenuse and is called angle C and is in upper case. The angle opposite the altitude is angle A and the angle opposite the base is angle B.

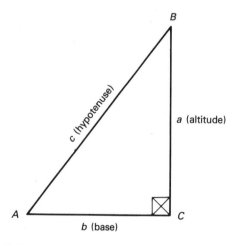

FIGURE 10–1   Standard reference right triangle

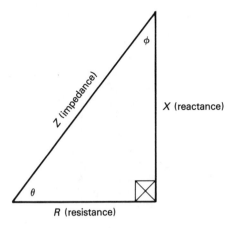

FIGURE 10-2   Impedance triangle

10-2 THE RIGHT TRIANGLE IN ELECTRONICS. In electronics the right triangle is labeled X, R, and Z, where: the altitude is side X; the base is side R; and the hypotenuse is side Z. This triangle is illustrated in Figure 10-2 and is referred to as the impedance triangle. The sides are described by the letters R, X, and Z (upper case), and the angles are theta (θ) and phi (ϕ). The angle theta is the angle between R and Z, whereas phi is between X and Z.

where X is the reactance measured in ohms
         R is the resistance measured in ohms
         Z is the impedance measured in ohms

These values represent opposition to alternating current, where X is the opposition caused by a capacitance or an inductance; R is the opposition caused by a resistance; and Z is the total opposition in an alternating-current circuit. These topics are discussed in more detail in later chapters.

10-3 PYTHAGOREAN THEOREM. Over two thousand years ago the Greek mathematician Pythagoras determined that the sum of the squares of the altitude and the base equals the square of the hypotenuse. This relation is illustrated in Figure 10-3. The triangle measures three units on the base, four units on the altitude, and five units on the hypotenuse. It can be seen that squaring the base and adding it to the square of the altitude equals the square of the hypotenuse.

$$\begin{aligned} \text{altitude} &= 4 \text{ and } 4^2 = 16 \\ \text{base} &= 3 \text{ and } 3^2 = +9 \\ \hline \text{hypotenuse} &= 5 \text{ and } 5^2 = 25 \end{aligned}$$

In a standard right triangle this is summarized: $c^2 = a^2 + b^2$; in the impedance triangle it is $Z^2 = R^2 + X^2$. Transposing the equation results in:

$$Z = \sqrt{X^2 + R^2}$$
$$R = \sqrt{Z^2 - X^2}$$
$$X = \sqrt{Z^2 - R^2}$$

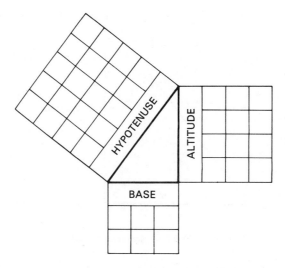

FIGURE 10-3   Pythagorean theorem

Example A
Given a right triangle with the values $X = 15 \, \Omega$ and $R = 20 \, \Omega$, find Z.

Draw and label the right triangle (Figure 10-4).

Select the correct form of the Pythagorean Theorem; substitute and solve for Z.

$$Z = \sqrt{X^2 + R^2}$$
$$Z = \sqrt{15^2 + 20^2}$$
$$Z = \sqrt{225 + 400}$$
$$Z = \sqrt{625}$$
$$Z = 25 \, \Omega$$

Most electronic calculators are programmed to work the Pythagorean relation. The calculator's instruction manual should be consulted for further explanation.

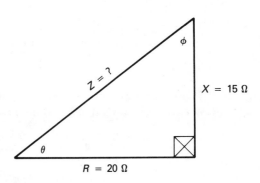

FIGURE 10-4   Triangle for Example A

EXERCISE 10-1 Using the Pythagorean Theorem, calculate the missing component.

1. $R = 22\ \Omega$ $X = 47\ \Omega$ $Z =$ __51.894__

$Z = \sqrt{X^2 + R^2}$
$R = \sqrt{Z^2 - X^2}$
$X = \sqrt{Z^2 - R^2}$

2. $X = 330\ \Omega$ $R = 520\ \Omega$ $Z =$ __615.87__

3. $Z = 10\ k\Omega$ $R = 4.7\ k\Omega$ $X =$ __8.826 kΩ__

4. $Z = 8.2\ M\Omega$ $X = 5\ M\Omega$ $R =$ __6.499 mΩ__

5. $R = 0.12\ \Omega$ $X = 0.24\ \Omega$ $Z =$ __.2683__

6. $Z = 150\ k\Omega$ $X = 47\ k\Omega$ $R =$ __142.446 kΩ__

$R = \sqrt{Z^2 - X^2}$

10-4 TRIGONOMETRIC FUNCTIONS. The sides of any right triangle form ratios that are constant for any given angle. There are six ratios, but only three are used in this text. Figure 10-5 illustrates a right triangle, defining the altitude and base as opposite and adjacent sides. Figure 10-5a shows the sides that are opposite and adjacent the angle theta (θ); Figure 10-5b shows the sides that are opposite and adjacent the angle phi (ϕ).

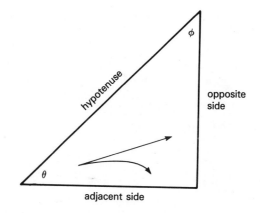

FIGURE 10–5a   Opposite and adjacent sides for theta

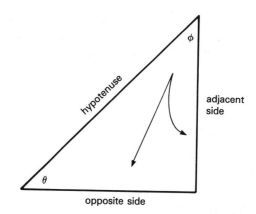

FIGURE 10–5b   Opposite and adjacent sides for phi

The ratio of the sides used in this text are defined as follows.

sine ratio:
$$\sin \theta = \frac{\text{opposite side}}{\text{hypotenuse}}$$

cosine ratio:
$$\cos \theta = \frac{\text{adjacent side}}{\text{hypotenuse}}$$

tangent ratio:
$$\tan \theta = \frac{\text{opposite side}}{\text{adjacent side}}$$

The ratios apply to both angle theta and phi, but the positions of the opposite and adjacent sides reverse when the other angle is used.

Similar triangles are those triangles whose angles (θ and φ) are equal but whose sides may be of different lengths. Figure 10–6 illustrates similar triangles, where the sine, cosine, and tangent ratios are the same regardless of the lengths of the sides.

Chapter 10
Right-Angle Trigonometry

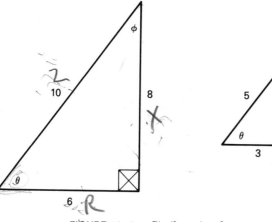

FIGURE 10–6   Similar triangles

Compare the ratios of each triangle.

           Large triangle   *Small triangle*

$$\sin \theta = \frac{\text{opp}}{\text{hyp}} \qquad \sin \theta = \frac{\text{opp}}{\text{hyp}}$$

$$\sin \theta = \frac{8}{10} \qquad \sin \theta = \frac{4}{5}$$

$$\sin \theta = 0.8000 \qquad \sin \theta = 0.8000$$

$$\cos \theta = \frac{\text{adj}}{\text{hyp}} \qquad \cos \theta = \frac{\text{adj}}{\text{hyp}}$$

$$\cos \theta = \frac{6}{10} \qquad \cos \theta = \frac{3}{5}$$

$$\cos \theta = 0.6000 \qquad \cos \theta = 0.6000$$

$$\tan \theta = \frac{\text{opp}}{\text{adj}} \qquad \tan \theta = \frac{\text{opp}}{\text{adj}}$$

$$\tan \theta = \frac{8}{6} \qquad \tan \theta = \frac{4}{3}$$

$$\tan \theta = 1.333 \qquad \tan \theta = 1.333$$

10–5 DETERMINING TRIGONOMETRIC RATIOS. For the value of each possible angle of theta, a ratio can be determined. Tables have been prepared listing these ratios for angles between 0° and 90°. With the advent of the electronic calculator, however, it is not necessary (it is also time-consuming) to use tables. Tables or calculators can express angles in either degrees or degrees, minutes, and seconds. In this text and in electronics, angles are expressed in degrees and fractions of degrees.

A calculator of this type has keys marked sin, cos, and tan. Enter the angle and press the desired function.

Find the sine, cosine, and tangent ratios for the angle 53.13°.

enter 53.13 in the calculator and depress the "sin" key

        read 0.79999 or 0.8000

enter 53.13 and depress "cos"

$$\text{read } 0.6000$$

enter 53.13 and depress "tan"

$$\text{read } 1.333$$

These are the same ratios used in Section 10–4, and the angle theta is 53.13° for both triangles. The sum of the angles theta and phi is equal to 90°. If either angle is known, the other can be determined.

where θ = 53.13°

$$\theta + \phi = 90°$$

$$53.13° + \phi = 90°$$

$$\phi = 90° - 53.13°$$

$$\phi = 36.87°$$

10–6 DETERMINING INVERSE FUNCTIONS. If the ratio of the sides of a right triangle are known, the angles theta or phi can be determined by use of the calculator. The inverse function may be called the "arc" function on some calculators. Refer to the instruction manual to determine how the calculator is programmed. The term "arc" means to find an angle whose ratio is known.

the arcsin 0.707 = (find the angle whose sin ratio is 0.707)

enter 0.707 and depress the "inv" and "sin" keys

read: 45°

If the "arc" key is used, it may be necessary to use the "2nd" function key. Some calculators use keys marked $\sin^{-1}$, $\cos^{-1}$, or $\tan^{-1}$ when the second function key is used. It is not a difficult operation; refer to the calculator's instruction manual.

EXERCISE 10–2 Using a calculator, find the ratios or the angles as required.
1. Find the sin, cos, and tan. Use four significant figures.

		sin	cos	tan
a.	62°	.8829	.4695	1.880
b.	43.8°	.6921	.7217	.9589
c.	8.6°	.1495	.9875	.1512
d.	87.9°	.9993	.0366	27.27
e.	45°	.7071	.7071	1

2. Find an angle whose sin is:
   a. 0.7071 = 44.99 = 45°
   b. 0.9291 = ~~75.72~~ 68.29°
   c. 0.7604 = 49.499
   d. 0.9792 = 78.293
   e. 0.2368 = 13.697

3. Find an angle whose cos is:
   a. 0.7071 = 45
   b. 0.5707 = 55.20
   c. 0.9952 = 5.6160
   d. 0.2706 = 74.30
   e. 0.9461 = 18.897

4. Find an angle whose tan is:
   a. 0.7071 = 35.26
   b. 0.5774 = 30.00
   c. 16.35 = 86.50
   d. 1.477 = 55.90
   e. 0.0875 = 5.0006

$\theta + \phi = 90°$

5. If $\phi = 36.4°$, find the cos of θ. = 53.6°   .5934

6. If $\theta = 72.4°$, find tan ϕ. 17.6°   .3172

7. If $\phi = 21.7°$, find sin θ. 68.3°   .9291

8. Find the angle θ if the tan ϕ = 0.6346.   57.6007°
   32.3992

10–7 THE SINE RATIO. Use the sine ratio to find the side of a right triangle.

   Example B Given a standard right triangle or impedance triangle, find angle ϕ and side Z.

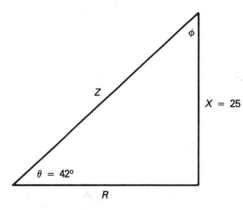

θ = 42°

X = 25

FIGURE 10-7   Right triangle for Example B

$$\phi = 90° - \theta$$

$$\phi = 90° - 42° = 48°$$

$$\sin \theta = \frac{X}{Z}$$

$$\sin 42° = \frac{25}{Z}$$

$$Z = \frac{25}{\sin 42°} \quad \text{(taken from a table or calculator: } \sin 42° = 0.6691\text{)}$$

$$Z = \frac{25}{0.6691}$$

$$Z = 37.36$$

EXERCISE 10-3   Using the sine ratio, find the missing value. Draw and label the impedance triangle for each problem.

1.   θ = 63°   X = 550 Ω   Z = 617.279

2.   θ = 25.5°   Z = 1020 Ω   X = 439.12

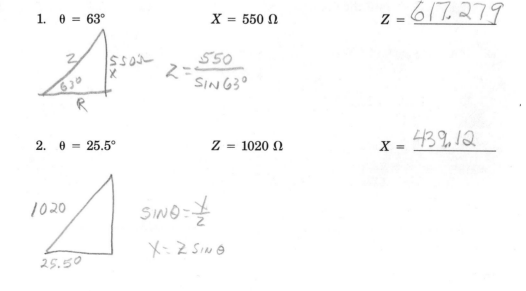

3. φ = 15.6°   R = 22 kΩ   Z = __81.81K__   **137**

θ = 74.4°
$\cos\theta = \frac{A}{H}$ $Z = \frac{A}{\cos\theta}$

Chapter 10
Right-Angle Trigonometry

4. φ = 82.4°   Z = 150 kΩ   R = __148,682 KΩ__

θ = 7.6°
$\sin\phi = \frac{O}{H} = \frac{R}{Z}$ $R = Z\sin\phi$

5. θ = 45°   Z = 1.41 Ω   R = __.997 Ω__

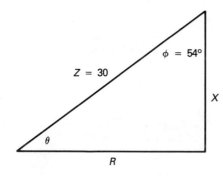
$\sin\theta = \frac{X}{Z}$
$\cos\theta = \frac{A}{H} = \frac{R}{Z}$ $R = Z\cos\theta$

6. θ = 53.1°   X = 120 Ω   Z = __150.059__

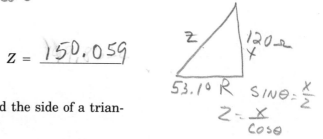
$\sin\theta = \frac{X}{Z}$
$Z = \frac{X}{\cos\theta}$

10-8 THE COSINE RATIO. Use the cosine ratio to find the side of a triangle.

Example C Given a standard right triangle or impedance triangle, find angle θ and side R.

φ = 54°

Z = 30

FIGURE 10-8 Right triangle for Example C

$$\theta = 90° - \phi$$

$$\theta = 90° - 54° = 36°$$

$$\cos\theta = \frac{R}{Z}$$

Part 2
Mathematics for DC and AC Circuits

$$\cos 36° = \frac{R}{30}$$

$$R = \cos 36°(30)$$

$$R = 0.8090(30) \quad \text{(from a table or calculator, } \cos 36° = 0.8090\text{)}$$

$$R = 24.27$$

EXERCISE 10-4 Use the cosine ratio to find the missing component. Draw and label the impedance triangle for each problem.

$\cos \theta = \frac{ADJ}{HYP}$

$\cos \theta = \frac{R}{Z}$

1. $\theta = 30°$ $R = 360\ \Omega$ $Z = \underline{415.69}$

   $Z = \frac{R}{\cos \theta}$

2. $\theta = 19.8°$ $Z = 2500\ \Omega$ $R = \underline{2352.2}$

   $R = Z \cos \theta$

$\cos \phi = \frac{X}{Z}$

3. $\phi = 53.1°$ $X = 64\ k\Omega$ $Z = \underline{106.592\ k\Omega}$

   $Z = \frac{X}{\cos \phi}$

4. $\phi = 66.5°$ $Z = 3.3\ k\Omega$ $X = \underline{1315.87}$

   $X = Z \cos \phi$

5. $\theta = 87.6°$ $R = 82\ M\Omega$ $Z = \underline{1958\ M\Omega}$

   $Z = \frac{R}{\cos \theta}$ 1,958,178,385

6. $\theta = 33.6°$ $R = 0.75\ \Omega$ $Z = \underline{.9\ \Omega}$

   $Z = \frac{R}{\cos \theta}$

10–9 USING THE TANGENT RATIO.
Use the tangent ratio to find the side of a right triangle.

Example D Given a standard right triangle, find angle φ and side R.

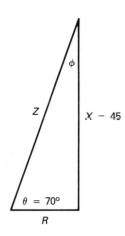

$\theta = 70°$

$X = 45$

FIGURE 10–9 Right triangle for Example D

$$\phi = 90° - \theta$$

$$\phi = 90° - 70° = 20°$$

$$\tan \theta = \frac{X}{R}$$

$$\tan 70° = \frac{45}{R}$$

$$R = \frac{45}{\tan 70°}$$

$$R = \frac{45}{2.75} \quad \text{(from a table or calculator, } \tan 70° = 2.75\text{)}$$

$$R = 16.36$$

EXERCISE 10–5 Using the tangent ratio, find the missing component. Draw the impedance triangle and label the sides for each problem.

$\tan \theta = \frac{O}{A} = \frac{X}{R}$

$\tan \phi = \frac{R}{X}$

1. $\theta = 35°$ $X = 470 \, \Omega$ $R = \underline{671.22}$

$\tan \theta = \frac{X}{R}$ $R = \frac{X}{\tan \theta}$

Part 2
Mathematics for DC and AC Circuits

$\tan\theta = \frac{X}{R}$

$\tan\phi = \frac{R}{X}$

2. $\theta = 8.2°$ $R = 2200\ \Omega$ $X = \underline{317.02}$

   $X = R\tan\theta$

3. $\phi = 83.4°$ $X = 18\ \Omega$ $R = \underline{156\ \Omega}$

   $R = \dfrac{X\tan\phi}{\tan\theta}$

4. $\phi = 63.2°$ $R = 3.3\ M\Omega$ $X = \underline{1.67\ M\Omega}$

   $X = \dfrac{R}{\tan\phi}$

5. $\theta = 45°$ $R = 1000\ k\Omega$ $X = \underline{1\ M\Omega}$

   $X = \dfrac{R\tan\theta}{\tan\theta}$

6. $\theta = 53.1°$ $X = 0.86\ \Omega$ $R = \underline{.6457\ \Omega}$

   $R = \dfrac{X\tan\theta}{\tan\theta}$

10-10 USING THE SINE, COSINE, OR TANGENT RATIO. Use the sine, cosine, or tangent ratio to determine the acute angles of a right triangle.

Example E Given a standard right triangle or impedance triangle, find angle θ with sine ratio.

$\sin\theta = \dfrac{\text{opp}}{\text{hyp}}$

$\sin\theta = \dfrac{12}{16}$

$\sin\theta = 0.75$

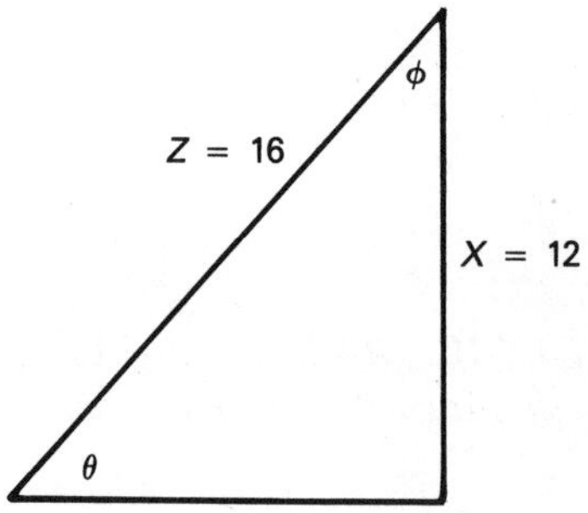

FIGURE 10-10   Right triangle for Example E

From the calculator, using inverse sin or arc sin, the angle can be determined by:

θ = 48.6°

φ = 90° − θ

φ = 90° − 48.6° = 41.4°

Example F Given a standard right triangle, find angle θ with cosine ratio.

$\cos \theta = \dfrac{\text{adj}}{\text{hyp}}$

$\cos \theta = \dfrac{9}{16}$

cos θ = 0.5625

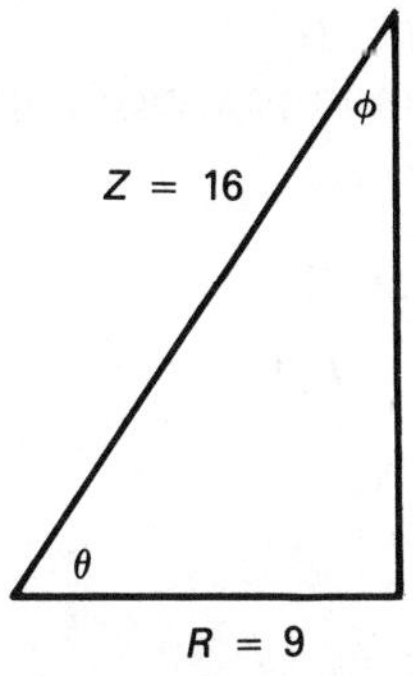

FIGURE 10–11   Right triangle for Example F

From the calculator, using inverse cosine or arc cosine, the angle can be determined by:

θ = 55.8°

φ = 90° − θ

φ = 90° − 55.8° = 34.2°

Example G Given a standard right triangle or impedance triangle, find angle θ with tangent ratio.

$\tan \theta = \dfrac{\text{opp}}{\text{adj}}$

$\tan \theta = \dfrac{17}{5}$

tan θ = 3.4

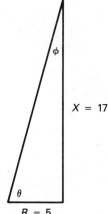

FIGURE 10–12   Right triangle for Example G

From the calculator, using inverse tangent or arc tangent, the angle can be determined by:

$\theta = 73.6°$

$\phi = 90° - 73.6° = 16.4°$

The same relationship can be shown for the angle ϕ.

EXERCISE 10-6 Using the sine, cosine, or tangent ratio, find angle ϕ or θ. Draw the impedance triangle and label the sides.

1. $Z = 560 \ \Omega$ $R = 320 \ \Omega$ $\theta = \underline{55.15°}$

   $\cos\theta = \dfrac{R}{Z}$

2. $X = 150 \ \Omega$ $Z = 820 \ \Omega$ $\theta = \underline{10.54°}$

   $\sin\theta = \dfrac{X}{Z}$

3. $X = 680 \ \Omega$ $R = 470 \ \Omega$ $\theta = \underline{55.35°}$

   $\tan\theta = \dfrac{X}{R}$

4. X = 3.3 kΩ   Z = 10,000 Ω   φ = __70.73°__

$\cos\theta = \frac{X}{Z}$

143

Chapter 10
Right-Angle Trigonometry

5. R = 0.25 Ω   X = 1.7 Ω   φ = __8.365°__

$\tan\phi = \frac{R}{X}$

6. Z = 8.2 kΩ   R = 5700 Ω   θ = __45.96°__

$\cos\theta = \frac{R}{Z}$

EVALUATION EXERCISE Draw an impedance triangle for each problem and find the missing component, using trig functions. Do not use the Pythagorean Theorem.

1. R = 1250 Ω   θ = 44.5°   Z = __1752.54 Ω__

X = __1228.37 Ω__

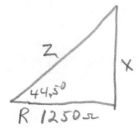

$\cos\theta = \frac{R}{Z}$ $\tan\theta = \frac{X}{R}$

$Z = \frac{R}{\cos\theta}$ $X = R\tan\theta$

2. Z = 680 Ω   θ = 8.4°   X = __99.34 Ω__

R = __672.71 Ω__

$\cos\theta = \frac{R}{Z}$ $\sin\theta = \frac{X}{Z}$

$R = Z\cos\theta$ $X = Z\sin\theta$

3. $X = 4.7\ \text{M}\Omega$ $\phi = 33.8°$ $Z = \underline{5.66\ \text{M}\Omega}$

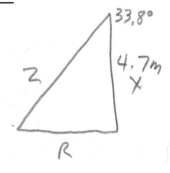

$\cos\phi = \dfrac{X}{Z}$
$Z = \dfrac{X}{\cos\phi}$
$\tan\phi = \dfrac{R}{X}$
$R = X\tan\phi$

$R = \underline{3.15\ \text{M}\Omega}$

4. $R = 22\ \Omega$ $X = 25\ \Omega$ $\theta = \underline{48.65°}$

$\tan\theta = \dfrac{X}{R}$
$\tan\theta = 1.136$
$\theta = 48.65°$

$\cos\theta = \dfrac{R}{Z}$
$Z = \dfrac{R}{\cos\theta}$
$Z =$

$Z = \underline{33.30\ \Omega}$

5. $Z = 15\ \text{k}\Omega$ $X = 6500\ \Omega$ $\theta = \underline{25.68°}$

$\sin\theta = \dfrac{X}{Z}$
$\sin\theta = .43333$
$\cos\theta = \dfrac{R}{Z}$
$R = Z\cos\theta$

$R = \underline{13{,}518\ \Omega}$

6. $R = 0.260\ \Omega$ $Z = 0.75\ \Omega$ $\theta = \underline{69.72°}$

$\cos\theta = \dfrac{R}{Z}$
$\sin\theta = \dfrac{X}{Z}$
$X = Z\sin\theta$

$X = \underline{.70\ \Omega}$

7. $\theta = 82.4°$ $\qquad R = 220\ \Omega \qquad$ $Z = \underline{1663\ \Omega}$

$\cos\theta = \dfrac{R}{Z}$ $\qquad \tan\theta = \dfrac{X}{R} \qquad$ $X = \underline{1648.8\ \Omega}$

$Z = \dfrac{R}{\cos\theta}$ $\qquad X = R\tan\theta$

8. $\phi = 14.5°$ $\qquad X = 3.3\ \text{k}\Omega \qquad$ $Z = \underline{3408.57\ \Omega}$

$\cos\phi = \dfrac{X}{Z}$ $\qquad \tan\phi = \dfrac{R}{X} \qquad$ $R = \underline{853.43\ \Omega}$

$Z = \dfrac{X}{\cos\phi}$ $\qquad R = X\tan\phi$

CHAPTER 11

Angles

Objectives

After completing this chapter you will be able to:
- Identify the measurement of angles in both degrees and radians
- Identify Cartesian coordinates
- Identify the four-quadrant system
- Locate values of x and y using Cartesian coordinates and the four-quadrant system
- Identify polar coordinants
- Generate angles in the four quadrants
- Identify the algebraic signs of sine, cosine, and tangent in each of the four quadrants
- Determine the position angle theta and the function angle in each quadrant
- Determine the inverse function of sine, cosine, and tangent in each quadrant

In Chapter 10, ratios and angles were studied as they apply to the right triangle. The angles were limited to values 90° or less. This chapter expands the subject of functions of any angle; in it, the rectangular and four-quadrant system of measurement is introduced.

11–1 ANGLE MEASUREMENT. Angles are commonly measured in degrees. Most students are familiar with this unit. In science and mathematics angles are measured in both degrees and radians (rad). The following definitions should be noted.

- *Degree:* $\frac{1}{360}$ part of the circumference of a circle; there are 360 degrees in a circle.

- *Radian:* approximately $\frac{1}{6.28}$ part of the circumference of a circle; there are 6.28 radians in a circle.

When a distance equal to the radius of the circle is measured along the circumference of the same circle, it encompasses an angle equal to 1 rad. Since the radius of a circle can be placed approximately 6.28 times on the circumference, there are approximately 6.28 rad in a circle. In terms of π this is 2π rads (see Figure 11–1).

Example A Angles can be measured in degrees or radians, as shown in Figure 11–2.

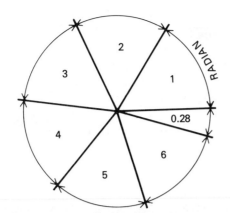

FIGURE 11–1   Radian measure

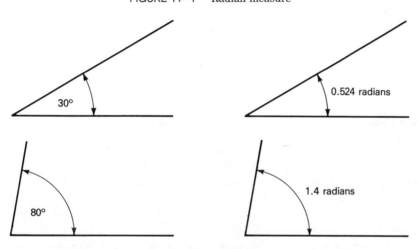

FIGURE 11–2   Angles measured in degrees and radians

11–2 CONVERTING RADIANS AND DEGREES. It is sometimes necessary to convert from radians to degrees or from degrees to radians.

$$2\pi \text{ rads} = 360°$$

$$\frac{2\pi \text{ rads}}{2\pi} = \frac{360°}{2\pi} \quad \text{(divide by } 2\pi\text{)}$$

$$\text{rads} = \frac{180°}{\pi}$$

Conversions   To convert radians to degrees:

$$\text{multiply by } \left(\frac{180°}{\pi}\right) \text{ or } (57.3°)$$

To convert degrees to radians:

$$\text{multiply by } \left(\frac{\pi}{180°}\right) \text{ or divide by } (57.3°)$$

(180° divided by π equals 57.3° which is usually sufficient for most calculations in electronics)

Example B

Change 75° to radians.

$$75° \div 57.3° = 1.31 \text{ rad}$$

Change 0.625 rad to degrees.

$$0.625 \times 57.3° = 35.8°$$

11-3 DEGREES, MINUTES, SECONDS, AND DECIMAL ANGLES. In measuring angles some practitioners use the degree as a basic unit of measure and further break the degree down into minutes and seconds. An angle can be expressed as 32 degrees 45 minutes 25 seconds (32° 45′ 25″). In electronics work the degree is further broken down into decimal units, such as 35.8°. It is the decimal unit that is used in this text.

EXERCISE 11-1 Convert the following angles to radian or degree measurements as indicated.

1. 65° to __1.1344__ rad
2. 132° to __2.3038__ rad
3. 240° to __4.1888__ rad
4. 180° to __3.1416__ rad
5. 1.6 rad to __91.67__ deg
6. 2π rad to __180__ deg
7. 2.4 rad to __137.5099__ deg
8. 820° to __14.3116__ rad
9. 7.5 rad to __429.7183__ deg
10. 1640° to __28.6233__ rad

11-4 CARTESIAN OR RECTANGULAR COORDINATES. Algebra problems can be solved by graphing solutions as seen in solving simultaneous equations. Two number lines that intersect at zero are used to divide the plane into four quadrants, as shown in Figure 11-3. The quadrants are numbered in the counterclockwise direction.

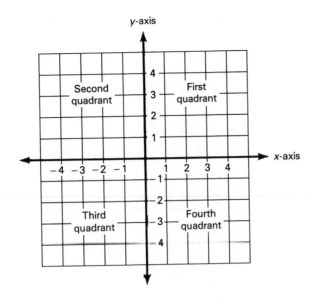

FIGURE 11-3   Four-quadrant system

11-5 LOCATING VALUES OF x AND y. Points determined by the values of x and y can be located in any of the four quadrants. Points are located by counting off the values of x and y in the positive or the negative direction. Figure 11-4 illustrates how the points $-4, 2$ are located. The value of x is listed first, and x is equal to -4. To locate this point, count four places in the negative direction on the x axis; then count two places in the positive direction on the y-axis.

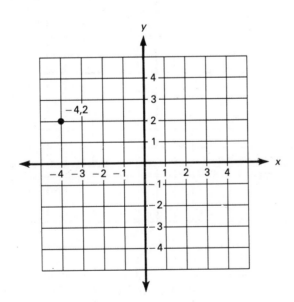

FIGURE 11-4   Plot of points $-4, 2$

Example C Plot the points $(-4, 7)$, $(-2, 4)$, $(1, 2)$, $(5, 0)$, $(3, -2)$, $(-5, -1)$, $(-2, -3)$.

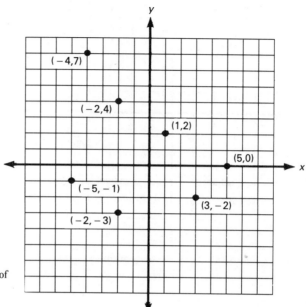

FIGURE 11–5 Plot of points for Example C

EXERCISE 11–2 Locate the following x, y points in Figure 11–6: $(7, -3)$, $(-4, -8)$, $(6, 4)$, $(-8, 5)$, $(-2, 6)$.

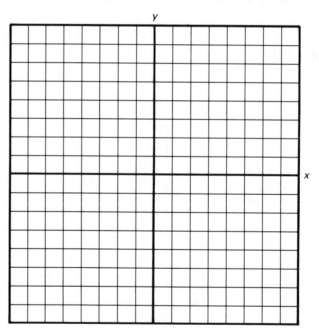

FIGURE 11–6 Graph for Exercise 11–2

11–6 POLAR COORDINATES. Points have been located by using values determined by the x and y axes. This system can be modified, and angles can be added to the x and y number lines, as shown in Figure 11–7. The angles also rotate, increasing in the counterclockwise direction. The point $(6, 7)$ can be located by using the rectangular coordinates 6 and 7, or it can be located by using the angle θ and the length of the line (r). The line (r) is called a radius vector; it is always a positive value. Locating a point using this method is called a polar coordinate. It can be expressed in a notation such as $15\underline{/30°}$, which means $r = 15$ at the position of 30°.

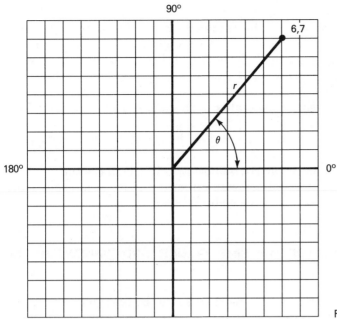

FIGURE 11–7 Polar coordinate, quadrant

11–7 GENERATION OF ANGLES.
Angles can be generated in this system and can be measured either in degrees or radians. Figure 11–8 illustrates angles generated in the different quadrants. The radius vector (r) has been rotated into each quadrant. The distance that r has been rotated is measured by the angle θ.

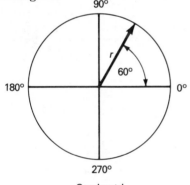

Quadrant I
θ = 60° or 1.05 rad

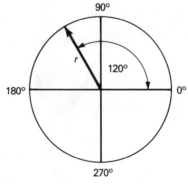

Quadrant II
θ = 120° or 2.09 rad

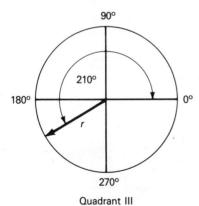

Quadrant III
θ = 210° or 3.67 rad

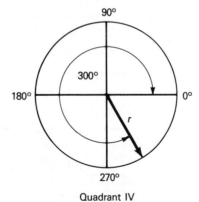

Quadrant IV
θ = 300° or 5.24 rad

FIGURE 11–8 Positive angle rotation

11–8 NEGATIVE ANGLES.

Angles generated in Figure 11–8 are all positive angles and were generated in the counterclockwise direction. Angles generated in the clockwise direction are negative angles. As on a basic number line, positive and negative simply indicate a direction from zero. To summarize:

1. Positive angles are generated in the counterclockwise direction.

2. Negative angles are generated in the clockwise direction.

An angle of $-45°$ is in the fourth quadrant, $-120°$ is in the third, and so on. The position of r can be expressed with either a positive or a negative angle. A rotating radius (r) at 315° can also be at $-45°$.

11–9 ALGEBRAIC SIGNS OF THE SINE, COSINE, AND TANGENT IN THE QUADRANTS.

First Quadrant An angle theta (θ) is generated into the first quadrant in Figure 11–9. A perpendicular is dropped to the (X) axis, as in Figure 11–10, forming a right triangle that includes the values of x and y. In this case the position angle and the function angle of the right triangle are the same angle. In the first quadrant (0° to 90°) the sine, cosine, and tangent ratios are all positive.

NOTE: The rotating radius (r) is positive in all quadrants.

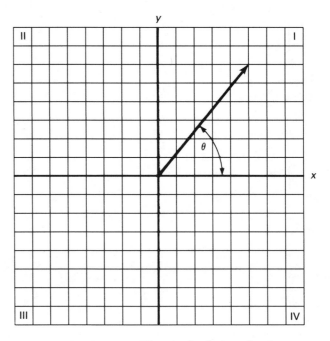

FIGURE 11–9   Theta in the first quadrant

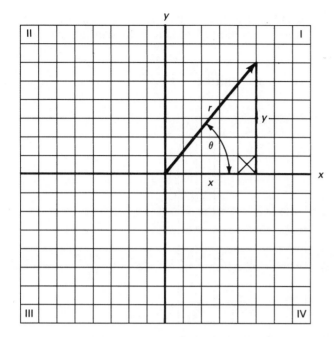

$$\sin \theta = \frac{y}{r} = +\sin \theta$$

$$\cos \theta = \frac{x}{r} = +\cos \theta$$

$$\tan \theta = \frac{y}{x} = +\tan \theta$$

FIGURE 11–10 Algebraic signs in the first quadrant

Second Quadrant In the second quadrant the position angle θ is between 90° and 180°. The function angle of the right triangle containing the values of x and y is the complement of θ. This angle will be referred to as θ' (theta prime).

$$\theta' = 180° - \theta$$

The sine, cosine, and tangent of θ are the same ratios as that of θ'. However, if θ is used to obtain these ratios, a calculator will also give the correct algebraic sign, as shown in Figure 11–11.

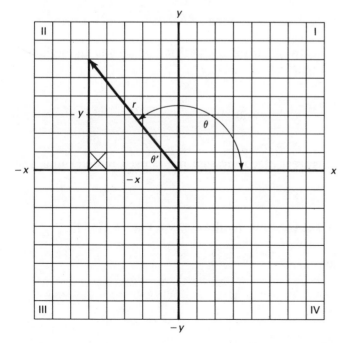

$$\sin \theta = \frac{y}{r} = +\sin \theta$$

$$\cos \theta = \frac{-x}{r} = -\cos \theta$$

$$\tan \theta = \frac{y}{-x} = -\tan \theta$$

FIGURE 11–11 Algebraic signs in the second quadrant

Third Quadrant In the third quadrant θ is at a position between 180° and 270° (see Figure 11–12). The function angle (θ′) of the right triangle can be determined by:

$$\theta' = \theta - 180°$$

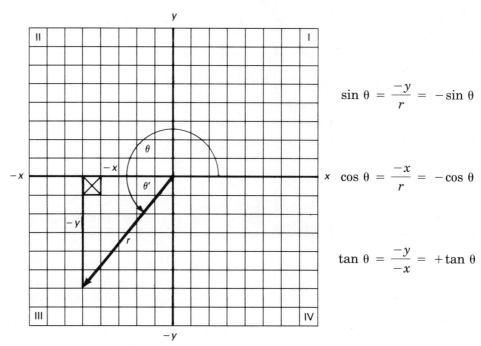

FIGURE 11–12   Algebraic signs in the third quadrant

A calculator will give the correct algebraic sign if θ is used to determine the ratio rather than θ′.

Fourth Quadrant In the fourth quadrant the position angle θ is an angle between 270° and 360°. See Figure 11–13.

$$\theta' = 360° - \theta$$

A calculator provides the correct algebraic sign if θ is used to determine the ratio.

For algebraic signs in each of the four quadrants, see Figure 11–14.

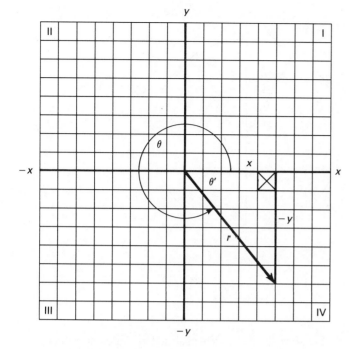

FIGURE 11–13 Algebraic signs in the fourth quadrant

$$\sin \theta = \frac{-y}{r} = -\sin \theta$$

$$\cos \theta = \frac{x}{r} = +\cos \theta$$

$$\tan \theta = \frac{-y}{x} = -\tan \theta$$

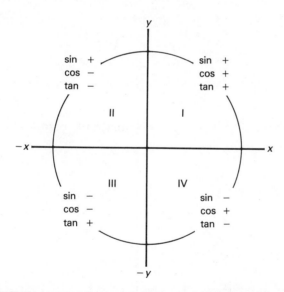

FIGURE 11–14 Summary of signs in the four quadrants

Example D Find the cosine of 120°, expressing its algebraic sign, ratio, quadrant, and the function angle theta prime (θ'). From the calculator, the algebraic sign is negative and the ratio is 0.8660. From calculation,

$$\theta' = 180° - \theta$$
$$\theta' = 180° - 120°$$
$$\theta' = 30°$$

From rotation, 120° is in the second quadrant (Figure 11–15).

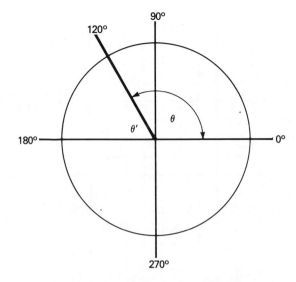

FIGURE 11-15 120° rotation for Example D

Example E Find the sine of 250°, expressing the algebraic sign, the ratio, the quadrant, and the function angle theta prime (θ'). From the calculator, the algebraic sign is negative and the ratio is 0.9397. From calculation,

$$\theta' = \theta - 180°$$
$$\theta' = 250° - 180°$$
$$\theta' = 70°$$

From rotation, 250° is in the third quadrant (Figure 11-16).

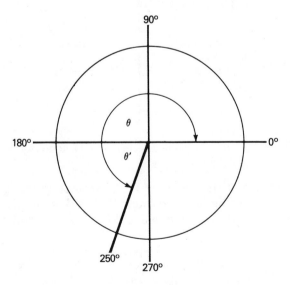

FIGURE 11-16 250° rotation for Example E

Example F Find the cosine of 325°, expressing the algebraic sign, ratio, quadrant, and function angle theta prime (θ'). From the calculator,

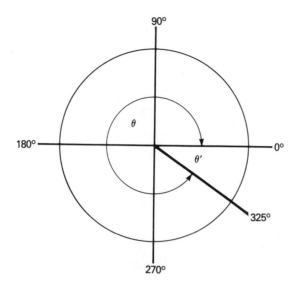

FIGURE 11–17   325° rotation for Example F

the algebraic sign is positive and the ratio is 0.8192. From calculation,

$$\theta' = 360° - \theta$$
$$\theta' = 360° - 325°$$
$$\theta' = 35°$$

From rotation, 325° is in the fourth quadrant (Figure 11–17).

The function angle theta prime and the position angle theta will have the same ratio. In Example D the cosine ratio of 120° is the same as the cosine ratio of 30°, or 0.8660. When determined on the calculator, the cosine of 120° is negative. The cosine of 30° is positive because the calculator sees 30° in the first quadrant. The function angle, theta prime is needed to determine inverse functions.

EXERCISE 11–3 Determine the algebraic sign, the ratio, quadrant, and theta prime for each.

Angle	Algebraic sign	Ratio	Quadrant	θ'
sin 118°	+	.8829	2	62°
tan 35°	+	.7002	1	35°
cos 342.5°	+	.9537	4	17.5°
sin 254°	−	.9613	3	16°

158
Part 2
Mathematics for DC
and AC Circuits

	SIGN	RATIO	QUAD	θ'
143° cos 2.5 rad	−	−.8011	2	37°
sin 6.8°	+	.1184	1	6.8°
−306° tan − 5.35 rad	+	1.3498	1	54°
37° sin 863°	+	.6018	1	37°
85° cos −1525°	+	.08715	4	85°
733° 13° tan 12.8 rad	+	.2379	1	13°
sin −26.5°	−	−.44619	4	−26.5
cos −270°	+	0	1	270
sin 45°	+	.7071	1	45°
807°.95 tan π/2 rad	+	14.1014	1	87.95
−138° cos −942°	−	−.7431	3	−42°

11–10 INVERSE FUNCTIONS. Although a calculator can provide the sine, cosine, and tangent ratio of any angle, it cannot always provide the position angle when using inverse functions. That position should be determined by observing the values of x and y used to find the ratio. If the ratio used to find the tangent function were $-x$ and $-y$, then:

$$\tan \theta = \frac{-y}{-x}$$

and if $\tan \theta = 0.8333$, find the position of θ. Arc tan $0.8333 = 39.8°$, but $39.8°$ is in the first quadrant. By inspecting the values of x and y it can be seen that both are negative. This can only mean the position is somewhere in the third quadrant. (See Figure 11–18.)

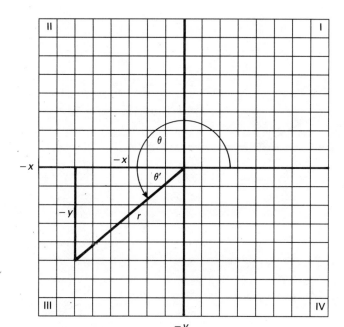

FIGURE 11-18   Inverse functions in the third quadrant

The angle 39.8° is the function angle (θ'); the position angle is:

$$\theta = 180° + \theta'$$
$$\theta = 180° + 39.8°$$
$$\theta = 219.8° \qquad \text{(third quadrant)}$$

Example G

Given
$$\cos \theta = -0.8941$$
$$x = -3, y = 1.5$$

Find θ' and θ.

The inverse cosine of −0.8941, taken from the calculator is 153.4°. Since x is negative and y is positive, θ is in the second quadrant and is 153.4° (see Figure 11-19).

$$\theta' = 180° - 153.4°$$
$$\theta° = 26.6°$$

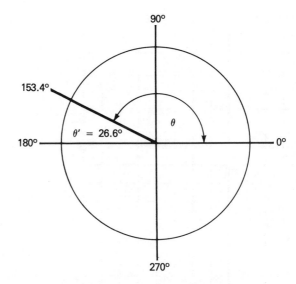

FIGURE 11-19   Inverse function in the second quadrant

Example H

Given
$$\tan x = 1.6$$
$$x = -5, y = -8$$

Find θ' and θ.

The inverse tangent of 1.6, taken from the calculator, is the angle 58°. Since x is negative and y is negative, the position angle must be in the third quadrant. The angle given on the calculator, 58°, is the function angle theta prime (see Figure 11-20).

$$\theta = 180° + \theta$$
$$\theta = 180° + 58°$$
$$\theta = 238°$$

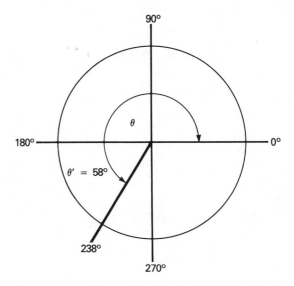

FIGURE 11-20   Inverse function in the third quadrant

EXERCISE 11-4 Determine the position angle θ and the function angle θ′ for the ratios.

Chapter 11
Angles

1. $\tan \theta = -2.79$
   $x = 4.2 \quad y = -12$

   Q4

   θ′ = __70.28__   θ = __-70.28__

2. $\sin \theta = -0.5769$
   $x = 10.6 \quad y = -7.5$

   Q4

   θ′ = __55__   θ = __124.41__

3. $\cos \theta = -0.5652$
   $x = -3.25 \quad y = 4.74$

   Q2

   θ′ = __55.58__   θ = __124.4__

4. $\tan \theta = 1.846$
   $x = -84.5 \quad y = -156$

   Q3

   θ′ = __61.56__   θ = __118.44__

5. $\cos \theta = -0.2727$
   $x = -0.15 \quad y = 0.529$

   Q2

   θ′ = __74.18__   θ = __105.82__

6. $\sin \theta = 0.833$
   $x = -1225 \quad y = 1844$

   Q2

   θ′ = __56.41__   θ = __123.59__

162

*Part 2
Mathematics for DC
and AC Circuits*

7. $\sin\theta = -0.9429$ $\theta' = \underline{70.54}$ $\theta = \underline{-70.54}$
   $x = 6.82$ $y = -19.3$
   Q 4

8. $\tan\theta = 2.156$ $\theta' = \underline{65.12}$ $\theta = \underline{65.12}$
   $x = 8.65$ $y = 18.65$
   Q 1

9. $\tan\theta = -0.6818$ $\theta' = \underline{34.29}$ $\theta = \underline{-34.29}$
   $x = 24.2$ $y = -16.5$
   Q 4

10. $\cos\theta = -0.8320$ $\theta' = \underline{33.69}$ $\theta = \underline{-146.3°}$
    $x = -15{,}000$ $y = -10{,}000$
    Q 3

EVALUATION EXERCISE Using a calculator, determine the algebraic sign and ratio for each angle in Exercises 1 through 8. Also determine the quadrant and the function angle theta prime.

	Algebraic sign and ratio (four significant figures)	Quadrant	θ'
1. $\cos 85°$	$+.0872$	1	$85°$
2. $\tan 247°$	$+2.356$	3	$67°$
3. $\sin -45°$	$-.7071$	4	$45°$
4. $\cos 0.754$ rad *43.20°*	$+.7290$	1	$43.20°$
5. $\tan 835°$	-2.145	4	$65°$
6. $\sin -135°$	$-.7071$	3	$45°$
7. $\tan 5.72$ rad *327.73°*	$-.6314$	4	32.27
8. $\cos -225°$	$-.7071$	2	$45°$

Determine the position angle theta and theta prime for each ratio in Exercises 9 through 16.

9. $\tan -1.447$ $\theta = \underline{-55.35°}$ $\theta' = \underline{55.35°}$
   $x = 3.8$ $y = -5.5$
   Q 4

10. $\cos -0.829$ $\theta = \underline{145.99°}$ $\theta' = \underline{34°}$
    $x = -126$ $y = 85$ $146°$
    Q 2

```
        90
        y
  sin+        sin+
  cos-        cos+
  tan-        tan+
       2    1
180 -x ─────┼───── x 0
       3    4
  sin-        sin-
  cos-        cos+
  tan+        tan-
        -y
        270
```

11. sin −0.3939 θ = 203.2° θ′ = 23.2°
    x = −42   y = −18
    Q3

12. sin 0.9040 θ = 64.69° θ′ = 64.69°
    x = 3.65   y = 7.72
    Q1

13. cos 0.7071 θ = 315° θ′ = 45°
    x = 35   y = −35
    Q4

14. tan 0.6598 θ = 213.42° θ′ = 33.42°
    x = −6.82   y = −4.5
    Q3

15. sin 0.9097 θ = 114.54 θ′ = 65.46°
    x = −586   y = 1284
    Q2

16. cos −0.8846 θ = 152.20° θ′ = 27.8°
    x = −16.5   y = 8.70
    Q2

CHAPTER 12

Simultaneous Equations and Determinants

Objectives

After completing this chapter you will be able to:
- Graph a linear equation
- Solve two simultaneous equations by graphing
- Solve two simultaneous equations by addition and subtraction
- Solve two simultaneous equations by substitution
- Solve two simultaneous equations with second-order determinants
- Solve three simultaneous equations with third-order determinants

Many circuits used in electronics contain more than one power source, which results in problems with more than one unknown. Bridge circuits are also analyzed with more than one unknown. The use of simultaneous equations provides a method of analyzing these circuits.

12-1 GRAPHING A LINEAR EQUATION. Equations can be graphed using x and y coordinate points.

Graph the equation $x + 2y = 8$. Let x arbitrarily equal any convenient number; then solve the equation for y.

x	4	2	0	-2	-4
y	2	3	4	5	6

$$x + 2y = 8$$
$$4 + 2y = 8$$
$$2y = 8 - 4$$
$$2y = 4$$
$$y = 2$$

Plot the x and y points and connect them. This equation results in a straight line, as will all first-degree equations with two unknowns (see Figure 12-1). A first-degree equation is one in which the exponent of the unknown is 1. The values of x and y anywhere on this line will satisfy the equation. Only two points are needed to determine a straight line; so a linear equation can be graphed with two x and y values.

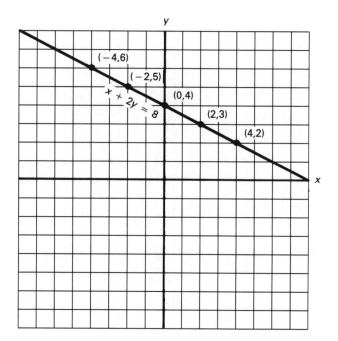

FIGURE 12-1   Graph of $x + 2y = 8$

12-2 GRAPHING TWO LINEAR EQUATIONS. The equations $2x - y = 6$ and $3x + y = 4$ are plotted by first determining two points at which x and y intersect.

$$2x - y = 6 \qquad \begin{array}{c|c|c} x & 2 & 0 \\ \hline y & -2 & -6 \end{array}$$

$$3x + y = 4 \qquad \begin{array}{c|c|c} x & 0 & 2 \\ \hline y & 4 & -2 \end{array}$$

The equations are then plotted as shown in Figure 12-2. The equations cross at the coordinate points 2, -2. At this point both equations have the same solution. This operation is referred to as solving equations simultaneously. Certain electronic circuits contain two unknowns and must be solved simultaneously. The solutions can be graphed, but that would be time-consuming and cumbersome for most applications. Sections 12-3 and 12-5 show how simultaneous equations provide a mathematical solution.

12-3 SOLVING TWO EQUATIONS SIMULTANEOUSLY BY ADDITION OR SUBTRACTION. The solution to two simultaneous equations can be found by adding or subtracting the two equations. First, one term in each equation must be identical. If the identical terms have unlike algebraic signs, they can be added, and that term drops out of each equation, leaving only one unknown in the equation. If the identical terms have like signs, then the equations should be subtracted. To mathematically solve the equations graphed in Section 12-2, first place them in the same order. Since they already have identical terms with unlike signs, they can be added, which causes the y term to drop out.

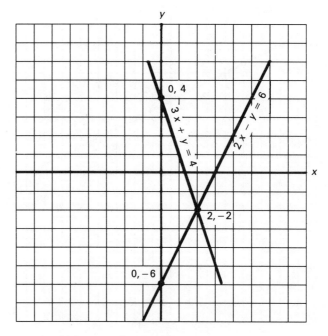

FIGURE 12-2   Graph of $2x - y = 6$ and $3x + y = 4$

$$\begin{aligned} 2x - y &= 6 \\ (+)\quad 3x + y &= 4 \\ \hline 5x &= 10 \\ x &= 2 \end{aligned}$$

To find y, substitute the value of x in either of the two original equations.

$$\begin{aligned} 3x + y &= 4 \\ 3(2) + y &= 4 \\ y &= 4 - 6 \\ y &= -2 \end{aligned}$$

The coordinate points check with the graph of Figure 12-2, $x = 2$ and $y = -2$.

If the terms are not identical, they can be made identical by multiplying one or both equations by constants that will make identical terms in each equation.

$$\begin{aligned} 3x + 4y &= 5 \\ 2x + 6y &= 2 \end{aligned}$$

Multiply the first equation by 2 and the second equation by 3

$$\begin{aligned} 2(3x + 4y = 5) &= 6x + 8y = 10 \\ 3(2x + 6y = 2) &= 6x + 18y = 6 \end{aligned}$$

Subract the equations:

$$\begin{aligned} 6x + 8y &= 10 \\ (-)\quad -6x - 18y &= -6 \\ \hline -10y &= 4 \\ y &= -0.4 \end{aligned}$$

Determine x:

$$3x + 4y = 5$$
$$3x + 4(-0.4) = 5$$
$$3x = 5 + 1.6$$
$$3x = 6.6$$
$$x = 2.2$$

Example A Solve the simultaneous equations:

$$5a + 5b = 7$$
$$2a + 5b = 1$$

$$(-) \quad \begin{array}{r} 5a + 5b = 7 \\ 2a + 5b = 1 \\ \hline 3a = 6 \\ a = 2 \end{array}$$

$$2a + 5b = 1$$
$$2(2) + 5b = 1$$
$$5b = 1 - 4$$
$$5b = -3$$
$$b = -0.6$$

Example B Solve the simultaneous equations:

$$2x + 4y = 12$$
$$x + y = 2$$

$$2x + 4y = 12$$
$$-2(x + y = 2)$$

$$(+) \quad \begin{array}{r} 2x + 4y = 12 \\ -2x - 2y = -4 \\ \hline 2y = 8 \\ y = 4 \end{array}$$

$$x + y = 2$$
$$x + 4 = 2$$
$$x = -2$$

12–4 SOLVING SIMULTANEOUS EQUATIONS BY SUBSTITUTION. Although addition and subtraction are the most commonly used methods, substitution is also used to solve equations.

Use substitution to solve the equations in the graph in Figure 12–2.

$$2x - y = 6$$
$$3x + y = 4$$

Solve one of the equations for either unknown. In this problem the first equation is solved for y:

Chapter 12
Simultaneous Equations
and Determinants

$$2x - y = 6$$
$$y = 2x - 6$$

Substitute $2x - 6$ in the second equation to get the value of y:

$$3x + (2x - 6) = 4$$
$$3x + 2x - 6 = 4$$
$$5x = 10$$
$$x = 2$$

The value of y can be determined by substituting 2 in either of the original equations:

$$2x - y = 6$$
$$2(2) - y = 6$$
$$-y = 6 - 4$$
$$y = -2$$

EXERCISE 12–1 Solve each simultaneous equation for the two unknowns by the addition or the subtraction method.

1. $3A + 5B = 51$ $A = $ _____

    $6A - B = 36$ $B = $ _____

2. $3x + 2y = 61$ $x = $ _____

    $x - y = 2$ $y = $ _____

3. $v - 6\mu = 8$ $v = $ _____

    $\mu + v = 1$ $\mu = $ _____

4. $3x - 3y = 3$ $x = $ _____
   $5x - y = 13$ $y = $ _____

169

Chapter 12
*Simultaneous Equations
and Determinants*

5. $3I + I_2 = 11$ $I = $ _____
   $3I_2 - I = 13$ $I_2 = $ _____

6. $4x - 6y = -6$ $x = $ _____
   $12y - 5x = 30$ $y = $ _____

7. $6x + 12y = 12$ $x = $ _____
   $12x + 16y = 18$ $y = $ _____

8. $3I - 5V = 9$ $I = $ _____
   $6I - 12 = 4V$ $V = $ _____

9. $4E + 5e = 3$ $E = $ _____

   $7 + 2E = -2e$ $e = $ _____

10. $3R + 5r = -6$ $R = $ _____

    $9 = (-8r - 6R)$ $r = $ _____

Example C

Solve simultaneous equations by substitution.

$$5a + 5b = 7$$
$$2a + 5b = 1$$

$$2a + 5b = 1$$
$$5b = 1 - 2a$$
$$b = \frac{1 - 2a}{5}$$

$$5a + \cancel{5}\left(\frac{1 - 2a}{\cancel{5}}\right) = 7$$
$$5a + 1 - 2a = 7$$
$$3a = 7 - 1$$
$$3a = 6$$
$$a = 2$$

Find b by substituting a in one of the original equations:

$$2a + 5b = 1$$
$$2(2) + 5b = 1$$
$$5b = 1 - 4$$
$$5b = -3$$
$$b = -\frac{3}{5}$$

EXERCISE 12-2 Solve the simultaneous equations, using the substitution method.

1. $2I + 3i = -4$ $\qquad I = $ _____

    $6i = 4I + 2$ $\qquad i = $ _____

2. $4R = 15 - 3r$ $\qquad R = $ _____

    $10 = 8R - 2r$ $\qquad r = $ _____

3. $3B + 2 = 5b$ $\qquad B = $ _____

    $4b - 12B - 18 = 0$ $\qquad b = $ _____

4. $6R - 16 = 18E$ $\qquad R = $ _____

    $14 - 2R = 3E$ $\qquad E = $ _____

5. $6B - 10b = 8$ $\qquad B = $ _____

    $B - 2b = 3$ $\qquad b = $ _____

6. $8I - 20 = -3i$ $\qquad I = $ _____

    $2i - 14 = -10I$ $\qquad i = $ _____

7. $10E - 6e = 12$ $E = $ _____

   $6E - 4 = e$ $e = $ _____

8. $4n - 24 = -6m$ $m = $ _____

   $8m - 22 = -2n$ $n = $ _____

12–5 SECOND-ORDER DETERMINANTS. Solving two equations by simultaneous solution is effective in electronic circuits having two unknowns. Another method is to use determinants. The term *second-order determinants* refers to solving equations with two unknowns; the term *third-order determinants* to solving equations with three unknowns; fourth-order to those with four unknowns; and so on.

In the two equations given below, a and b are the coefficients and c is the constant. The equations are of the same type that were solved using the simultaneous solution.

$$a_1x + b_1y = c_1$$

$$a_2x + b_2y = c_2$$

Using a simultaneous solution, the equations are solved for x and y. The result is an equation for x and y in the terms of the coefficients and the constant. The coefficients and constant are known values when solving practical problems.

$$x = \frac{b_2c_1 - b_1c_2}{a_1b_2 - a_2b_1}$$

$$y = \frac{a_1c_2 - a_2c_1}{a_1b_2 - a_2b_1}$$

NOTE: the denominators are the same in both equations.

12–6 EVALUATING A MATRIX FOR SECOND-ORDER DETERMINANTS. The equations shown in Section 12–5 can be used to solve for x and y. A convenient method is to arrange the coefficients and constants in columns referred to as an "array" or "matrix." The procedure for determining the solution of the arrangement is that of multiplication. By definition, a determinant is "a numerical value assigned to a square matrix." The coefficients and constants are arranged in a 2-by-2 square matrix, as shown below. Multiplication is then used to evaluate the matrix; by definition, the procedure is called positive and negative multiplication.

$$\begin{vmatrix} a_1 & b_1 \\ a_2 & b_2 \end{vmatrix}$$

In positive multiplication, multiply, using the algebraic sign across the diagonal downward, as indicated by the arrow.

$$\begin{vmatrix} a_1 & b_1 \\ a_2 & b_2 \end{vmatrix} = (a_1 b_2)$$

In negative multiplication, multiply, using the algebraic sign across the diagonal upward, as indicated by the arrow. The result will be negative.

$$\begin{vmatrix} a_1 & b_1 \\ a_2 & b_2 \end{vmatrix} = -(a_2 b_1)$$

The results are then collected according to algebraic sign.

$$(a_1 b_2) - (a_2 b_1)$$

Example D Evaluate the matrix by positive and negative multiplication.

$$\begin{vmatrix} 2 & 3 \\ 1 & 5 \end{vmatrix}$$

positive multiplication   negative multiplication

$$\begin{vmatrix} 2 & 3 \\ 1 & 5 \end{vmatrix} = (10) \qquad \begin{vmatrix} 2 & 3 \\ 1 & 5 \end{vmatrix} = -(3)$$

then

$$(10) - (3) = 7$$

Example E Evaluate the matrix given.

$$\begin{vmatrix} 4 & 12 \\ -3 & -1 \end{vmatrix} = (-4) \qquad \begin{vmatrix} 4 & 12 \\ -3 & -1 \end{vmatrix} = -(-36)$$

$$(-4) - (-36) = 32$$

EXERCISE 12-3 Evaluate the following matrix, using positive and negative multiplication.

1. $\begin{vmatrix} 4 & 3 \\ 5 & 2 \end{vmatrix}$ _____

2. $\begin{vmatrix} 6 & 9 \\ -2 & 3 \end{vmatrix}$ _____

3. $\begin{vmatrix} -12 & 4 \\ 10 & 6 \end{vmatrix}$

4. $\begin{vmatrix} 1 & 24 \\ -1 & 6 \end{vmatrix}$

5. $\begin{vmatrix} 12 & -4 \\ -2 & -1 \end{vmatrix}$

6. $\begin{vmatrix} -3 & -5 \\ -1 & -7 \end{vmatrix}$

7. $\begin{vmatrix} 12 & 30 \\ 22 & 16 \end{vmatrix}$

8. $\begin{vmatrix} -12 & -10 \\ 5 & 12 \end{vmatrix}$

12-7 SOLVING EQUATIONS WITH SECOND-ORDER DETERMINANTS.
The general equations in Section 12-5 are given below in matrix or determinant form. It can be seen that the denominators in both equations are identical.

$$x = \frac{b_2 c_1 - b_1 c_2}{a_1 b_2 - a_2 b_1}$$

$$x = \frac{\begin{vmatrix} b_2 & c_2 \\ b_1 & c_1 \end{vmatrix}}{\begin{vmatrix} a_1 & b_1 \\ a_2 & b_2 \end{vmatrix}}$$

$$y = \frac{a_1 c_2 - a_2 c_1}{a_1 b_2 - a_2 b_1}$$

$$y = \frac{\begin{vmatrix} a_1 & c_1 \\ a_2 & c_2 \end{vmatrix}}{\begin{vmatrix} a_1 & b_1 \\ a_2 & b_2 \end{vmatrix}}$$

The above determinant solution can be applied to simultaneous equations having two unknowns, as in the equations:

$$2x - y = 6 \quad \text{and} \quad 3x + y = 4$$

In the equations

$$a_1 = 2 \qquad a_2 = 3$$
$$b_1 = -1 \qquad b_2 = 1$$
$$c_1 = 6 \qquad c_2 = 4$$

The solution for x is

$$x = \frac{\begin{vmatrix} 1 & 4 \\ -1 & 6 \end{vmatrix}}{\begin{vmatrix} 2 & -1 \\ 3 & +1 \end{vmatrix}} = \frac{6 - (-4)}{2 - (-3)} = \frac{10}{5} = 2$$

The solution for y is

$$y = \frac{\begin{vmatrix} 2 & 6 \\ 3 & 4 \end{vmatrix}}{\begin{vmatrix} 2 & -1 \\ 3 & 1 \end{vmatrix}} = \frac{8 - (18)}{2 - (-3)} = \frac{-10}{5} = -2$$

In the equations, $x = 2$ and $y - 2$.

Example F Solve the two equations for the unknowns:

$$6m + 4n = 24$$
$$8m + 2n = 22$$

$$m = \frac{\begin{vmatrix} 2 & 22 \\ 4 & 24 \end{vmatrix}}{\begin{vmatrix} 6 & 4 \\ 8 & 2 \end{vmatrix}} = \frac{48 - (88)}{12 - (32)} = \frac{-40}{-20} = 2$$

$$n = \frac{\begin{vmatrix} 6 & 24 \\ 8 & 22 \end{vmatrix}}{-20} = \frac{132 - (192)}{-20} = \frac{-60}{-20} = 3$$

$$m = 2 \quad \text{and} \quad n = 3$$

EXERCISE 12–4 Solve the equations for the unknowns, using second-order determinants.

1. $3a - 3y = 3$
   $5a - y = 13$

   $a = $ _____

   $b = $ _____

2. $3I_1 + I_2 = 11$
   $3I_1 - I_2 = 13$

   $I_1 = $ _____

   $I_2 = $ _____

3. $4e - 6E = -6$
   $-5e + 12E = 30$

   $e =$ _____
   $E =$ _____

4. $4R + 5r = 3$
   $2R + 2r = -7$

   $R =$ _____
   $r =$ _____

5. $4B = 15 - 3A$
   $10 = 8B - 2A$

   $B =$ _____
   $A =$ _____

6. $4Z_1 - 2Z_2 = 8$
   $3Z_1 + 6Z_2 = 10$

   $Z_1 =$ _____
   $Z_2 =$ _____

7. $5a - 2b = -5$
   $3a + 5b = 12$

   $a =$ _____
   $b =$ _____

8. $6x - 10y = 24$
   $4x - 12 = -8$

   $x = \underline{\hspace{2cm}}$

   $y = \underline{\hspace{2cm}}$

12-8 THIRD-ORDER DETERMINANTS.

When three unknowns appear in an equation, the equation can be solved by simultaneous solution; but that usually becomes cumbersome and time-consuming. Third-order determinants can be used for solutions to equations with three unknowns. Fourth-order determinants or greater are not needed to solve most network problems required by technicians, and thus will not be considered in this text.

The matrix for a third-order determinant is 3 by 3. It is solved by positive and negative multiplication. To simplify the procedure of this multiplication, the first two columns are repeated at the right of the matrix, as seen below.

3-by-3 matrix

$$\begin{vmatrix} 2 & 3 & 1 \\ 4 & 2 & 3 \\ 1 & 2 & 4 \end{vmatrix}$$

3-by-3 matrix with two columns repeated

$$\begin{vmatrix} 2 & 3 & 1 & 2 & 3 \\ 4 & 2 & 3 & 4 & 2 \\ 1 & 2 & 4 & 1 & 2 \end{vmatrix}$$

The determinant is then solved in the following manner.

positive multiplication:

$$\begin{vmatrix} 2 & 3 & 1 & 2 & 3 \\ 4 & 2 & 3 & 4 & 2 \\ 1 & 2 & 4 & 1 & 2 \end{vmatrix} = 16 \quad \begin{vmatrix} 2 & 3 & 1 & 2 & 3 \\ 4 & 2 & 3 & 4 & 2 \\ 1 & 2 & 4 & 1 & 2 \end{vmatrix} = 9 \quad \begin{vmatrix} 2 & 3 & 1 & 2 & 3 \\ 4 & 2 & 3 & 4 & 2 \\ 1 & 2 & 4 & 1 & 2 \end{vmatrix} = 8$$

$$(16) + (9) + (8) = (33)$$

negative multiplication:

$$\begin{vmatrix} 2 & 3 & 1 & 2 & 3 \\ 4 & 2 & 3 & 4 & 2 \\ 1 & 2 & 4 & 1 & 2 \end{vmatrix} = 2 \quad \begin{vmatrix} 2 & 3 & 1 & 2 & 3 \\ 4 & 2 & 3 & 4 & 2 \\ 1 & 2 & 4 & 1 & 2 \end{vmatrix} = 12 \quad \begin{vmatrix} 2 & 3 & 1 & 2 & 3 \\ 4 & 2 & 3 & 4 & 2 \\ 1 & 2 & 4 & 1 & 2 \end{vmatrix} = 48$$

$$(2) + (12) + (48) = -(62)$$

and $(33) - (62) = -29$

Example G Solve the determinant given.

$$\begin{vmatrix} 3 & 2 & 1 \\ 4 & -3 & 2 \\ -4 & 5 & 2 \end{vmatrix} \quad \begin{vmatrix} 3 & 2 & 1 & 3 & 2 \\ 4 & -3 & 2 & 4 & -3 \\ -4 & 5 & 2 & -4 & 5 \end{vmatrix}$$

$(3)(-3)(2) + (2)(2)(-4) + (1)(4)(5) = (-14)$
$(-4)(-3)(1) + (5)(2)(3) + (2)(4)(2) = -(58)$

$(-14) - (58) = -72$

EXERCISE 12-5 Solve the following matrix for third-order determinants.

1. $\begin{vmatrix} 3 & 7 & 4 \\ 6 & 5 & 2 \\ 1 & 4 & 2 \end{vmatrix}$

2. $\begin{vmatrix} -2 & 5 & 7 \\ 4 & -4 & 3 \\ 2 & 5 & -1 \end{vmatrix}$

3. $\begin{vmatrix} 7 & -12 & 4 \\ 1 & 10 & 3 \\ -3 & 5 & 4 \end{vmatrix}$

4. $\begin{vmatrix} -12 & 5 & 6 \\ 8 & 2 & -3 \\ 20 & 6 & 14 \end{vmatrix}$

5. $\begin{vmatrix} 7 & 6 & -9 \\ -3 & 4 & 2 \\ 5 & -2 & 1 \end{vmatrix}$

6. $\begin{vmatrix} 12 & 4 & 6 \\ -10 & 2 & 3 \\ -2 & -4 & 2 \end{vmatrix}$

12-9 SOLVING EQUATIONS WITH THIRD-ORDER DETERMINANTS.

Network problems in electronics sometimes develop into simultaneous equations having three unknowns. Three unknowns in simultaneous equations can be solved using third-order determinants. The matrix is developed in the same manner as a second-order matrix. Standard equations are given to show the development of the 3-by-3 matrix.

$$a_1 x + b_1 y + c_1 z = d_1$$
$$a_2 x + b_2 y + c_2 z = d_2$$
$$a_3 x + b_3 y + c_3 z = d_3$$

The denominator is the same for each matrix.

$$\begin{vmatrix} a_1 & b_1 & c_1 \\ a_2 & b_2 & c_2 \\ a_3 & b_3 & c_3 \end{vmatrix}$$

The matrix for each numerator of the unknowns is developed by replacing the coefficient of the unknown with the constant, or d.

$$x = \frac{\begin{vmatrix} d_1 & b_1 & c_1 \\ d_2 & b_2 & c_2 \\ d_3 & b_3 & c_3 \end{vmatrix}}{\text{denominator}}$$

$$y = \frac{\begin{vmatrix} a_1 & d_1 & c_1 \\ a_2 & d_2 & c_2 \\ a_3 & d_3 & c_3 \end{vmatrix}}{\text{denominator}}$$

$$z = \frac{\begin{vmatrix} a_1 & b_1 & d_1 \\ a_2 & b_2 & d_2 \\ a_3 & b_3 & d_3 \end{vmatrix}}{\text{denominator}}$$

Example H Solve the simultaneous equations, using third-order determinants.

$$2x + 2y + z = 4$$
$$x + 3y + 4z = 6$$
$$4x + y + 3z = 8$$

$$x = \frac{\begin{vmatrix} 4 & 2 & 1 \\ 6 & 3 & 4 \\ 8 & 1 & 3 \end{vmatrix} \begin{matrix} 4 & 2 \\ 6 & 3 \\ 8 & 1 \end{matrix}}{\begin{vmatrix} 2 & 2 & 1 \\ 1 & 3 & 4 \\ 4 & 1 & 3 \end{vmatrix} \begin{matrix} 2 & 2 \\ 1 & 3 \\ 4 & 1 \end{matrix}}$$

$(36) + (64) + (6) - [(24) + (16) + (36)] =$
$(106) - (76) = 30$

$(18) + (32) + (1) - [(12) + (8) + (6)] =$
$(51) - (26) = 25$

$$x = \frac{30}{25} = 1.2$$

$$y = \frac{\begin{vmatrix} 2 & 4 & 1 \\ 1 & 6 & 4 \\ 4 & 8 & 3 \end{vmatrix} \begin{matrix} 2 & 4 \\ 1 & 6 \\ 4 & 8 \end{matrix}}{25}$$

$(36) + (64) + (8) - [(24) + (64) + (12)] =$
$(108) - (100) = 8$

$$y = \frac{8}{25} = 0.32$$

$$z = \frac{\begin{vmatrix} 2 & 2 & 4 \\ 1 & 3 & 6 \\ 4 & 1 & 8 \end{vmatrix} \begin{matrix} 2 & 2 \\ 1 & 3 \\ 4 & 1 \end{matrix}}{25}$$

$(48) + (48) + (4) - [(48) + (12) + (16)] =$
$(100) - (76) = 24$

$$z = \frac{24}{25} = 0.96$$

EXERCISE 12–6 Solve the simultaneous equations, using third-order determinants.

1. $x + 2y + z = 7$
   $x + 3y + 4z = 14$
   $2x + y + 2z = 2$

   $x = $ _____
   $y = $ _____
   $z = $ _____

2. $2a - 3b + 4c = 12$
   $a - b + 2c = 6$
   $4a - 4b + c = 10$

   $a = $ _____
   $b = $ _____
   $c = $ _____

3. $2I + 4i - 3e = 16$
   $I - 12i + 4e = 14$
   $8I - 5i + e = 28$

   $I = $ _____
   $i = $ _____
   $e = $ _____

4. $4R - 3r - Z = -1$
   $10R - 4r + 4Z = 5$
   $-11R - 6r + 2Z = 1$

   $R =$ _____
   $r =$ _____
   $Z =$ _____

5. $12M - 2m - I = 12 - 4I$
   $-6M + 10m - 7I = 6$
   $2M + 4m - 3I = 10m + 16$

   $M =$ _____
   $m =$ _____
   $I =$ _____

6. $5s - 6t + 2\mu = 10$
   $-s + 12t + 8\mu = 15$
   $10s - 5t - 18\mu = 10\mu - 18$

   $s =$ _____
   $t =$ _____
   $\mu =$ _____

EVALUATION EXERCISE Solve the simultaneous equations in Exercises 1 through 4, using addition, subtraction, or substitution.

1. $4a - 3b = 5$
   $a + b = 12$

   $a =$ _____
   $b =$ _____

2. $3x + 2y = 12$
   $4x - 3y = 6$

   $x =$ _____
   $y =$ _____

Chapter 12
Simultaneous Equations
and Determinants

3. $R + 3r = 11$
   $7r + 4R = 20$

   $R = $ _____

   $r = $ _____

4. $12V + 4v = 14$
   $5V - 2v = 8$

   $V = $ _____

   $v = $ _____

Solve the equations in Exercises 5 through 10, using determinants.

5. $2a + 5b = 3$
   $2a + 2b = -7$

   $a = $ _____

   $b = $ _____

6. $5x - 2y = -5$
   $3x + 5y = 12$

   $x = $ _____

   $y = $ _____

7. $4c + 3B = 15$
   $2B + 10 = 8c$

   $B = $ _____

   $c = $ _____

8. $2x - 3y + 4z = 12$
   $4x - 4y + z = 10$
   $x - y + 2z = 6$

   $x =$ _____

   $y =$ _____

   $z =$ _____

9. $a + 2b - c = 6$
   $a + 3b + 3c = 12$
   $2a + b + 2c = 2$

   $a =$ _____

   $b =$ _____

   $c =$ _____

10. $R + 2r - s = 4$
    $2R - r + 4s = 3$
    $3R + 4r - 2s = 6$

    $R =$ _____

    $r =$ _____

    $s =$ _____

Chapter 12
Simultaneous Equations and Determinants

CHAPTER 13

Networks

Objectives

After completing this chapter you will be able to:
- Apply Kirchhoff's Law to circuits containing two power sources
- Develop loop equations for dc circuits with two voltage loops
- Develop loop equations for dc circuits with three voltage loops
- Solve simultaneous loop equations

Some electrical circuits or networks have more than one power source. One method of solving networks is the application of Kirchhoff's Law, which results in simultaneous equations. The equations can be solved simultaneously with algebraic manipulation or by using determinants.

13-1 APPLICATION OF SIMULTANEOUS EQUATIONS. Some circuits contain more than one power source, as shown in Figure 13-1. Circuits of this type can be solved by using loop equations and applying the principles of simultaneous equations to finish the solution.

There are as many methods of applying loop equations to circuits as there are authors of textbooks. Any method will work; but to avoid confusion the method used in this text is the conventional and widely used one that follows. Given the circuit in Figure 13-1, find the current I_a and I_b.

1. Simplify the circuit by combining any series or parallel components (Figure 13-2).
2. There are three possible loops in which current can flow in the circuit. At this point a direction of electron flow is assumed in the counterclockwise direction, without regard to polarity of sources (Figure 13-3).

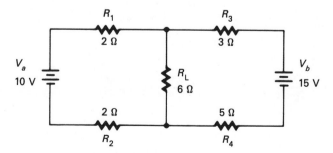

FIGURE 13-1   Circuit with two power sources

184

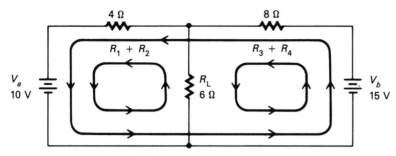

FIGURE 13–2   Three possible loops

3. Choose two of the three loops. In this problem the two internal loops have been chosen and have been identified with the currents I_a and I_b, as shown in Figure 13–3. Assign a polarity to each component based on the assumed direction of electron flow (negative to positive), as shown in Figure 13–3.
4. Kirchhoff's Law states that the algebraic sum of the voltage drops around a closed path of current taken with proper signs is equal to zero:

$$V_1 + V_2 + V_3 + \text{etc.} = 0$$

"Proper sign" is defined as the first algebraic sign (+ or −) encountered when developing loops. Each component was assigned a polarity when the direction of current was assumed.
5. The voltage across each component will be an Ohm's Law function ($V = IR$). In Figure 13–4, the voltage across each component will be the value of the given resistance times the current through the resistor. The voltage across the 4 Ω resistor taken with the proper sign is $(-4I_a)$. Following the assumed current (I_b), the voltage across the 8 Ω resistor is $(-8I_b)$. (See Figure 13–4.)
6. The voltage across the 6 Ω resistor is determined by 6 Ω and both I_a and I_b.

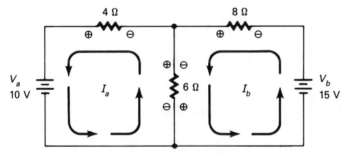

FIGURE 13–3   Voltage polarities for each component

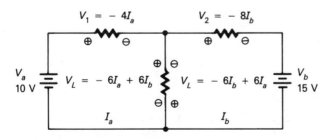

FIGURE 13–4   Voltage equations for each component

loop I_a $\quad 10 - 6I_a + 6I_b - 4I_a = 0$

$$-10I_a + 6I_b = -10 \text{ V}$$

loop I_b $\quad -15 - 8I_b - 6I_b + 6I_a = 0$

$$6I_a - 14I_b = 15 \text{ V}$$

7. Each loop is collected independently of the other:

Loop using I_a

$V_a - V_L - V_1 = 0$ (Kirchhoff's Law; collected with proper signs)

$V_a = 10$ V

$V_1 = -4I_a$

$V_L = -6I_a$ and $+6I_b$

Since both currents are present in R_L, the voltage is determined by I_a and I_b. The sign for I_a is negative, and I_b is positive.

Collecting the voltages in terms of *IR*:

$$10 - 6I_a + 6I_b - 4I_a = 0$$

Simplify and put in standard equation form:

$$-10I_a + 6I_b = -10$$

Loop using I_b

$$V_b - V_2 - V_L = 0$$

$-15 - 8I_b - 6I_b + 6I_a = 0$ ($6I_a$ is positive because, when collecting the I_b loop, the sign I_a is +.)

Simplify and put in standard form:

$$6I_a - 14I_b = 15$$

$-10I_a + 6I_b = -10$

$6I_a - 14I_b = 15$ (both equations are in standard form)

Solving for I_a and I_b

$$3(-10I_a + 6I_b = -10)$$

$$5(6I_a - 14I_b = 15)$$

$$18I_b = -30$$
$$\frac{-70I_b = 75}{-52I_b = 45}$$

$$I_b = -0.865 \text{ A} = -865 \text{ mA}$$

$$-10I_a + 6(-0.865) = -10$$

$$-10I_a = -10 + 5.19$$

$$-10I_a = -4.81$$

$$I_a = 0.481 \text{ A} = 481 \text{ mA}$$

NOTE: Because I_b is a negative value, the electron flow was assumed to be in the opposite direction. The value is correct, but the direction of flow is reversed. When further analyzing the circuit, the current should be used in the proper direction.

SUMMARY:
- Assume all possible paths of electron flow. Conventional current may be used if desired.

- The assumptions should all be in the same direction, either counter-clockwise or clockwise.

- Assign polarities to the components in accordance with the assumed electron flow.

- Collect the voltage drops for each loop and put equations in standard form.

- Solve for the currents by solving the equations simultaneously.

Example A Find I_a, I_b, and I_L.

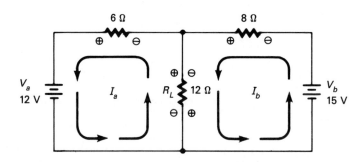

FIGURE 13–5   Circuit for Example A

loop A $12 - 18I_a + 12I_b = 0$ $-18I_a + 12I_b = -12$

loop B $-15 - 20I_b + 12I_a = 0$ $12I_a - 20I_b = 15$

$$5(-18I_a + 12I_b = -12)$$

$$3(12I_a - 20I_b = 15)$$

$$-90I_a = -60$$

$$\underline{36I_a = 45}$$

$$-54I_a = -15$$

$$I_a = 0.277 \text{ A} = 277 \text{ mA}$$

$$12I_a - 20I_b = 15$$

$$12(0.277) - 20I_b = 15$$

$$-20I_b = 15 - 3.32$$

$$-20I_b = 11.68$$

$$I_b = -0.584 \text{ A} = 584 \text{ mA}$$

Finding the Current I_L in Figure 13–5 Because current I_b is a negative value, it is actually flowing in the opposite direction, and the current $I_L = I_a + I_b$.

$$I_L = 0.277 \text{ A} + 0.584 \text{ A}$$

$$I_L = 0.861 \text{ A} = 861 \text{ mA}$$

13–2 REVERSING ONE POWER SUPPLY. The procedure is the same in reversing one power supply, but one polarity changes, as shown in Figure 13–6.

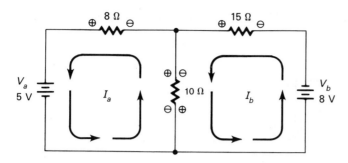

FIGURE 13-6   Circuit with reversed power supply

loop I_a $5 - 18I_a + 10I_b = 0$ and $-18I_a + 10I_b = -5$

loop I_b $8 - 25I_b + 10I_a = 0$ and $10I_a - 25I_b = -8$

$$5(-18I_a + 10I_b = -5)$$

$$2(10I_a - 25I_b = -8)$$

$$-90I_a = -25$$

$$\frac{20I_a = -16}{-70I_a = -41}$$

$$I_a = 0.586 \text{ A} = 586 \text{ mA}$$

$$10I_a - 25I_b = -8$$

$$10(0.586) - 25I_b = -8$$

$$-25I_b = -8 - 5.86$$

$$-25I_b = -13.86$$

$$I_b = 0.554 \text{ A} = 554 \text{ mA}$$

$$I_L = I_a - I_b$$

$$I_L = 586 \text{ mA} - 554 \text{ mA}$$

$$I_L = 32 \text{ mA}$$

NOTE: Since I_a and I_b are both positive, the assumption of electron flow is correct, and these currents are opposing.

EXERCISE 13–1 Using loop equations solve the following for the unknown values indicated.

1.
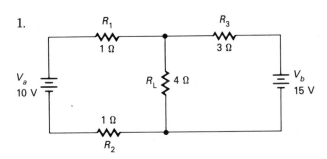

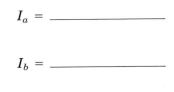

$I_a = $ _____

$I_b = $ _____

$I_L = $ _____

FIGURE 13–7   Circuit for Exercise 13–1 (1)

2.
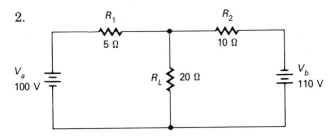

$I_a = $ _____

$I_b = $ _____

$I_L = $ _____

FIGURE 13–8   Circuit for Exercise 13–1 (2)

3.
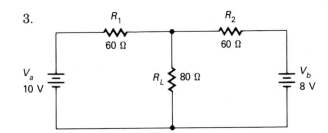

$I_a = $ _____

$I_b = $ _____

$I_L = $ _____

FIGURE 13–9   Circuit for Exercise 13–1 (3)

4.
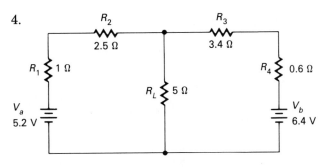

$I_a = $ _____

$I_b = $ _____

$I_L = $ _____

FIGURE 13–10   Circuit for Exercise 13–1 (4)

5.

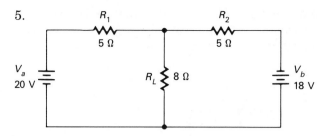

FIGURE 13-11   Circuit for Exercise 13-1 (5)

$I_a =$ _____

$I_b =$ _____

$I_L =$ _____

6.

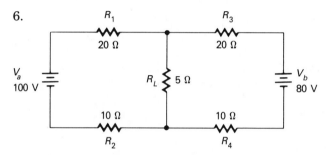

FIGURE 13-12   Circuit for Exercise 13-1 (6)

$I_a =$ _____

$I_b =$ _____

$I_L =$ _____

7.

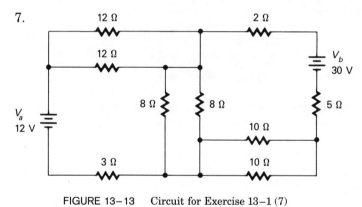

FIGURE 13-13   Circuit for Exercise 13-1 (7)

$I_a =$ _____

$I_b =$ _____

8.

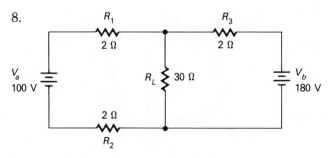

FIGURE 13-14   Circuit for Exercise 13-1 (8)

$I_L =$ _____

13-3 CIRCUITS WITH THREE LOOPS. Although the method for developing loops is the same, it is easier to solve the equations by using determinants. Using this method, determine the currents in the circuit given in Figure 13–15.

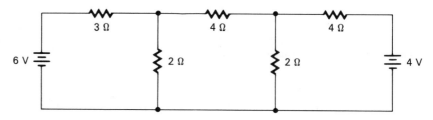

FIGURE 13–15   Circuit with three loops

1. Assume a counterclockwise flow of current in each loop, and assign polarities for each component (Figure 13–16).

2. Develop equations for each loop:

   I_a: $6V - 2I_a + 2I_b - 3I_a = 0$

   I_b: $-2I_b + 2I_a - 2I_b + 2I_c - 4I_b = 0$

   I_c: $-4V - 4I_c - 2I_c + 2I_b = 0$

   Collect terms:

   $$-5I_a + 2I_b + 0 = -6 \text{ V}$$
   $$2I_a - 8I_b + 2I_c = 0$$
   $$0 + 2I_b - 6I_c = 4 \text{ V}$$

3. Develop the matrix for I_a and solve by using determinants:

$$I_a = \frac{\begin{array}{ccc|cc} -6 & 2 & 0 & -6 & 2 \\ 0 & -8 & 2 & 0 & -8 \\ 4 & 2 & -6 & 4 & 2 \end{array}}{\begin{array}{ccc|cc} -5 & 2 & 0 & -5 & 2 \\ 2 & -8 & 2 & 2 & -8 \\ 0 & 2 & -6 & 0 & 2 \end{array}}$$

$(-288 + 16 + 0) - (0 - 24 + 0) =$
$(-272) - (-24) = \mathbf{-248}$

$(-240 + 0 + 0) - (0 - 20 - 24) =$
$(-240) - (-44) = \mathbf{-196}$

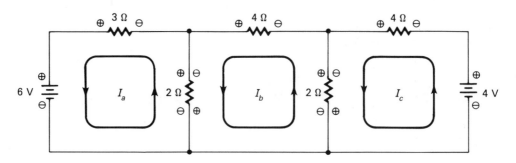

FIGURE 13–16   Loops and voltage polarities

$$I_a = \frac{-248}{-196}$$
$$I_a = 1.27 \text{ A}$$

$$I_b = \frac{\begin{vmatrix} -5 & -6 & 0 \\ 2 & 0 & 2 \\ 0 & 4 & -6 \end{vmatrix} \begin{matrix} -5 & -6 \\ 2 & 0 \\ 0 & 4 \end{matrix}}{-196} \quad \begin{matrix} (0 + 0 + 0) - (0 - 40 + 72) = \\ (0) - (32) = \mathbf{-32} \end{matrix}$$

$$I_b = \frac{-32}{-196}$$
$$I_b = 0.163 \text{ A} = 163 \text{ mA}$$

$$I_c = \frac{\begin{vmatrix} -5 & 2 & -6 \\ 2 & -8 & 0 \\ 0 & 2 & 4 \end{vmatrix} \begin{matrix} -5 & 2 \\ 2 & -8 \\ 0 & 2 \end{matrix}}{-196} \quad \begin{matrix} (160 + 0 - 24) - (0 + 0 + 16) = \\ (136) - (16) = \mathbf{120} \end{matrix}$$

$$I_c = \frac{120}{-196}$$
$$I_c = -0.612 \text{ A} = 612 \text{ mA}$$

EXERCISE 13–2 Use determinants to solve the following exercises.

1. Find currents I_a and I_b in the circuit in Figure 13–17.

    $I_a =$ _____
    $I_b =$ _____

2. Find currents I_a and I_b in the circuit in Figure 13–17 if all the resistances were changed to 5Ω.

    $I_a =$ _____
    $I_b =$ _____

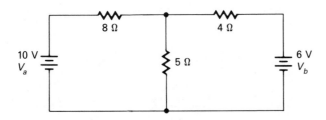

FIGURE 13–17 Circuit for Exercise 13–2 (1)

3. Find the currents I_a, I_b, and I_c for the circuit in Figure 13–18.

$I_a =$ _____
$I_b =$ _____
$I_c =$ _____

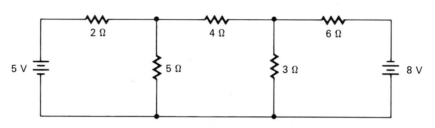

FIGURE 13–18   Circuit for Exercise 13–2 (3)

4. Repeat Exercise 3, with all the resistors equal to 6Ω.

$I_a =$ _____
$I_b =$ _____
$I_c =$ _____

5. Find currents I_a, I_b, and I_c for the circuit given in Figure 13–19.

$I_a =$ _____
$I_b =$ _____
$I_c =$ _____

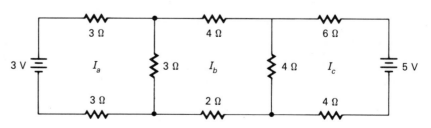

FIGURE 13–19   Circuit for Exercise 13–2 (5)

EVALUATION EXERCISE

1. Find the current through the load in the circuit in Figure 13–20, using loop equations.

   Chapter 13 Networks

   $I_L = $ _____

   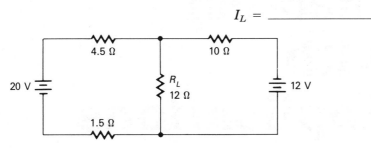

   FIGURE 13–20 Circuit for Evaluation Exercise (1)

2. What would be the current through the load in Exercise 1 if an 8 Ω resistor were in series with the load? Use loop equations to determine the answer.

   $I_L = $ _____

3. Find the current through the load in the circuit in Figure 13–21, using determinants.

   $I_L = $ _____

   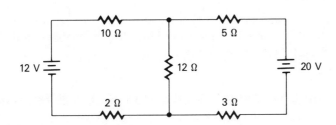

   FIGURE 13–21 Circuit for Evaluation Exercise (3)

4. Find currents I_a, I_b, and I_c in the circuit in Figure 13–22. Use determinants to find the solution.

   $I_a = $ _____
   $I_b = $ _____
   $I_c = $ _____

   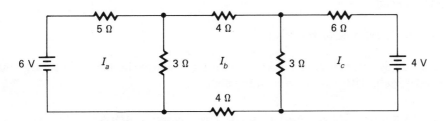

   FIGURE 13–22 Circuit for Evaluation Exercise (4)

CHAPTER 14
Thevenin's Theorem with Applications

Objectives

After completing this chapter you will be able to:
- Develop Thevenin's Theorem
- Solve for load currents with one power source
- Solve for load currents with two power sources
- Develop Norton's Theorem
- Solve for load currents with one power source
- Solve for load currents with two power sources

Electronic circuits, especially the type used in Chapter 13, are often analyzed by using Thevenin's equivalent circuits. This chapter utilizes that principle to analyze the circuits used in Chapter 13.

14–1 THEVENIN'S THEOREM. Thevenin, a French engineer, stated a practical theorem which, when applied to a circuit, basically states that the current through a load can be determined if the load resistance, Thevenin's internal equivalent resistance, and the open circuit voltage are known. Thevenin's Theorem can be stated as an equation

$$I_L = \frac{V_{open}}{R_L + R_{Th}}$$

where:

V_{open} is the voltage measured across the output terminals of a supply with the load removed.

R_L is the resistance of the load.

R_{Th} is the resistance of the circuit looking back into the circuit from the output terminals with the power source shorted.

14–2 THEVENIZING A CIRCUIT WITH ONE POWER SOURCE. Given the circuit in Figure 14–1, find the current through the load. The load (R_L) is

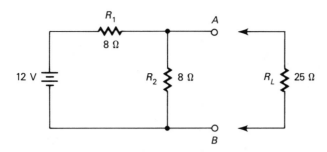

FIGURE 14-1   Thevenizing with one power source

placed across points A to B. The load current can be calculated if the voltage measured from points A to B can be determined and the resistance measured from A to B with the power source shorted.

$$I_L = \frac{V_{open}}{R_L + R_{Th}}$$

V_open (Open-circuit Voltage)   The open-circuit voltage is the voltage drop across R_2 with the load (R_L) removed (Figure 14–2).

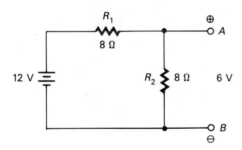

FIGURE 14-2   Open-circuit voltage

By Ohm's Law, calculation of this voltage is 6 V.

$$I_T = \frac{12 \text{ V}}{16 \text{ }\Omega} = 0.75 \text{ A}$$

$$V_{R_2} = V_{open} = 0.75 \text{ A}(8 \text{ }\Omega) = 6 \text{ V}$$

Thevenin's Equivalent Resistance (R_{Th})   If the power source were shorted and the resistance measured from points A to B, the resistors R_1 and R_2 would be in a parallel combination, as they are in Figure 14–3a. Two 8 Ω resistors in parallel result in $R_{Th} = 4$ Ω, Figure 14–3b.

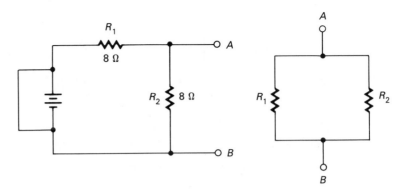

FIGURE 14-3a   Thevenin's equivalent resistance

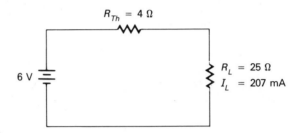

FIGURE 14-3b   Thevenin's equivalent circuit

Solution by Thevenin's Equation

$$I_L = \frac{V_{open}}{R_L + R_{Th}}$$

$$I_L = \frac{6}{25 + 4}$$

$$I_L = 207 \text{ mA}$$

Validation of Thevenin's Theorem   To verify Thevenin's Theorem calculate the load current by conventional methods (Figure 14-4).

$$R_T = R_1 + \frac{R_2 R_L}{R_2 + R_L} \qquad V_{R_2} = V_{R_L} = I_T\left(\frac{R_2 R_L}{R_2 + R_L}\right)$$

$$R_T = 8 + \frac{8(25)}{33} \qquad V_{R_2} = V_{R_L} = 0.851\left(\frac{8(25)}{8 + 25}\right)$$

$$R_T = 14.1 \text{ }\Omega \qquad V_{R_L} = 5.16 \text{ V}$$

$$I_T = \frac{12 \text{ V}}{14.1 \text{ }\Omega} = 0.851 \text{ A} \qquad I_{R_L} = \frac{5.16 \text{ V}}{25 \text{ }\Omega} = 206 \text{ mA}$$

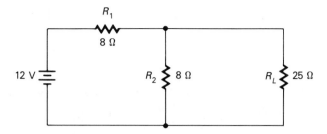

FIGURE 14-4 Finding I_L by Ohm's Law

14-3 THEVENIZING CIRCUITS WITH TWO POWER SOURCES.
Given the circuit in Figure 14-5a (the same circuit used in Chapter 13 to explain the use of simultaneous equations), simplify the circuit by collecting series components (Figure 14-5b).

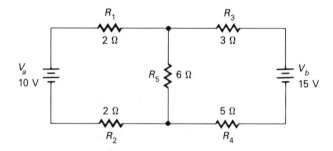

FIGURE 14-5a Thevenizing with two power supplies

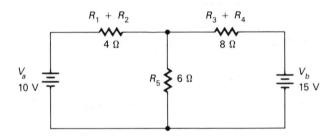

FIGURE 14-5b Simplified circuit of Figure 14-5a

Thevenize for the Current I_b Because current I_b flows through the 8 Ω resistor it is considered the load and is removed from the circuit (Figure 14-6). The position of V_b has been changed to simplify the circuit, but electrically it is in the same position.

$$I_L = \frac{V_{open}}{R_L + R_{Th}}$$

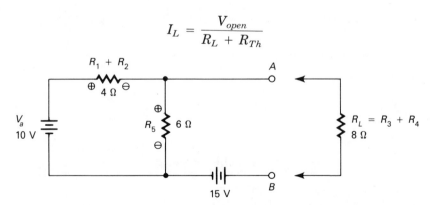

FIGURE 14-6 Thevenizing for current I_b

The Load Resistance R_L The 8 Ω resistor is removed from the circuit and becomes the load.

Open Circuit Voltage Determine the voltage across the 6 Ω resistor in Figure 14–6. This is a simple Ohm's Law calculation for the series circuit containing the 6 Ω resistor, 4 Ω resistor, and the 10 V supply.

$$R_5 = \frac{10\text{ V}}{4\text{ Ω} + 6\text{ Ω}} \times 6 = 6\text{ V}$$

Collect the voltage with proper algebraic signs from points A to B through R_5 in Figure 14–7.

$$V_{open} = V_{A-B} = 6\text{ V} - 15\text{ V} = 9\text{ V}$$

$$V_{open} = 9\text{ V}$$

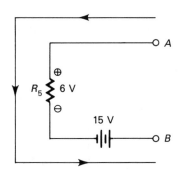

FIGURE 14–7   Determining V_{open}

Thevenin's Equivalent Resistance Thevenin's equivalent resistance is the resistance looking back into the circuit from points A to B with the power sources shorted. This can also be stated: if the resistance from points A to B is measured with an ohmmeter, and the power sources are shorted, the ohmmeter will read R_{Th} (Figure 14–8a).

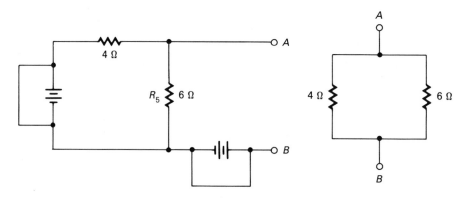

FIGURE 14–8a   Thevenin's equivalent resistance

The resistances are in parallel; the total resistance is:

$$R_{Th} = \frac{4(6)}{4 + 6}$$

$$R_{Th} = \frac{24}{10}$$

$$R_{Th} = 2.4 \, \Omega$$

Current I_b The current I_b and the Thevenin's equivalent circuit are the following (see Figure 14–8b):

$$I_L = I_b = \frac{V_{open}}{R_L + R_{Th}}$$

$$I_b = \frac{9 \text{ V}}{8 + 2.4}$$

$$I_b = 865 \text{ mA}$$

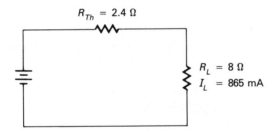

FIGURE 14–8b   Thevenin's equivalent circuit

Current I_a The current I_a can be determined by using simple voltage loops, since the voltage across the 8 Ω resistors is known; or Thevenin's can be applied to the circuit for I_a, in which case it would be applied in the same manner (Figures 14–9a and 14–9b).

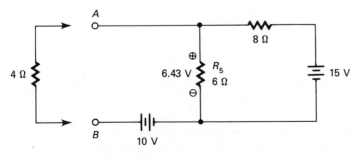

FIGURE 14–9a   Determining the current I_a

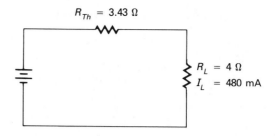

FIGURE 14–9b   Thevenin's equivalent circuit

$$I_a = I_L = \frac{V_{open}}{R_L + R_{Th}}$$

$$R_L = 4\ \Omega$$

$$V_{open} = 6.43\ \text{V} - 10\ \text{V} = -3.57\ \text{V}$$

$$R_{Th} = \frac{8(6)}{14} = 3.43\ \Omega$$

$$I_a = I_L = \frac{3.57}{4\ \Omega + 3.43\ \Omega}$$

$$I_a = 480\ \text{mA}$$

Current Through R_5 R_5 is the load and is removed from the circuit. The value of the load is 6 Ω (Figure 14–10).

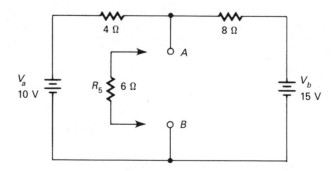

FIGURE 14–10   Determining the current through R_5

Open Circuit Voltage   The power supplies are in a series opposing connection, and the result across the 4 Ω and 8 Ω resistors is 5 V. The voltage across the 4 Ω and 8 Ω resistors is a simple series-circuit calculation.

$$V_{4\Omega} = \frac{5\text{ V}}{12} \times 4\ \Omega = 1.67\text{ V}$$

$$V_{8\Omega} = \frac{5\text{ V}}{12} \times 8\ \Omega = 3.33\text{ V}$$

Because V_b is the larger of the two voltages it controls the direction of current, and the electron flow is from V_b through V_a. The polarity is as indicated in Figure 14–11. Collect the voltage, with proper sign, from A to B, either through V_a or V_b; the result is the same.

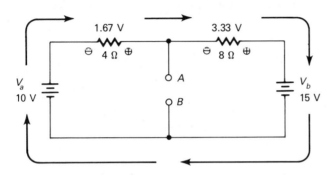

FIGURE 14–11   Open-circuit voltage

$$V_{open} = V_{a-b} = 1.67\text{ V} + 10\text{ V} \quad \text{(through } V_a\text{)}$$
$$V_{open} = 11.67\text{ V}$$

or

$$V_{open} = -3.33\text{ V} + 15\text{ V} \quad \text{(through } V_b\text{)}$$
$$V_{open} = 11.67\text{ V}$$

Thevenin's Equivalent Resistance   Again, the equivalent Thevenin's resistance is the parallel combination of the 4 Ω and 8 Ω resistance (Figures 14–12a and 14–12b).

$$R_{Th} = \frac{4 \times 8}{12}$$

$$R_{Th} = 2.67\ \Omega$$

$$I_L = I_b = \frac{V_{open}}{R_L + R_{Th}}$$

$$I_L = I_b = \frac{11.67\text{ V}}{6\ \Omega + 2.67\ \Omega}$$

$$I_b = 1.35\text{ A}$$

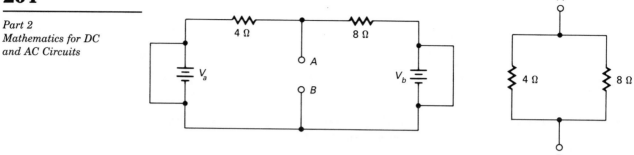

FIGURE 14–12a Thevenin's equivalent resistance

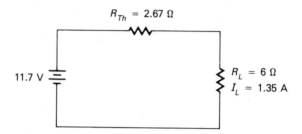

FIGURE 14–12b Thevenin's equivalent circuit

EXERCISE 14–1 Find the currents indicated in each problem, using Thevenin's Theorem.

1.

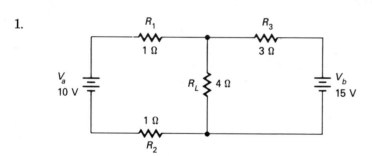

FIGURE 14–13 Circuit for Exercise 14–1 (1)

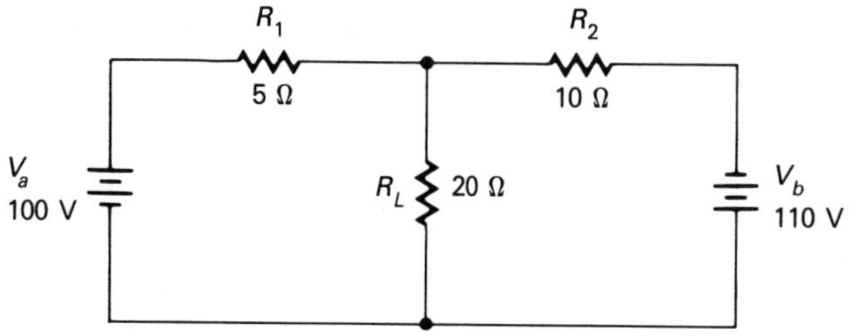

$I_a = $ _____

$I_b = $ _____

$I_L = $ _____

2.

FIGURE 14–14 Circuit for Exercise 14–1 (2)

$I_a = $ _____

$I_b = $ _____

$I_L = $ _____

3.

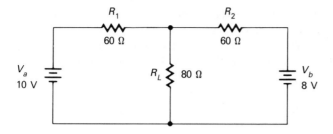

FIGURE 14–15   Circuit for Exercise 14–1 (3)

$I_a =$ _____

$I_b =$ _____

$I_L =$ _____

4.

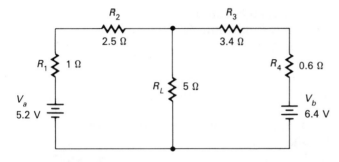

FIGURE 14–16   Circuit for Exercise 14–1 (4)

$I_a =$ _____

$I_b =$ _____

$I_L =$ _____

5.

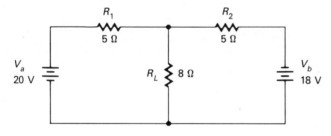

FIGURE 14–17   Circuit for Exercise 14–1 (5)

$I_a =$ _____

$I_b =$ _____

$I_L =$ _____

6.

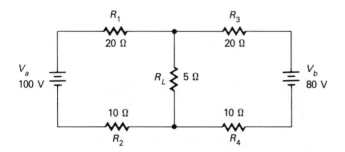

FIGURE 14–18   Circuit for Exercise 14–1 (6)

$I_a =$ _____

$I_b =$ _____

$I_L =$ _____

14-4 NORTON'S THEOREM.

Norton's and Thevenin's theorems are similar, but Norton's considers a generator a constant current device. The open circuit voltage and the Thevenin's equivalent resistance are calculated in the same manner. Norton's Theorem is often used to analyze such current devices as transistors.

Norton's Theorem states:

$$I_{Norton} = \frac{V_{open}}{R_{Thevenin}}$$

where: I_{Norton} is a constant current produced by a source.

V_{open} is the open circuit voltage of the source, calculated in the same manner as it is for Thevenin's.

$R_{Thevenin}$ is the internal resistance with the power source shorted.

The Thevenin's and Norton's equivalent circuits are compared in Figures 14-19a and 14-19b.

Find the load current in the circuit in Figure 14-20a, using Norton's equivalent circuit.

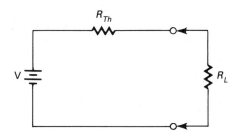

FIGURE 14-19a Thevenin's equivalent circuit

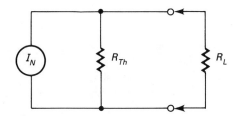

FIGURE 14-19b Norton's equivalent circuit

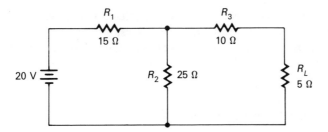

FIGURE 14–20a   Determining the current through R_L with Norton's Theorem

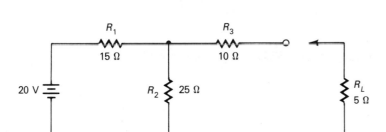

FIGURE 14–20b   Finding open-circuit voltage

Remove the load from the circuit as shown in Figure 14–20b and calculate the open circuit voltage. There is no voltage drop across R_3 because no current is flowing through it. The open circuit voltage is the voltage across R_2, which is an Ohm's Law calculation.

$$V_{open} = \left(\frac{20 \text{ V}}{40 \text{ }\Omega}\right)(25 \text{ }\Omega) = 12.5 \text{ V}$$

Thevenin's equivalent resistance, with the power source shorted, will be the parallel combination of R_1 and R_2 in series with R_3.

$$R_{Th} = \frac{(15)(25)}{15 + 25} + 10$$
$$R_{Th} = 19.4 \text{ }\Omega$$

Norton's equivalent circuit is shown in Figure 14–21; the constant current is calculated by Norton's Theorem.

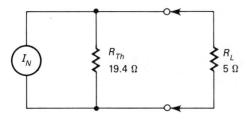

FIGURE 14–21   Norton's equivalent circuit

$$I_N = \frac{V_o}{R_{Th}}$$
$$I_N = \frac{12.5}{19.4}$$
$$I_N = 644 \text{ mA}$$

Norton's constant current source supplies a constant current of 644 mA divided between the Thevenin's equivalent resistance and the load. The load current can now be determined by conventional calculation of parallel circuits.

Calculate the parallel combination of R_{Th} and R_L

$$R_{par} = \frac{(19.4)(5)}{19.4 + 5}$$
$$R_{par} = 3.98 \ \Omega$$

Calculate the voltage across the parallel combination

$$V_{par} = (3.98 \ \Omega)(0.644 \ A)$$
$$V_{par} = 2.56 \ V$$

Calculate the load resistance by Ohm's Law

$$I_L = \frac{2.56 \ V}{5}$$
$$I_L = 512 \ mA$$

14–5 NORTONIZING CIRCUITS WITH TWO POWER SOURCES. Refer to Section 14–3, which is used to illustrate Thevenin's Theorem. In the procedure used to determine the current I_b, it was established that the open circuit voltage V_o, was 9 V and the Thevenin's equivalent resistance, R_{Th}, was 2.4 Ω. This resistance is used to determine the Norton's equivalent circuit and the constant current, as shown in Figure 14–22.

$$I_N = \frac{V_o}{R_{Th}}$$
$$I_N = \frac{9}{2.4}$$
$$I_N = 3.75 \ A$$

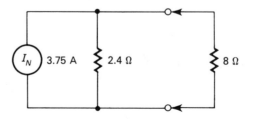

FIGURE 14–22 Norton's equivalent circuit for Section 14–5

Calculate I_b by a conventional parallel circuit method. Determine the parallel resistance of the circuit in Figure 14–22.

$$R_{par} = \frac{R_{Th} R_L}{R_{Th} + R_L}$$

$$R_{par} = \frac{(2.4)(8)}{2.4 + 8}$$

$$R_{par} = 1.85 \ \Omega$$

Calculate the voltage across the parallel combination:

$$V_{par} = I_N R_{par}$$
$$V_{par} = (3.75)(1.85)$$
$$V_{par} = 6.94 \ V$$

Calculate the load current by Ohm's Law. The load current is the current I_b determined in Section 14–3.

$$I_L = \frac{V_{par}}{R}$$
$$I_L = \frac{6.94}{8}$$
$$I_L = 867 \ mA$$

The current I_L can be compared with that calculated for I_b in Section 14–3. In that section I_b was 865 mA; where it was calculated to be 867 mA, well within tolerance. The currents I_a and I_b calculated in Section 14–3 can also be determined using the same procedure.

Ratio Method A ratio between currents and resistance can also be used to find the load current.

$$\frac{I_N}{I_L} = \frac{R_L}{R_{par}}$$

$$\frac{3.75}{I_L} = \frac{8}{1.85}$$

$$I_N = 867 \ mA$$

EXERCISE 14–2 Solve the problems, using Norton's Theorem.

1. The Thevenin's equivalent resistance of a circuit is 15.5 Ω, and the Norton's current is 550 mA. What is the load current through a 75 Ω load added to the circuit?

$$I_L = \underline{\hspace{2in}}$$

2. What is the current through a 50 Ω load if it is placed across a power source, the Thevenin's equivalent resistance of the source is 22.5 Ω, and the Norton's constant current is 875 mA?

$I_L =$ _____

3. Calculate the load current I_L in the circuit shown in Figure 14–23.

$I_L =$ _____

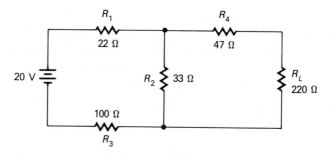

FIGURE 14–23 Circuit for Exercise 14–2 (3)

4. Repeat Exercise 3, with $R_1 = 68$ Ω and $R_L = 150$ Ω.

$I_L =$ _____

5. Determine the current from the source V_a in the circuit in Figure 14–24.

$I_a =$ _____

6. Determine the current through the load in the circuit given in Figure 14–24.

$I_L = $ _____

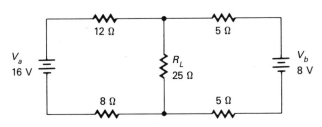

FIGURE 14–24   Circuit for Exercise 14–2 (5)

EVALUATION EXERCISE   Solve, using Thevenin's Theorem.

1. Find the load current in the circuit given in Figure 14–25.

$I_L = $ _____

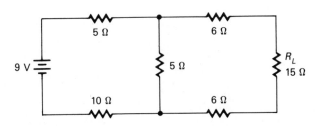

FIGURE 14–25   Circuit for Evaluation Exercise (1)

2. In Figure 14–26 determine the current from the source V_b.

$I_b = $ _____

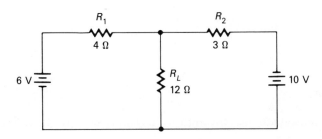

FIGURE 14–26   Circuit for Evaluation Exercise (2)

3. In the circuit given in Figure 14–26, change R_1 to 15 Ω and calculate the current through the load.

$I_L = $ _____

Solve Exercises 4 through 6, using Norton's Theorem.

4. Determine the current through R_1, using the circuit in Figure 14–26, with the following changes: $R_1 = 22\ \Omega$, $R_2 = 12\ \Omega$, $R_L = 20\ \Omega$.

$I_1 = $ _____

5. Find the current through R_2 in the circuit given in Exercise 4.

$I_2 = $ _____

6. Determine the current through R_L in the circuit given in Exercise 4.

$I_L = $ _____

CHAPTER 15

The Bridge

Objectives

After completing this chapter you will be able to:
- Identify a balanced bridge circuit
- Solve a bridge circuit for the unbalanced current, using loop equations
- Solve a bridge circuit for unbalanced current, using Thevenin's Theorem
- Solve a bridge circuit for all branch currents, using a combination of loop equations and Thevenin's Theorem

A common circuit used in measuring instruments, test equipment, and power supplies is the bridge circuit, commonly referred to as a wheatstone bridge. The characteristics and operation of this circuit can be observed by studying the current flow in the circuit.

15–1 BALANCED BRIDGE. The circuit in Figure 15–1 is in the balanced condition. By this it is meant that the voltage between points A and B is 0 V. Since there is no potential between A and B, no current flows through the meter. The circuit, in effect, is a simple series parallel circuit, as shown in Figure 15–2, where the meter is effectively out of the circuit. The currents can be calculated by an Ohm's Law relationship.

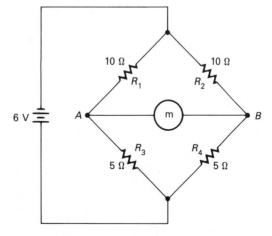

FIGURE 15–1   The bridge circuit

213

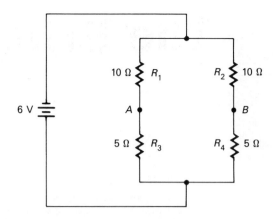

FIGURE 15-2 The balanced bridge condition

15-2 DETERMINING THE BALANCED CONDITION. The resistances in a balanced bridge are in the relationship:

$$\frac{R_1}{R_3} = \frac{R_2}{R_4} \quad \text{or} \quad \frac{R_1}{R_2} = \frac{R_3}{R_4}$$

by cross-multiplication

$$R_1 R_4 = R_2 R_3$$

In the bridge circuit, the bridge is balanced if the products of the resistors opposite each other are equal (see Figure 15-3).

$$R_1 R_4 = R_2 R_3$$
$$(10)(5) = (5)(10)$$
$$50 \ \Omega = 50 \ \Omega \text{ (the bridge is balanced)}$$

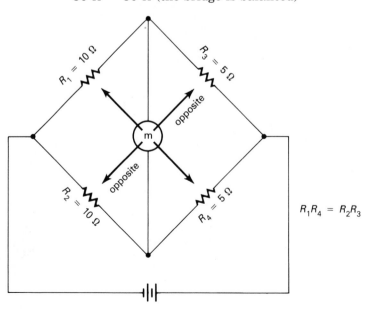

FIGURE 15-3 Balanced bridge ratio

EXERCISE 15-1 Determine whether the bridges are balanced. Refer to Figure 15-3, where the resistors have the values given (values are in ohms).

	R_1	R_2	R_3	R_4	
1.	300 Ω	150 Ω	100 Ω	50 Ω	_____
2.	250 Ω	500 Ω	500 Ω	25 Ω	_____
3.	6 kΩ	9 kΩ	2000 Ω	3000 Ω	_____
4.	3.3 MΩ	330 kΩ	4 kΩ	4 MΩ	_____
5.	0.5 Ω	0.05 Ω	2.5 Ω	0.25 Ω	_____

15–3 THE UNBALANCED BRIDGE. In the unbalanced bridge electrons flow from points A to B or B to A, depending on the polarity of the unbalance. Three loop equations can be developed, as indicated in Figure 15–4.

Equation 1: Electrons leave negative terminal of power source and flow through R_1 and R_2.

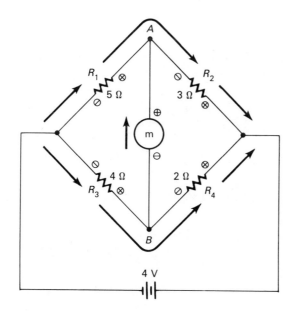

FIGURE 15–4 Current paths in the unbalanced bridge

Equation 2: Electrons leave negative terminal of power source and flow through R_3 and R_4.

Equation 3: Electrons could flow through the meter from points A to B or B to A, depending on the polarity. In determining the equations either path can be used. In this example the assumption is from points B to A, setting up the polarity indicated in Figure 15–4.

Collecting the Loops To keep the unknown values of current to a minimum, the voltage loops are collected in terms of I_1, I_3, and I_M, as shown in Figure 15–5.

Three voltage loops are then collected:

equation 1 $\quad 4 - 5I_1 - 3(I_1 + I_m) = 0$

equation 2 $\quad 4 - 4I_3 - 2(I_3 - I_m) = 0$

equation 3 $\quad 4 - 4I_3 - 10I_m - 3(I_1 + I_m) = 0$

The equations are expanded

$$4 - 5I_1 - 3I_1 - 3I_m = 0$$

$$4 - 4I_3 - 2I_3 + 2I_m = 0$$

$$4 - 4I_3 - 10I_m - 3I_1 - 3I_m = 0$$

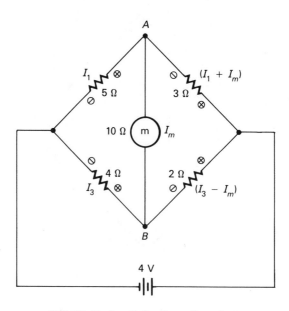

FIGURE 15–5 Collecting voltage loops

Given in standard form are:

equation 1 $\qquad -8I_1 + 0 - 3I_m = -4$

equation 2 $\qquad 0 - 6I_3 + 2I_m = -4$

equation 3 $\qquad -3I_1 - 4I_3 - 13I_m = -4$

Solving Equations Simultaneously One of the equations resulted in three unknowns; the other two equations have two unknowns. In most cases the meter current is the value that is to be determined. Use the first and third equations and eliminate I_1.

$$3(-8I_1 + 0 - 3I_m = -4)$$

$$-8(-3I_1 - 4I_3 - 13I_m = -4)$$

$$\begin{array}{r} 0 + -9I_m = -12 \\ 32I_3 + 104I_m = 32 \\ \hline 32I_3 + 95I_m = 20 \end{array} \qquad \text{(resulting equation)}$$

Use the resulting equation and the remaining equation (equation 2); eliminate I_3 and solve for I_m:

$$3(32I_3 + 95I_m = 20) \qquad \text{(resulting equation)}$$

$$16(-6I_3 + 2I_m = -4) \qquad \text{(equation 2)}$$

$$\begin{array}{r} 285I_m = 60 \\ 32I_m = -64 \\ \hline 317I_m = -4 \end{array}$$

$$I_m = 12.6 \text{ mA}$$

15-4 SOLVING BRIDGES WITH DETERMINANTS. Because three equations develop from the bridge, it may be easier to use determinants. The equations are developed in the same manner; then the solution is found using determinants. It is easier to work with positive numbers. Since all but one sign in the three equations in Section 15-3 is negative, the signs of the equations have been changed.

Equations used in Section 15-3 in standard form, with the signs changed.

$$8I_1 + 0 + 3I_m = 4$$
$$0 + 6I_3 - 2I_m = 4$$
$$3I_1 + 4I_3 + 13I_m = 4$$

Develop the matrix and solve for I_m.

$$I_m = \frac{\begin{vmatrix} 8 & 0 & 4 \\ 0 & 6 & 4 \\ 3 & 4 & 4 \end{vmatrix} \begin{matrix} 8 & 0 \\ 0 & 6 \\ 3 & 4 \end{matrix}}{\begin{vmatrix} 8 & 0 & 3 \\ 0 & 6 & -2 \\ 3 & 4 & 13 \end{vmatrix} \begin{matrix} 8 & 0 \\ 0 & 6 \\ 3 & 4 \end{matrix}}$$

$(192 + 0 + 0) - (72 + 128 + 0) =$
$(192) - (200) = -8$

$(624 + 0 + 0) - (54 - 64 + 0) =$
$(624) - (-10) = 634$

$I_m = \dfrac{-8}{634}$

$I_m = -12.6$ mA

The answer is negative because the signs in the equations were changed.

Currents I_1 and I_3 can also be found with determinants, but it is easier to substitute I_m in equations 1 and 2 and solve for the values. A combination of loop equation and determinants can be helpful in solving bridge circuits.

EXERCISE 15-2 Find I_m, I_1, and I_3.

1.

$R_1 = 16 \; \Omega$

$R_2 = 30 \; \Omega$

$R_3 = 10 \; \Omega$

$R_4 = 5 \; \Omega$

$R_m = 25 \; \Omega$

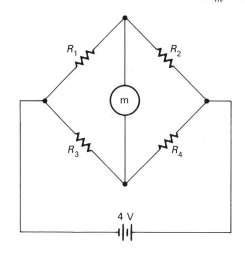

FIGURE 15-6 Circuit for Exercise 15-2 (1)

$I_m = $ _____

$I_1 = $ _____

$I_3 = $ _____

2.

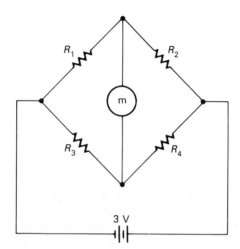

FIGURE 15-7   Circuit for Exercise 15-2 (2)

$R_1 = 6 \text{ k}\Omega$

$R_2 = 3 \text{ k}\Omega$

$R_3 = 1 \text{ k}\Omega$

$R_4 = 5 \text{ k}\Omega$

$R_m = 2 \text{ k}\Omega$

$I_m = $ _____

3.

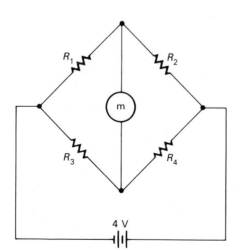

FIGURE 15-8   Circuit for Exercise 15-2 (3)

$R_1 = 2 \text{ }\Omega$

$R_2 = 4 \text{ }\Omega$

$R_3 = 12 \text{ }\Omega$

$R_4 = 10 \text{ }\Omega$

$R_m = 100 \text{ }\Omega$

$I_m = $ _____

4.

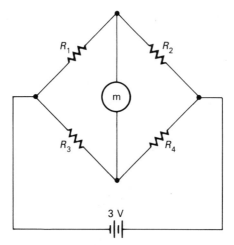

FIGURE 15-9 Circuit for Exercise 15-2 (4)

$R_1 = 6\ \Omega$

$R_2 = 10\ \Omega$

$R_3 = 3\ \Omega$

$R_4 = 25\ \Omega$

$R_m = 20\ \Omega$

$I_m =$ _____

$I_2 =$ _____

$I_4 =$ _____

15-5 THEVENIZING A BRIDGE. The current flowing through the meter or from points A to B in Figure 15-10 can be found by using Thevenin's Theorem. Thevenin's Theorem states that the current through a load can be determined if we know about the load, the open circuit voltage, the resistance of the load, and the Thevenin's equivalent resistance.

$$I_L = \frac{V_{open}}{R_L + R_{Th}}$$

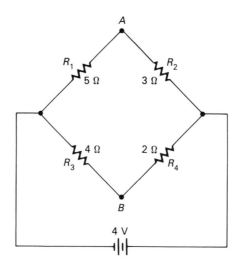

FIGURE 15-10 Removing the meter

Determining the Current Through the Meter in Figure 15–10 Because the meter becomes the load, it is removed from the circuit (Figure 15–10). The circuit then becomes a simple series parallel circuit.

Finding V Open Determine the voltage drop across R_1 and R_3 by Ohm's Law.

$$V_1 = \frac{4 \text{ V}}{8 \text{ }\Omega} \times 5 \text{ }\Omega = 2.500 \text{ V}$$

$$V_3 = \frac{4 \text{ V}}{6 \text{ }\Omega} \times 4 \text{ }\Omega = 2.666 \text{ V}$$

V_{open} is the difference between V_1 and V_3:

$$V_{open} = 2.666 \text{ V} - 2.500 \text{ V} = 0.166 \text{ V}$$

Finding R_{Th} With the power supply shorted, the resistance configuration between points A and B is shown (Figure 15–11).

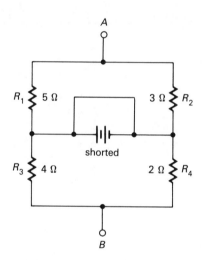

FIGURE 15–11   Finding R_{Th}

$$R_{Th} = \frac{R_1 R_2}{R_1 + R_2} + \frac{R_3 R_4}{R_3 + R_4}$$

$$R_{Th} = \frac{5(3)}{5 + 3} + \frac{4(2)}{4 + 2}$$

$$R_{Th} = \frac{15}{8} + \frac{8}{6}$$

$$R_{Th} = 1.88 + 1.33 = 3.21 \text{ }\Omega$$

Then

$$I_L = \frac{V_{open}}{R_L + R_{Th}}$$

$$I_L = \frac{0.166 \text{ V}}{10 \text{ }\Omega + 3.21 \text{ }\Omega}$$

$$I_L = 12.6 \text{ mA}$$

The value of I_m computed for this circuit using the loop equation was also 12.6 mA.

15-6 CURRENT THROUGH THE RESISTORS IN THE LEGS. Current in these resistors can also be Thevenized, but the circuit is slightly more complex. Generally it is easier to use a combination of Thevenin's Theorem and loops.

Given the circuit in Figure 15-12, find I_m, I_1, I_2, I_3, I_4.

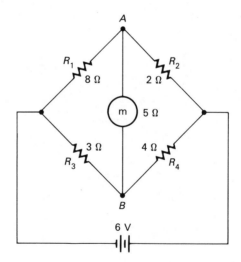

FIGURE 15-12   An unbalanced bridge

Finding I_m by Thevenin's Theorem V_{open} is the difference between V_1 and V_3:

$$V_1 = \frac{6 \text{ V}}{10 \text{ }\Omega} \times 8 \text{ }\Omega = 4.8 \text{ V}$$

$$V_3 = \frac{6 \text{ V}}{7 \text{ }\Omega} \times 3 \text{ }\Omega = 2.57 \text{ V}$$

$$V_{open} = 4.8 \text{ V} - 2.57 \text{ V} = 2.23 \text{ V}$$

Thevenin's Equivalent Resistance:

$$R_{Th} = \frac{R_1 R_2}{R_1 + R_2} + \frac{R_3 R_4}{R_3 + R_4}$$

$$R_{Th} = \frac{8(2)}{8 + 2} + \frac{3(4)}{3 + 4}$$

$$R_{Th} = 1.6 + 1.71 = 3.31 \ \Omega$$

$$I_m = I_L = \frac{V_{open}}{R_L + R_{Th}}$$

$$I_m = \frac{2.23 \text{ V}}{5 \ \Omega + 3.31 \ \Omega}$$

$$I_m = 0.268 \text{ A}$$

Finding I_1, I_2, I_3, I_4 Determine the actual direction of electron flow in the circuit by observing the polarity of the voltage from points A to B in Figure 15–13. Since the voltage across R_1 is greater than the voltage across R_3, point A is more positive than point B and electrons flow from B to A.

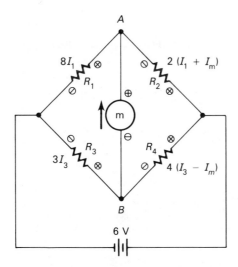

FIGURE 15–13   Finding currents in the bridge

Finding I_1 Collect the top loop, substitute I_m (0.268 A), and solve for I_1:

$$-10I_1 - 2I_m = -6$$

$$-10I_1 - 2(0.268 \text{ A}) = -6$$

$$-10I_1 = -6 + 0.536$$

$$-10I_1 = -5.464$$

$$I_1 = 546 \text{ mA}$$

Finding I_3 Collect the lower loop, substitute I_m, and solve for I_3:

$$-7I_3 + 4I_m = -6$$

$$-7I_3 + 4(0.268) = -6$$

$$-7I_3 = -6 - 1.07$$

$$-7I_3 = -7.07$$

$$I_3 = 1.01 \text{ A}$$

Finding I_2

$$I_2 = I_1 + I_m$$

$$I_2 = 546 \text{ mA} + 268 \text{ mA}$$

$$I_2 = 814 \text{ mA}$$

Finding I_4

$$I_4 = I_3 - I_m$$

$$I_4 = 1.01 \text{ A} - 0.268 \text{ A}$$

$$I_4 = 742 \text{ mA}$$

EXERCISE 15-3

1. Find I_m by Thevenin's Theorem.

$R_1 = 1\ \Omega$

$R_2 = 3\ \Omega$

$R_3 = 4\ \Omega$

$R_4 = 6\ \Omega$

$R_m = 40\ \Omega$

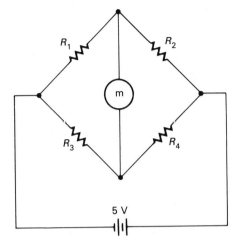

FIGURE 15-14 Circuit for Exercise 15-3 (1)

$I_m =$ _____

2. Find I_m, I_2, and I_4 by a combination of Thevenin's Theorem and loop equations.

$R_1 = 2\ \Omega$

$R_2 = 10\ \Omega$

$R_3 = 5\ \Omega$

$R_4 = 4\ \Omega$

$R_m = 50\ \Omega$

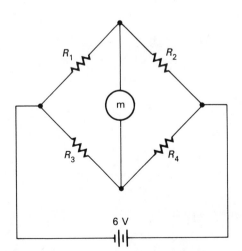

FIGURE 15-15 Circuit for Exercise 15-3 (2)

$I_m =$ _____

$I_2 =$ _____

$I_4 =$ _____

3. Find I_m by Thevenin's Theorem.

$R_1 = 10\ \Omega$

$R_2 = 2\ \Omega$

$R_3 = 10\ \Omega$

$R_4 = 8\ \Omega$

$R_m = 1\ \Omega$

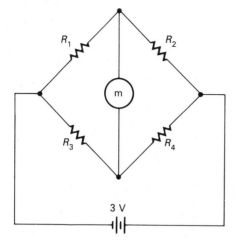

FIGURE 15–16   Circuit for Exercise 15–3 (3) $I_m =$ _____

4. Find I_m, I_2, and I_4 by a combination of Thevenin's Theorem and loop equations.

$R_1 = 12\ \Omega$

$R_2 = 10\ \Omega$

$R_3 = 12\ \Omega$

$R_4 = 5\ \Omega$

$R_m = 50\ \Omega$

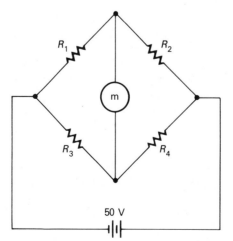

FIGURE 15–17   Circuit for Exercise 15–3 (4)

$I_m =$ _____

$I_2 =$ _____

$I_4 =$ _____

EVALUATION EXERCISE

1. Find the current through the meter, using loop equations, in the circuit in Figure 15–18.

   Chapter 15
   The Bridge

   $I_m =$ _____

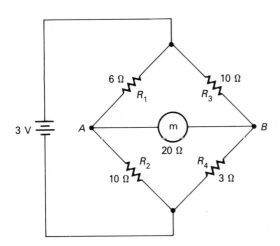

FIGURE 15–18   Circuit for Evaluation Exercise (1)

2. Find the currents I_2 and I_4 in the circuit in Figure 15–18.

   $I_2 =$ _____
   $I_4 =$ _____

3. In the circuit in Figure 15–18 change the resistance values to: $R_1 = 8\ \Omega$, $R_3 = 12\ \Omega$, $R_2 = 5\ \Omega$, $R_4 = 4\ \Omega$, and find the meter current, using Thevenin's Theorem.

   $I_m =$ _____

4. Find the current through R_1 and R_3 in the circuit used in Exercise 3.

$I_1 = $ _____

$I_2 = $ _____

CHAPTER 16
Principles of Vector Algebra

Objectives

After completing this chapter you will be able to:
- Compare scalar and vector quantities
- Graphically add vectors at right angles
- Graphically add vectors at other than right angles
- Determine x and y components of vectors
- Compute the resultant vector mathematically from two forces
- Compute the resultant vector mathematically from multiple forces

Measurements are made in such units as feet, inches, meters, grams, quarts, and liters. These magnitudes are measured with a scale and are called scalar quantities. Forces that act on a body are measured by using vector quantities. A vector is a quantity that has both magnitude and direction.

A person pulling a weight attached to a rope is exerting a force or magnitude measured in a scalar unit that represents a force such as a pound or a kilogram. Because the force is exerted in a measurable direction, it is considered a vector quantity, having both direction and magnitude.

16–1 VECTOR REPRESENTATION. A scalar quantity is represented graphically by a scale (Figure 16–1a). It is similar to a measurement made on a ruler. The same measurement—which has both magnitude and direction—is represented graphically with an arrow (Figure 16–1b).

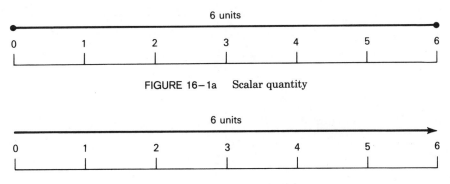

FIGURE 16–1a   Scalar quantity

FIGURE 16–1b   Vector quantity

229

16–2 A SINGLE VECTOR FORCE. Figure 16–2a–d gives examples of vector quantities acting in various directions. They are shown within the four-quadrant system used in trigonometry. Vector quantities can be expressed in polar form, as in Figure 16–2a–d.

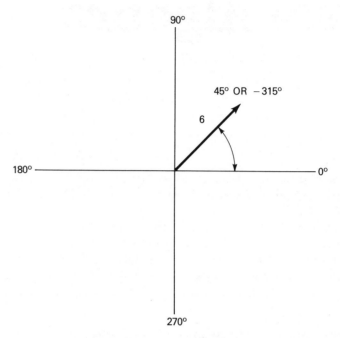

FIGURE 16–2a 6/45° or 6/−315°

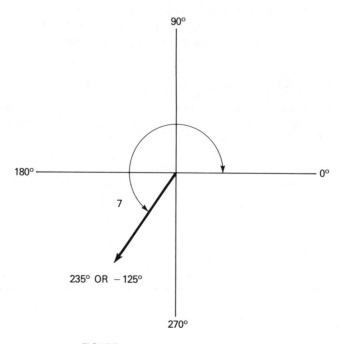

FIGURE 16–2b 7/235° or 7/−125°

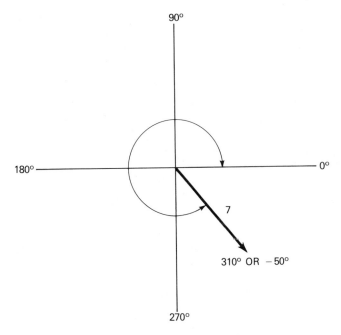

FIGURE 16-2c $7\underline{/310°}$ or $7\underline{/-50°}$

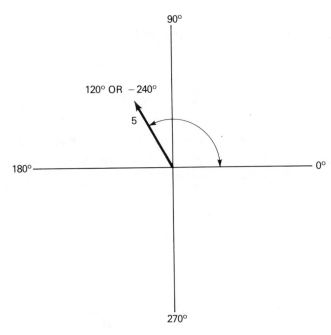

FIGURE 16-2d $5\underline{/120°}$ or $5\underline{/-240°}$

Vectors can also be expressed by using the negative angle of rotation.

$$16\text{-}2a \text{ can also be } 6\underline{/-315°}$$
$$16\text{-}2b \text{ can also be } 7\underline{/-125°}$$
$$16\text{-}2c \text{ can also be } 7\underline{/-50°}$$
$$16\text{-}2d \text{ can also be } 5\underline{/-240°}$$

16-3 MULTIPLE VECTORS. Several forces may act on a given point at one time; a graphic example is given in Figure 16-3. Forces may act on more than one plane, although the forces illustrated in Figure 16-3 are acting on the same plane.

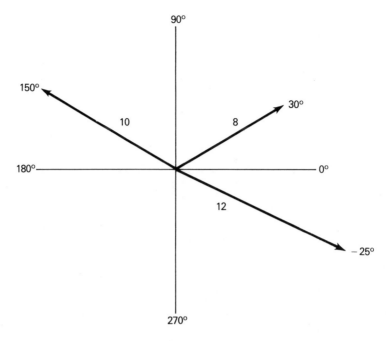

FIGURE 16-3   Multiple vectors acting on one plane

In polar form the forces are expressed:

$$\begin{array}{lll} 8/30° & \text{or} & 8/-330° \\ 10/150° & \text{or} & 10/-210° \\ 12/-25° & \text{or} & 12/335° \end{array}$$

16-4 GRAPHICALLY ADDING TWO VECTORS AT RIGHT ANGLES. Vectors can be graphically added, and a resultant can be determined, by measuring the length or magnitude of resultant and position angles. This method is known as "completing the parallelogram." Figure 16-4 illustrates this procedure.

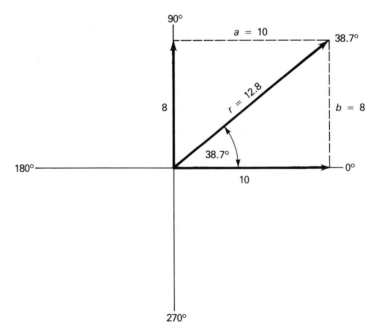

FIGURE 16-4   Adding vectors at right angles

Step 1: A parallelogram is completed by drawing side *a* parallel to the opposite side—"10." Side *b* is drawn parallel to the opposite side—"8."

Step 2: The resultant vector—*r*—is the diagonal of the parallelogram; in this case the parallelogram is a rectangle.

Step 3: The length and position of the resultant can be measured.

The results of the preceding analysis will be as accurate as the drawings and the accuracy of the measurements. The graphical-analysis operation is expressed:

$$10\underline{/0°} + 8\underline{/90°} = 12.8\underline{/38.7°}$$

Figures 16–5a–d also illustrate graphical addition of vectors at right angles to each other.

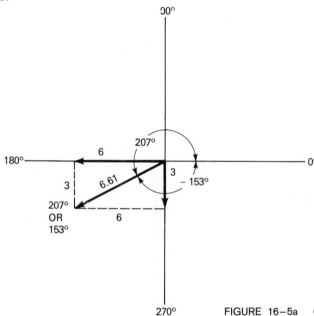

FIGURE 16–5a $6\underline{/180°} + 3\underline{/270°} = 6.61\underline{/207°}$ or $6.61\underline{/-153°}$

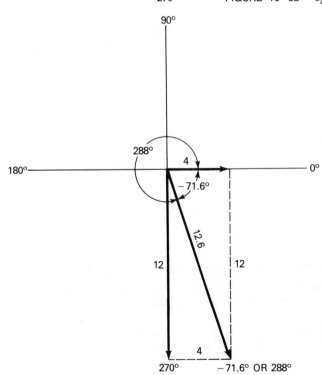

FIGURE 16–5b $4\underline{/0°} + 12\underline{/270°} = 12.6\underline{/-71.6°}$ or $12.6\underline{/288°}$

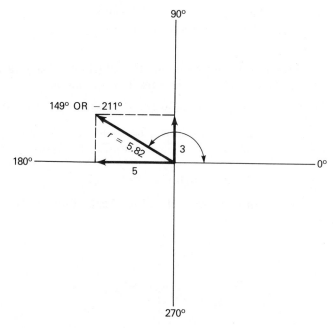

FIGURE 16–5c $3\underline{/90°} + 5\underline{/180°} = 5.82\underline{/149°}$ or $5.82\underline{/-211°}$

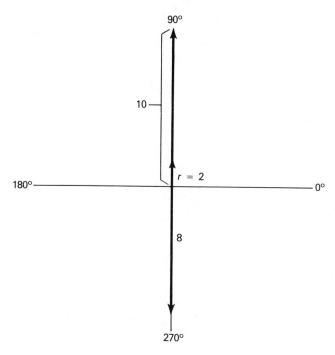

FIGURE 16–5d $10\underline{/90°} + 8\underline{/270°} = 2\underline{/90°}$ or $2\underline{/-270°}$

16–5 GRAPHICALLY ADDING VECTORS AT OTHER THAN 90°. The procedure of completing the parallelogram used for vectors at 90° is also used for vectors at other than 90°. Figures 16–6a–d illustrate this procedure.

Step 1: Draw side "5" at 0° and side "6" at 60°.

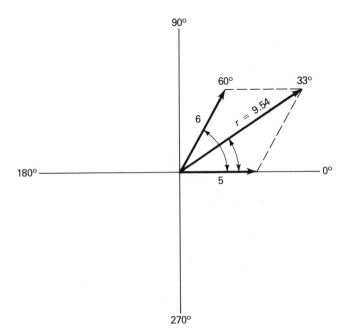

FIGURE 16-6a $5\underline{/0°} + 6\underline{/60°} = 9.54\underline{/33°}$

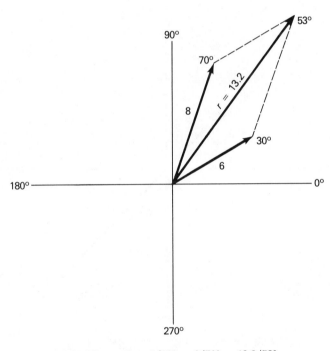

FIGURE 16-6b $6\underline{/30°} + 8\underline{/70°} = 13.2\underline{/53°}$

Step 2: Draw side *a* parallel to the side labeled "5," and draw side *b* parallel to the side labeled "6."

Step 3: Complete the parallelogram; measure the position angle the length of the resultant.

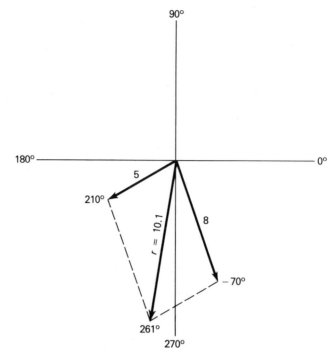

FIGURE 16-6c $5\underline{/210°} + 8\underline{/-70°} = 10.1\underline{/261°}$

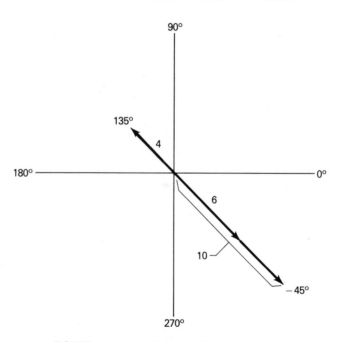

FIGURE 16-6d $4\underline{/135°} + 10\underline{/-45°} = 6\underline{/-45°}$

16-6 GRAPHICALLY ADDING MORE THAN TWO VECTORS. In any given set of vectors, a pair of vectors can be added first. Three vectors are given in Figure 16-7a. The vectors $8\underline{/30°}$ and $10\underline{/-80°}$ were arbitrarily selected to be added first. When these two vectors have been added, the resultant is added to the remaining vector (Figure 16-7b). If there were a fourth vector, the resultant of this addition would be added to the fourth vector.

$$8\underline{/30°} + 10\underline{/-80°} + 4\underline{/235°} = 11.2\underline{/-55°}$$

Chapter 16
Principles of
Vector Algebra

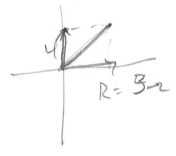

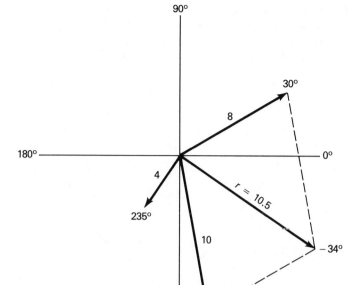

FIGURE 16–7a $8\underline{/30°} + 10\underline{/-80°} = 20.5\underline{/-34°}$

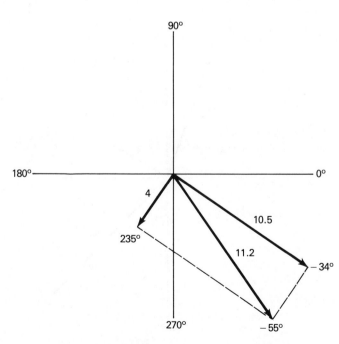

FIGURE 16–7b $10.5\underline{/-34°} + 4\underline{/235°} = 11.2\underline{/-55°}$

EXERCISE 16–1 Graphically determine the resultant vectors in Figures 16–8a–d. Extreme accuracy is not necessary. The final resultant will be close to the correct position if a straightedge is used. Nor is it necessary to determine the magnitude or the position angle; that will be proved in a later exercise.

1.

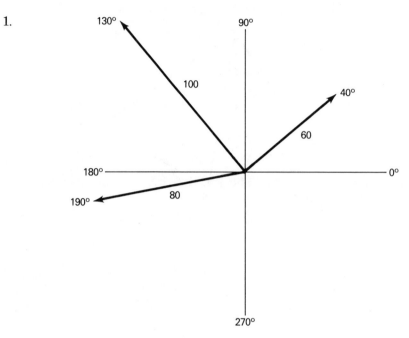

FIGURE 16–8a   Vectors for Exercise 16–1 (1)

2.

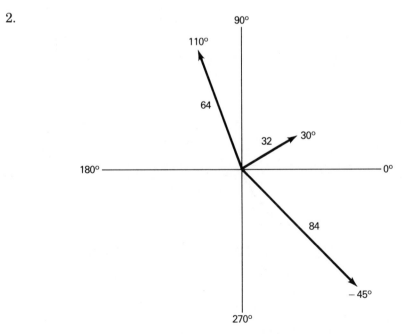

FIGURE 16–8b   Vectors for Exercise 16–1 (2)

3.

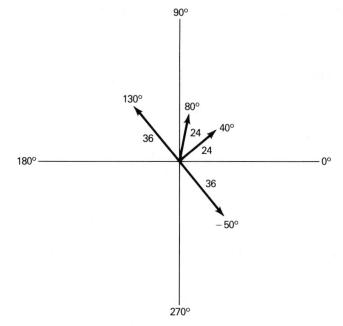

FIGURE 16–8c   Vectors for Exercise 16–1 (3)

4.

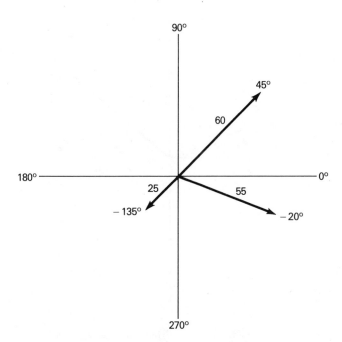

FIGURE 16–8d   Vectors for Exercise 16–1 (4)

16-7 DETERMINING X AND Y COMPONENTS OF VECTORS. When a vector is plotted in any quadrant, it has an x and a y component that can be determined in the manner shown in Example A.

Example A   Given the vector $10\underline{/40°}$

First, position the vector in the correct quadrant, as shown in Figure 16-9.

> Step 1: Drop a perpendicular from the end of the vector to the x axis, to form a right triangle.
>
> Step 2: In the right triangle the values of x and y can be calculated using the sine and cosine functions.

Find the value of x:

$$\cos 40° = \frac{x}{10}$$

$$x = \cos 40° \, (10)$$
$$x = 7.66$$

Find the value of y:

$$\sin 40° = \frac{y}{10}$$

$$y = \sin 40° \, (10)$$
$$y = 6.43$$

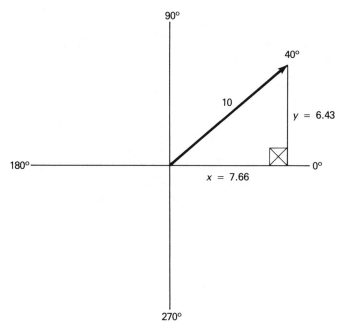

FIGURE 16-9   The vector $10\underline{/40°}$

A vector notation in polar form can be converted to x and y values by using the equations:

$$x = \cos \theta \, (r)$$

where r is the rotating radius or the vector magnitude and theta is the position angle of the vector

$$y = \sin \theta \, (r)$$

The x and y values of a vector are also called its "rectangular coordinates." Example 16–8 illustrates vectors in polar form, converted to x and y values or to rectangular coordinates.

Example B polar form

$18.5/25°$	$x = 16.8$	$y = 7.82$
$235/-75°$	$x = 60.8$	$y = 227$
$125/255°$	$x = -32.3$	$y = -121$
$30/180°$	$x = -30$	$y = 0$

EXERCISE 16–2 Compute the x and y components of the following given vectors.

	x	y
1. $42/81.2°$		
2. $1.92/40°$		
3. $72/180°$		
4. $500/270°$		
5. $36/90°$		
6. $108/-10.9°$		
7. $1600/106.5°$		
8. $364/285.1°$		
9. $61.2/221.4°$		
10. $40.9/-116.5°$		
11. $0.654/330.7°$		
12. $0.5/284.6°$		

16-8 COMPUTING THE RESULTANT MATHEMATICALLY. When vectors are at right angles y can be found by inspection as shown in Example C. Figure 16–10 graphically illustrates the mathematical example of adding the vectors $8\underline{/0°}$ and $8\underline{/90°}$.

Example C
Add the given vectors: $8\underline{/0°}$ and $8\underline{/90°}$

Figure 16–10 graphically shows the resultant, r, at the position angle θ (the symbol for the position of a vector).

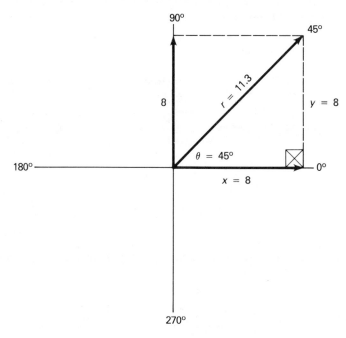

FIGURE 16–10 $8\underline{/0°} + 8\underline{/90°} = 11.3\underline{/45°}$

By right-triangle trigonometry the magnitude of the resultant vector, r, and the position angle, θ, can be found.

$$\tan \theta = \frac{y}{x}$$
$$\tan \theta = \frac{8}{8}$$
$$\theta = 45°$$

and
$$\sin \theta = \frac{y}{r}$$
$$r = \frac{y}{\sin \theta}$$
$$r = \frac{8}{\sin 45°}$$
$$r = 11.3\underline{/45°} \text{ (the resultant vector)}$$

Another mathematical solution for this operation is shown in Example D.

Example D

First determine the values of x and y; then sum the values of x and y.

	x	y
$8/0°$	8	0
$8/90°$	0	8
sum of xs and ys =	8	8

Theta can be determined by the equation:

$$\tan \theta = \text{sum of the } y\text{s divided by the sum of the } x\text{s}$$
$$\tan \theta = \frac{8}{8}$$
$$\theta = 45°$$

The resultant vector magnitude can be found by:

$$r = \text{sum of the } y\text{s divided by } \sin \theta$$

$$r = \frac{8}{\sin 45°}$$
$$r = 11.3$$

and $\qquad 8/0° + 8/90° = 11.3/45°$

While the system using the conventional right-angle trigonometry may seem a more simple solution at this time, it will be needed to solve vector problems for vectors other than 90°. The system illustrated in Example D should be used to master the system. Examples E and F also compare the graphical with the preferred mathematical solution.

Example E

Given vectors $25/0° + 15/90°$

	x	y
$25/0°$	25	0
$15/90°$	0	15
	25	15 (sum)

Chapter 16
Principles of Vector Algebra

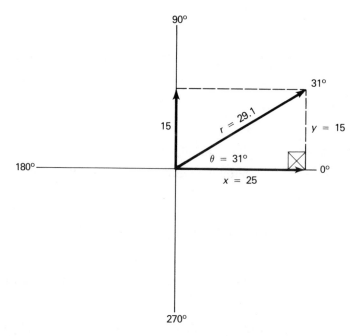

FIGURE 16–11 Graphical solution for Example E: $25\underline{/0°} + 15\underline{/90°} = 29.1\underline{/31°}$

Find theta:

$$\tan \theta = \frac{15}{25}$$
$$\theta = 31°$$

Find the magnitude of the resultant r:

$$r = \frac{15}{\sin 31°}$$
$$r = 29.1$$

The resultant vector is $29.1\underline{/31°}$.

Example F
Given $50\underline{/180°}$ and $30\underline{/-90°}$

	x	y
$50\underline{/180°}$	-50	0
$30\underline{/-90°}$	0	-30
	-50	-30 (sum)

Find theta:

$$\tan \theta = \frac{-30}{-50}$$
$$\theta = 211°$$

At this point the calculator will give an arc function of 31°; but remember that both x and y are negative and that the resultant must be in the third quadrant. The calculator shows the function angle theta prime, not theta.

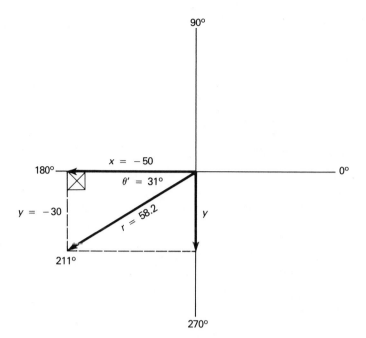

FIGURE 16-12 Graphical solution to Example F

$$\theta = 31° + 180° = 211°$$

$$r = \frac{-30}{\sin 211° \text{ (or } 31°)}$$

$$r = 58.2°$$

and $\quad 50\underline{/180°} + 30\underline{/-90°} = 58.2\underline{/211°}$

EXERCISE 16-3 Determine the resultant vector and draw a sketch of the position of the vectors and the resultant.

1. $65\underline{/0°} + 40\underline{/90°}$ _____

2. $10.5\underline{/0°} + 4.2\underline{/90°}$ _____

3. $120\underline{/90°} + 140\underline{/0°}$ _____

4. $320\underline{/90°} + 250\underline{/180°}$

5. $26.5\underline{/-90°} + 32\underline{/180°}$

6. $16\underline{/-270°} + 11.5\underline{/0°}$

7. $185\underline{/270°} + 186\underline{/90°}$

8. $0.65\underline{/-180°} + 1.25\underline{/-270°}$

9. $86\underline{/180°} + 38\underline{/360°}$

10. $3560\underline{/-90°} + 4520\underline{/180°}$

16–9 COMPUTING RESULTANT VECTORS AT OTHER THAN RIGHT ANGLES. The procedure that uses the sum of the xs and the sum of the ys is also used to find the resultant vectors when they are at other than right angles; Figure 16–13 graphically illustrates this procedure. A different method of plotting the vectors is shown. The first vector, $10\underline{/30°}$, is shown in the conventional position, whereas the second vector, $15\underline{/60°}$, is drawn at the end of the first vector. The resultant vector is drawn from the point of origin to the end of the second vector. The resultant is still the diagonal of the parallelogram. This method eliminates the need to draw the parallel sides and thus makes the explanation easier to follow. In Figure 16–12 the x and y components for $10\underline{/30°}$ are labeled x and y, but they are labeled x' and y' for the vector $15\underline{/60°}$. The sum of y components divided by the sum of the x components is the tangent of theta for the resultant.

the sum of the y components $= y + y'$
the sum of the x components $= x + x'$

$$\tan \theta_r = \frac{y + y'}{x + x'}$$

$10\underline{/30°}$	$x =$	8.66	$y = 5$
$15\underline{/60°}$	$x' =$	7.5	$y' = 13$
sum		16.16	18

$$\tan \theta_r = \frac{18}{16.16}$$
$$\theta_r = 48.1°$$

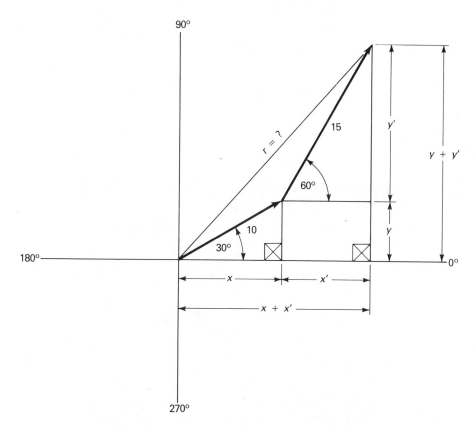

FIGURE 16–13 Vector addition other than right angles

The value of the resultant is determined by dividing the sum of the y components by the sine of θ_r.

$$r = \frac{18}{\sin 48.1°}$$
$$r = 24.2$$

The resultant vector is: $24.2\underline{/48.1°}$

Example G illustrates the method used to determine resultants of vectors at other than right angles. Figure 16–14 illustrates graphically the vectors and the resultant. A sketch of the addition is helpful when adding vectors.

Example G

Find the sum of $16\underline{/-40°}$ and $12\underline{/35°}$

	x	y
$16\underline{/-40°}$	12.3	-10.3
$12\underline{/35°}$	9.83	6.88
sum	22.13	-3.42

$$\tan \theta = \frac{y}{x}$$

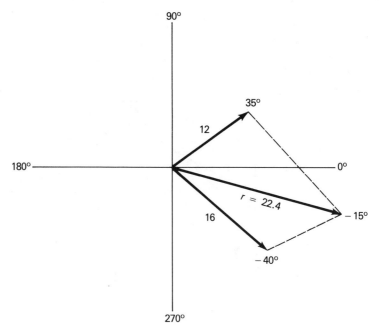

FIGURE 16–14   Graphical solution for Example G

$$\tan \theta = \frac{-3.42}{22.13}$$
$$\theta = -8.78°$$

$$r = \frac{y}{\sin \theta}$$
$$r = \frac{-3.42}{\sin -8.78°}$$
$$r = 22.4$$

The resultant vector is $22.4/\!-8.78°$.

EXERCISE 16-4 Determine the resultant vector in Exercises 1 through 6 and draw a sketch of the vectors and the resultants.

1. $75/32° + 50/65°$ _____

2. $180/\!-80° + 120/30°$ _____

3. $85/125° + 40/\!-140°$ _____

4. $22/210° + 34/\!-230°$ _____

249

Chapter 16
Principles of
Vector Algebra

5. $62\underline{/40°} + 100\underline{/-45°} + 72\underline{/150°}$

6. $16\underline{/90°} + 22\underline{/135°} + 30\underline{/222°}$

Exercises 7 through 10 are the same problems graphically sketched in Figures 16–8a–d. Mathematically prove the sketches are correct. Since they are sketches without values, the answer can only be approximated; but they will be close enough to determine the results. New sketches need not be drawn.

7. $60\underline{/40°}$

   $100\underline{/130°}$

   $80\underline{/190°}$

8. $32\underline{/30°}$

   $64\underline{/110°}$

   $84\underline{/-45°}$

9. $24/20°$

   $24/80°$

   $36/130°$

   $36/-50°$

10. $60/45°$

    $25/-135°$

    $55/-20°$

EVALUATION EXERCISE Graphically sketch the resultant vector for Exercises 1 and 2. It is not necessary to measure the length of the magnitude or the value of theta.

1.
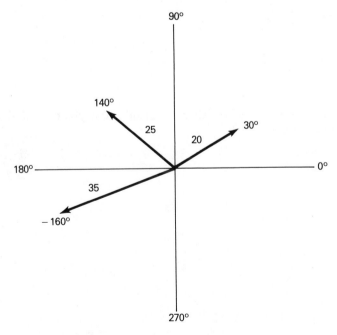

FIGURE 16-15 Vectors for Evaluation Exercise (1)

2.

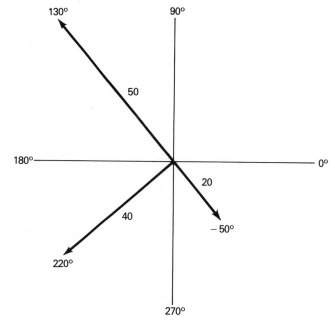

FIGURE 16–16   Vectors for Evaluation Exercise (2)

3. Determine mathematically the magnitude and the position of the resultant vector in Exercise 1.

4. Determine mathematically the magnitude and position of the resultant vector in Exercise 2.

5. Determine the resultant vector for the following given vectors and draw a sketch showing the addition and the resultant.

   $35/65°$

   $60/140°$

   $40/-25°$

6. Repeat Exercise 5 for the set of vectors.

   $12/135°$

   $15/200°$

   $12/-45°$

   $25/285°$

CHAPTER 17

Sine Waves

Objectives

After completing this chapter you will be able to:
- Understand the relationship between electromechanical and graphic generation of sine waves
- Determine instantaneous values of sine waves
- Identify phase shift between voltage and current sine waves
- Compare the cosine curve with sine curves
- Solve sine and cosine wave equations
- Graphically add sine waves
- Determine angular velocity
- Apply angular velocity to sine wave equations
- Understand harmonic generation
- Observe the effect of harmonic distortion

Sine waves can be generated by electromagnetic devices such as magnet pole generators. Mechanical generation of sine waves is limited to low frequencies, usually in heavy-power applications, for instance household appliances and industrial machinery. Electronic generators can produce frequencies millions of times greater than mechanical generators. In either case the resultant sine wave has the same shape and characteristics.

17–1 MECHANICAL SINE WAVE GENERATION. For ease of explanation a two-pole generator is used (see Figure 17–1). A magnetic field exists between the north and south poles. A conductor rotating in counterclockwise direction produces a current in the conductor. If the speed of the conductor is uniform, a sine wave current will be produced. A corresponding sine wave of voltage will also be produced across the conductor. Starting at position 1 in Figure 17–1, the sine wave is plotted, with the amplitude on the y axis and the time on the x axis.

1. At the instant the conductor is in position 1 it is moving parallel to the lines of force, and no current is produced in the conductor.

2. As the conductor is rotated between points 1 and 2 it begins to cut lines of force and begins to generate the sine wave.

3. In position 3 the conductor is cutting at a right angle (90°) to the lines of force, producing the maximum current flow in the conductor.

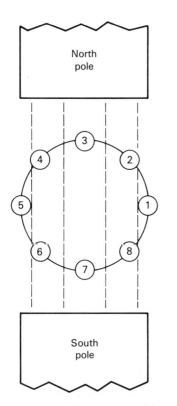

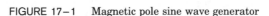

FIGURE 17-1 Magnetic pole sine wave generator

4. As the conductor rotates between points 3 and 4 it cuts the lines of force at other than a right angle, and less current is produced in the conductor.

5. At point 5 the conductor cuts parallel to the lines of force, and no current is produced in the conductor.

6. As the conductor rotates between points 5 and 6 it reverses the direction of rotation to the lines of force. Because it cuts the lines of force, it produces some current, but that current is produced in the opposite direction.

7. At position 7 the conductor cuts lines of force at a right angle, and maximum current is produced.

8. Between points 7 and 8 the conductor cuts at less than a right angle, and less current is produced.

9. When the conductor returns to point 9 it is again cutting parallel to the lines of force, and no current is produced.

The conductor rotates 360°, producing one sine wave; each revolution produces one sine wave. The rate of speed at which the conductor travels determines the number of sine waves produced in a given time and is called the "frequency" (measured in hertz [Hz]). A generator that produces one sine wave in 1 second has a frequency of 1 Hz. The number of sine waves produced in 1 second is also referred to as a "cycle." Sixty cycles or sine waves per second is a frequency of 60 Hz; 1000 cycles per second is 1 kilohertz (kHz), 1 million cycles per second is 1 megahertz (MHz). Mechanical generation of sine waves is generally limited to 400 Hz, whereas electronic generation is in millions of hertz.

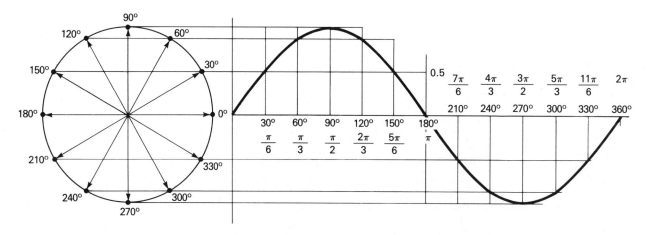

FIGURE 17-2 Graphical projection of a sine wave shown in degrees and radians

17-2 GRAPHICAL GENERATION OF SINE WAVES. A graph corresponding to the mechanical generation of a sine wave shown in Figure 17-1 is illustrated in Figure 17-2. Angles are generated through a 360° rotation and are graphically projected using time degrees on the horizontal axis and amplitude on the vertical axis. The angles generated by the rotating radius (r) are shown measured in degrees, but they are also measured in radians. Each angle shown in the projection is 30°. The first 30° angle is shown again in Figure 17-3; for the explanation the amplitude of the sine wave is 10 units. The rotating radius is always a positive value; its length is equal to the maximum value of the sine wave. In Figure 17-3 a perpendicular is dropped to the x axis showing the x and y components. The length of y is also the length or amplitude attained by the sine wave at the end of 30°. The sine function can be used to mathematically find this length,

$$\sin \theta = \frac{y}{r}$$

In general terms

$$y = r \sin \theta$$

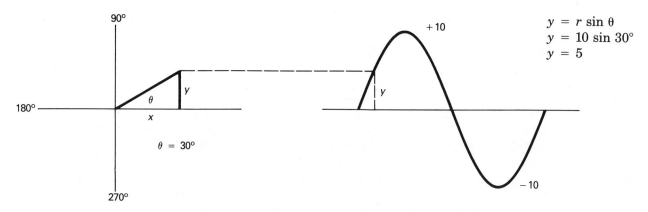

FIGURE 17-3 $y = r \sin \theta$

Each amplitude point on the sine curve can be determined by the equation $y = r \sin \theta$. Because the resultant curve varies by the sine of the generated angle, the result is called a "sine curve," or, when applied to voltage or current, a "sine wave."

17-3 VOLTAGE AND CURRENT CURVES. If the voltage applied to a circuit is sinusoidal the current will also have the same sine curve; but, of course, it will be measured in different units. Figure 17-4 illustrates a sine wave of current or voltage. Any point on the curve is an instantaneous value and is depicted by using lowercase symbols of either v or i. The maximum point is depicted by upper case V or I. There are several instantaneous points on the sine wave that need special definition.

- *Maximum or peak value* (I_p, I_m or V_p, V_m): there is a maximum value of amplitude on both the positive and the negative half of the sine wave. These values are also called the "peak value" of the sine wave.

- *Peak-to-peak value* (I_{p-p} or V_{p-p}): the value measured between the positive and negative peaks of a sine wave, or two times the maximum or peak value.

- *Average value* (I_a or V_a): the average value of a sine wave is computed over only half of the generation of either the positive or the negative

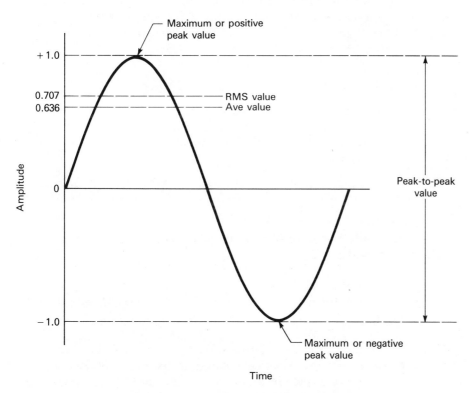

FIGURE 17-4 Sine wave of current or voltage

half of the sine wave. For every positive instantaneous value there is an equal negative value, and the average value for one full cycle of a sine wave would be zero.

When a series of instantaneous values are averaged (that is, every 10° for a half cycle), the average value will be 63.6% of the maximum value, referred to as the "0.636 point."

$$I_{ave} = 0.636\, I_{max} \qquad V_{ave} = 0.636\, V_{max}$$

- *RMS (root mean square) or effective value:* The rms, or effective value of the sine wave, is 70.7% of the maximum value. This value of current and voltage produces the same heating effect as an equivalent DC circuit.

  Unless otherwise stated, values of current and voltage are given in rms.

$$I_{rms} = 0.707\, I_{max} \qquad V_{rms} = 0.707\, V_{max}$$

17–4 INSTANTANEOUS VALUES OF CURRENT AND VOLTAGE. Because the sine curve is a function of the sine ratio, and $y = r \sin \theta$, a general equation for voltage and current can be written. The rotating radius at 90° and 270° is equal to the maximum value of the sine wave, and the following equations are valid.

$$i = I_{max} \sin \theta$$

where

i is the instantaneous value of current
I_{max} is the maximum or peak value of the sine wave
θ is the angle that measures the time of rotation, given in degrees or radians

$$v = V_{max} \sin \theta$$

where

v is the instantaneous value of voltage
V_{max} is the maximum or peak value of the sine wave
θ is the angle that measures the time of rotation, given in degrees or radians

Example A
Determine the instantaneous value of voltage at 40° if the peak value of the sine wave is 150 V.

$$v = V_{max} \sin \theta$$
$$v = 150 \sin 40° \qquad [150 (\sin 40°)] = [150 (0.6428)]$$
$$v = 96.4 \text{ V}$$

EXERCISE 17–1 Determine the instantaneous value of the following. In each case sketch one cycle of a sine curve and indicate the approximate point on the curve (see Example A).

1. $y = 50 \sin 35°$ _28.68_

2. $i = 150 \sin 42.4°$ mA _____

3. $v = 650 \sin 65.5°$ V _591.5 V_

4. $v = 8.9 \sin 45°$ mV _____

5. $i = 22 \sin -45°$ μA _−15.6 μA_

6. $y = 1250 \sin -120°$ _____

7. $i = 30 \sin 242°$ mA _−26.5 mA_

8. $i = 0.5 \sin 180°$ A _____

9. $v = 360 \sin 125°$ μV _294.9 μV_

10. $y = \sin 36°$

11. $i = 12 \sin 360°\ \mu A$

12. $v = 25 \sin -180°\ V$

17–5 PHASE SHIFT. Sine waves having an amplitude beginning at other than 0° are considered out of phase with that time axis. Generally, the sine wave beginning at zero on a given time base is called the "fundamental wave." Figure 17–5 shows a fundamental sine wave, v_1, and another sine wave, v_2, beginning at 45°. The second sine wave, v_2, reaches its peak at 135° or 45° after the fundamental, v_1. It would be lagging the fundamental by 45°.

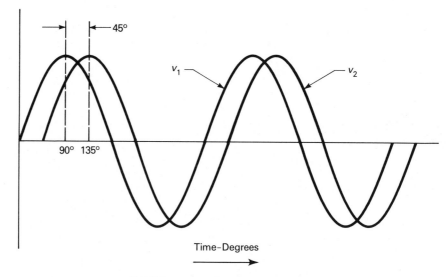

FIGURE 17–5 Sine wave lagging by 45°

The equations used to express the sine waves in Figure 17–5 are

$$v_1 = V_{max} \sin \theta$$

$$v_2 = V_{max} \sin (\theta - 45°)$$

Figure 17–6 shows two sine waves where v_2 is leading the fundamental by 45°. A convenient method for determining lead or lag is to compare the positive peak of the fundamental with the other sine wave. If, on a given time axis, the sine wave reaches its first positive peak *before* the fundamental peaks, it is leading; if it reaches its first positive peak *after* the fundamental peaks, it is lagging. In Figure 17–6, v_2 reaches its first positive peak at 45°, or 45° before the fundamental.

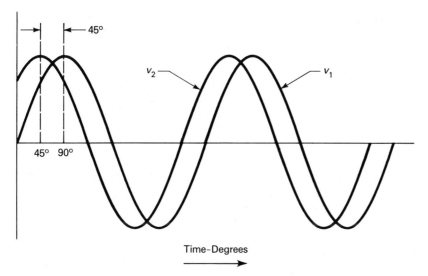

FIGURE 17-6 Sine wave leading by 45°

The equations used to express the sine waves in Figure 17-6 are

$$v_1 = V_{max} \sin \theta$$
$$v_2 = V_{max} \sin (\theta + 45°)$$

17-6 COSINE CURVE. A sine wave curve is plotted using the equation $y = r \sin \theta$. If the cosine function were used, $y = r \cos \theta$, the curve would be as illustrated in Figure 17-7(a-b). Figure 17-7a illustrates a leading cosine curve and Figure 17-7b illustrates a lagging cosine curve.

The equations for these curves can be written by expressing theta $\pm 90°$; the preferred method, however, is to use cosine.

leading curve equation	preferred equation
$v = V_{max} \sin (\theta + 90°)$	$v = V_{max} \cos \theta$
lagging curve equation	preferred equation
$v = V_{max} \sin (\theta - 90°)$	$v = -V_{max} \cos \theta$

When V_{max} is expressed with a negative sign, it is lagging by 90°, when expressed as a positive V_{max}, it is leading by 90°.

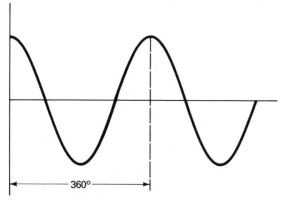

FIGURE 17-7a Leading cosine curve

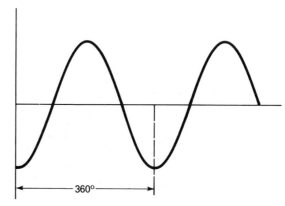

FIGURE 17-7b Lagging cosine curve

Example B

Determine the instantaneous value of current if the equation is $i = 50 \sin(\theta + 35°)$ mA and $\theta = 42°$.

$$i = 50 \sin(\theta + 35°) \text{ mA}$$
$$i = 50 \sin(42° + 35°)$$
$$i = 50 \sin 77°$$
$$i = 48.7 \text{ mA}$$

Determine the instantaneous value of voltage where $\theta = 30°$, $v = -35 \cos \theta$ V.

$$v = -35 \cos \theta \text{ V}$$
$$v = -35 \cos 30°$$
$$v = -30.3 \text{ V}$$

The negative sign in front of 35 indicates a lagging cosine curve, where 35 is the rotating radius and the rotating radius is always positive.

The equation $v = -30.3$ V can be calculated with $(\theta - 90°)$, and the result would be the same. The cosine of 30° is the same as the sine of 120°.

$$v = 35 \sin(\theta - 90°)$$
$$v = 35 \sin(30 - 90°)$$
$$v = 35 \sin -60°$$
$$v = -30.3 \text{ V}$$

An equation written with a negative sign in front of the sine function is 180° out of phase. The equation $v = -50 \sin \theta$ is 180° out of phase with the fundamental.

EXERCISE 17-2 Determine the instantaneous value of the sine curves.

1. $i = 30 \sin(\theta + 15°)$ mA, $\theta = 60°$ 29 mA

2. $i = 15 \sin(\theta - 65°)$ A, $\theta = 195°$

3. $v = 75 \sin(\theta - 85°)$ V, $\theta = 342°$ $\underline{-73.1 \text{ V}}$

263

Chapter 17
Sine Waves

4. $v = 150 \cos \theta$ mV, $\theta = 122°$ _____

5. $i = -750 \sin \theta$ µA, $\theta = 123°$ $\underline{-629 \text{ µA}}$

6. $y = 60 \cos \theta$, $\theta = 45°$ _____

7. $v = 80 \sin(\theta - 45°)$ mV, $\theta = 135°$ $\underline{80 \text{ V}}$

8. $i = 0.502 \sin(\theta - 115°)$ A, $\theta = 295°$ _____

9. $i = -1600 \cos \theta$ µA, $\theta = 125°$ $\underline{918 \text{ µA}}$

10. $v = -60 \sin \theta\ \mu V, \theta = -47°$

17-7 SINE WAVE ADDITION.
Figure 17-8 (a-d) illustrates the graphical addition of sine waves. Figure 17-8a shows two sine waves that are in phase and of the same frequency. They have different amplitudes; $v_1 = 100 \sin \theta$ and $v_2 = 60 \sin \theta$. The resultant voltage, $v_R = 160 \sin \theta$, is obtained by graphically adding instantaneous values of v_1 and v_2. If the instantaneous values were added every 10° an accurate sine wave would result. The amplitude of 160 is easily determined. The resultant equations for each of the drawings are mathematically proved in Chapter 18, "Phasors."

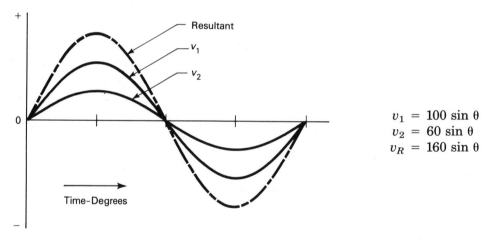

$v_1 = 100 \sin \theta$
$v_2 = 60 \sin \theta$
$v_R = 160 \sin \theta$

FIGURE 17-8a Sine waves in phase

Figure 17-8b shows two sine waves of the same frequency but that are out of phase. The sine wave v_1 is the fundamental and begins at zero on the time axis; the sine wave v_2 lags the fundamental by 90°. Because they are of equal amplitudes the resultant peaks halfway between the peaks of v_1 and v_2. The resultant sine wave peaks at 135° and lags the fundamental, v_1, by 45°. The amplitude and position of the resultant can be found graphically and are shown in Chapter 18.

Figure 17-8c shows a sine wave lagging by 90° but of less amplitude. The resultant does not peak halfway between the two; instead, it peaks closer to the one of the higher amplitude.

Figure 17-8d shows two sine waves with a phase shift of 180° and different amplitudes. For each positive and negative point on the fundamental there is a corresponding negative and positive point on v_2. The resultant is a sine wave in phase but at a lesser amplitude than either v_1 or v_2.

265

Chapter 17
Sine Waves

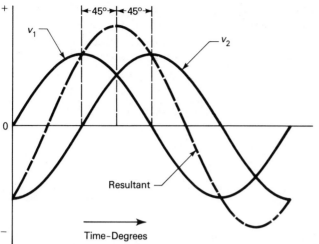

$v_1 = 100 \sin \theta$
$v_2 = -100 \cos \theta$
$V_R = 141 \sin (\theta - 45°)$

FIGURE 17–8b   Sine waves out of phase with the same amplitudes

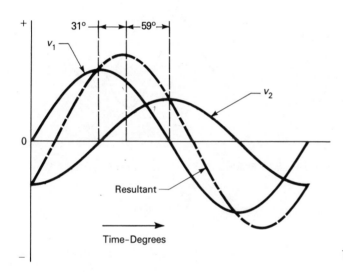

$v_1 = 100 \sin \theta$
$v_2 = -60 \cos \theta$
$V_R = 117 \sin (\theta - 31°)$

FIGURE 17–8c   Sine waves out of phase with different amplitudes

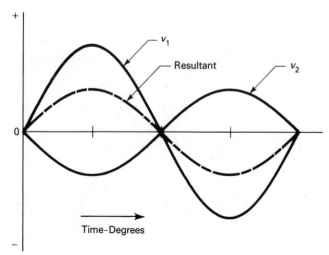

$v_1 = 100 \sin \theta$
$v_2 = -60 \sin \theta$
$V_R = 40 \sin \theta$

FIGURE 17–8d   Sine waves with a phase shift of 180°

EXERCISE 17–3 Graphically add the sine waves and sketch the resultant curve. Accuracy is not the purpose of these problems, so a straightedge or a ruler can be used to determine the instantaneous values. Points should be taken about every 45°. The work should result in a rough sketch of the resultant.

1.

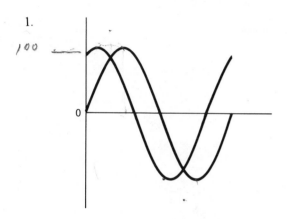

FIGURE 17–9a   Sine waves for Exercise 17–3 (1)

2.

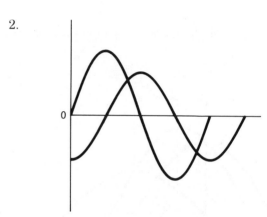

FIGURE 17–9b   Sine waves for Exercise 17–3 (2)

3.

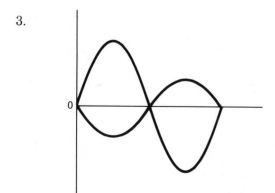

FIGURE 17–9c   Sine waves for Exercise 17–3 (3)

4.

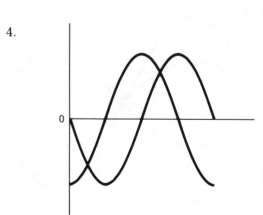

FIGURE 17–9d   Sine waves for Exercise 17–3 (4)

5.

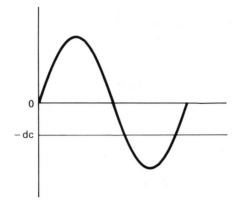

FIGURE 17-9e   Sine wave and dc voltage for Exercise 17-3(5)

17-8 ANGULAR VELOCITY. Angular velocity is directly proportional to frequency and is determined by the equation

$$\omega = 2\pi f$$

where   ω (omega) is the angular velocity measured in radians per second

f is the frequency measured in hertz

Example C
What is the angular velocity of a 60 Hz generator?

$$\omega = 2\pi f$$
$$\omega = 2\pi(60)$$
$$\omega = 377 \text{ rad/sec}$$

A common method of measuring the speed of a rotating shaft such as a motor is in revolutions per minute or second. The speed could just as easily be measured in radians per second. Since there are 2π radians in one revolution, a motor rotating at 60 rps is rotating at 377 radians per second.

The speed at which a sine wave is produced is its angular velocity and is measured in radians per second. Rather than measure the number of revolutions a generator is making to produce a certain frequency, the number of radians is measured.

The time required to produce one cycle of a frequency is called the "period" and is inversely proportional to the frequency, where $T = \frac{1}{f}$; T is the time required to produce one cycle, sec; and f is frequency, Hz.

17-9 SINE WAVES AND ANGULAR VELOCITY. To determine the instantaneous value of a voltage sine wave, the equation used is $v = V_m \sin \theta$. The angle theta is equal to ωt ($\theta = \omega t$); substituting ωt for θ, the equation becomes

$$v = V_m \sin \omega t$$

This equation is more commonly used to describe instantaneous values of voltage or current because it involves the frequency used to produce the voltage or current. It is often necessary to compare sine waves by "stopping" them at a given time and comparing their instantaneous values.

Example D
Determine the instantaneous value of voltage for a 60 Hz generator at the time of 2 msec, $V_m = 50$ V.

$v = V_m \sin \omega t$ $\omega = 2\pi f$
$v = 50 \sin (377 \text{ rad/sec})(0.002 \text{ sec})$ $\omega = 6.28(60)$
$v = 50 \sin 0.754 \text{ rad}$ $\omega = 377 \text{ rad/sec}$

At this point the sine ratio can be found by using radians, or radians can be changed to degrees:

$$(0.754 \text{ rad})(57.3°) = 43.2°$$

$v = 50 \sin 0.754 \text{ rad or } 43.2°$
$v = 34.2 \text{ V}$

Determine the instantaneous value of a sine wave at the end of 3 msec if the frequency is 100 Hz, $V_m = 60$ V, and it is leading by 30°.

$v = V_m \sin (\omega t + 30°)$
$v = 60 \sin (628)(0.003) + 30°$
$v = 60 \sin (1.884 \text{ rad} + 30°)$

Change radians to degrees before adding

$v = 60 \sin (108 + 30°)$
$v = 60 \sin 138°$
$v = 40.1 \text{ V}$

EXERCISE 17-4 Determine the instantaneous values of the sine waves for the given conditions.

1. $v = \sin \omega t$ V
   $f = 400$ Hz, $t = 4$ msec

   −.5848 V

   ωt = 2πf = 2513

2. $i = 35 \sin \omega t$ mA
   $f = 100$ Hz, $t = 5$ msec

Chapter 17
Sine Waves

3. $v = 750 \sin(\omega t + 45°)$ mV
   $f = 50$ Hz, 3 msec

   $\omega t = 2\pi f$
   $\omega t = 54$ RAD $= 3094°$
   $3094 + 45° = 3138.97°$

   −740.8 mV

4. $v = 120 \sin(\omega t - 52°)$ V
   $f = 60$ Hz, $t = 1.5$ msec

   −40.25

5. $i = 16 \sin(\omega t - 22°)$ mA
   $f = 1000$ Hz, $t = 450$ μsec

   $\omega t = 2\pi f$
   $= 2\pi(1000)$
   $= (6283 \text{ RAD})(450 \mu s)$
   $= 2.82735$ RAD
   $= 161.995° - 22 = 139.995°$

   −10.285 mA   10.3

6. $v = 250 \cos \omega t$ mV
   $f = 400$ Hz, $t = 800$ μsec

   _____

7. $v = -600 \sin \omega t$ μV
   $f = 10$ kHz, $t = 30$ μsec

   $\omega t = 2\pi f$
   $= 2\pi(10,000)(30 \mu s)$
   $= 1.885$ RAD
   $= 108°$

   −570.63 μV   −572 μV

8. $i = -12 \sin \omega t$ A
   $f = 8$ kHz, $t = 15$ μsec

   $\omega = 2\pi f$
   $= 2\pi(8\text{kHz})(15\text{μs})$
   $=$

150 mV

9. $v = 150 \cos \omega t$ mV
   $f = 400$ Hz, $t = 15$ μsec

   $\omega = 2\pi f$
   $= 2\pi(400)(15\text{μs})$
   $= .037699$ RAD
   $= 2.16°$

   $V = 150 \cos 2.16°$
   $=$

   149.89 mV

10. $i = 300 \sin(\omega t - 45°)$ V
    $f = 1$ kHz, $t = 0.05$ msec

17–10 HARMONICS. When a guitar string is plucked, it produces a sound of a fundamental frequency, as well as other frequencies called "harmonics." When sine waves are generated by electronic devices they generate other sine waves, also called harmonics. These relationships can be shown graphically by the addition of sine waves and their harmonics.

Harmonics are classified as even or odd. The second harmonic is twice the fundamental frequency and is called the "first even harmonic." The third harmonic is three times the fundamental frequency and is called the "first odd harmonic." Harmonics are usually produced at lower amplitudes than is the fundamental frequency.

Even harmonics	*Odd harmonics*
second (first even)	third (first odd)
fourth (second even)	fifth (second odd)
sixth (third even)	seventh (third odd)
eighth (fourth even)	ninth (fourth odd)
etc.	etc.

A harmonic is a multiple of a fundamental frequency. If the fundamental frequency were 100 Hz, the second harmonic would be 200 Hz, the third 300 Hz, and so on.

Fundamental frequency	First even harmonic	Second even harmonic	Third even harmonic
100 Hz	200 Hz	400 Hz	600 Hz
15 kHz	30 kHz	60 kHz	90 kHz
3 MHz	6 MHz	12 MHz	18 MHz

Fundamental frequency	First odd harmonic	Second odd harmonic	Third odd harmonic
100 Hz	300 Hz	500 Hz	700 Hz
15 kHz	45 kHz	75 kHz	105 kHz
3 MHz	9 MHz	15 MHz	21 MHz

The equation for a second harmonic sine wave is $y = \sin 2\omega t$, where the angular velocity omega is doubled, producing two cycles for the same time period of the fundamental.

Voltage, or current sine waves, indicating harmonics:

second harmonic $\quad v = V_m \sin 2\omega t$
third harmonic $\quad v = V_m \sin 3\omega t$
eighth harmonic $\quad i = I_m \sin 8\omega t$
ninth harmonic $\quad i = I_m \sin 9\omega t$
etc.

Figure 17–10 illustrates a fundamental sine wave and second and third harmonics. Generally, harmonics are produced at a lower amplitude than is the fundamental.

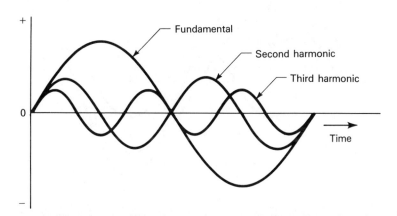

FIGURE 17–10   Fundamental with second and third harmonic

17–11 HARMONIC DISTORTION. Electronic devices producing sine-wave frequencies also produce harmonics. If the amplitude of the harmonic is great enough, and the harmonic and the fundamental exist electrically at the same point, the resultant wave shape is distorted. An oscilloscope measuring voltage at that point would detect a distorted wave shape or complex wave. Fundamental and second harmonics are shown in Figure 17–11. The resultant wave shape is observed on an oscilloscope. This is a common type of distortion that occurs in electronic devices. A trained technician observing this complex wave on the scope would have an indication of where to look for the problem. With the proper design techniques, the technician can keep this type of distortion to a minimum.

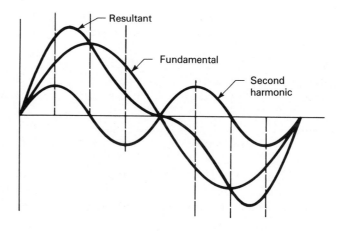

FIGURE 17-11   Fundamental and second harmonic

Complex waves can be analyzed using a spectrum analyzer; the fundamental and harmonics can also be observed. (They can also be analyzed mathematically using a Fourier series; but that is beyond the scope of this text.) The equation for the complex wave in Figure 17-11 is written:

$$v_R = V_1 \sin \omega t + V_2 \sin 2\omega t$$

Figures 17-12 and 17-13 illustrate common harmonic wave shapes.

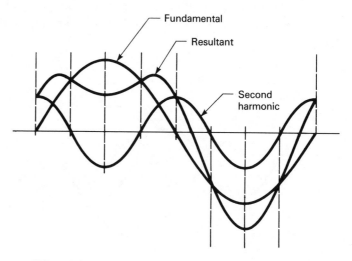

FIGURE 17-12   Effect of the second harmonic 90° out of phase $V_R = V_1 \sin \omega t + V_2 \cos 2\omega t$

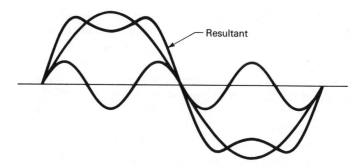

FIGURE 17-13   Effect of third harmonic in phase $V_R = V_1 \sin \omega t + V_2 \sin 3\omega t$

The square wave is an important complex wave used in electronic testing and it contains only in phase-odd harmonics. They contribute to shaping the square wave (see Figure 17–14a–c). A good square wave contains at least five odd harmonics. Figure 17–14b shows the result of the fundamental and the third harmonic (it begins to flatten the top and steepen the sides). By adding the fifth and seventh harmonics, the sides become steeper and the top flatter, Figure 17–14c. Analyzing a square wave with a spectrum analyzer will show the harmonic makeup of a square wave.

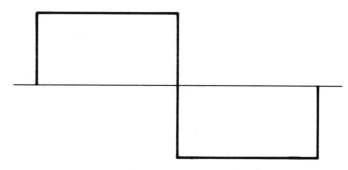

FIGURE 17–14a $v = V_M \sin \omega t + \dfrac{1}{3} V_M \sin 3\omega t + \dfrac{1}{5} V_M \sin 5\omega t + \dfrac{1}{7} V_M \sin 7\omega t +$ etc.

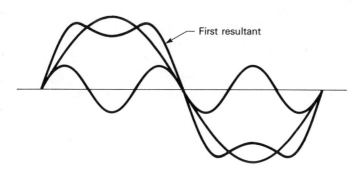

FIGURE 17–14b   Fundamental plus third harmonic

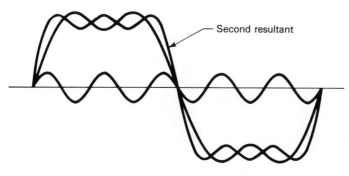

FIGURE 17–14c   First resultant with fifth harmonic

EVALUATION EXERCISE Determine the instantaneous value in Exercises 1 through 5. Provide a sketch of the sine wave, indicating the approximate instantaneous point.

1. $y = \sin(\theta + 25°)$
   $\theta = 15°$

   .64278

2. $v = 350 \sin(\theta - 45°)$ mV
   $\theta = 85°$

   224.975 mV

3. $i = 12.5 \sin(\theta - 36°)$ µA
   $\theta = 120°$

   12.43 µA

4. $i = -76 \cos\theta$ mA
   $\theta = 65°$

   −32.1189 mA

5. $v = -120 \sin\theta$ µV
   $\theta = 240°$

   103.92 µA

In Exercises 6 through 10 determine the instantaneous values of the sine waves for the given conditions.

6. $v = 450 \sin(\omega t + 60°)$ V
   $f = 50$ Hz, $t = 2$ msec
   $\omega = 2\pi f$
   $= 2\pi(50)(2ms)$
   $= .6283$ RAD $= 36° + 60° = 96°$

   <u>447.53 V</u>

7. $i = \cos \omega t$ mA
   $f = 400$ Hz, $t = 1.25$ msec
   $\omega = 2\pi f$
   $= 2\pi(400)(1.25ms)$
   $= 3.1415$ RAD $= 180°$

   <u>−1 mA</u>

8. $i = 52(\sin \omega t − 45°)$ μA
   $f = 10$ kHz, $t = 10$ μsec
   $\omega = 2\pi f$
   $= 2\pi(10,000)(10\mu s) − 45°$
   $= 36° − 45° = -9°$

   <u>−8.1345 μA</u>

9. $v = −25 \sin \omega t$ V
   $f = 60$ Hz, $t = 5$ msec
   $\omega = 2\pi f$
   $= 2\pi(60)(5ms)$
   $= 108°$

   <u>−23.776 V</u>

10. $i = 175(\sin \omega t + 15°)$ μA
    $f = 100$ Hz, $t = 0.15$ msec
    $\omega = 2\pi f$
    $= 2\pi(100)(.15ms) + 15°$
    $= 5.4° + 15° = 20.4°$

    <u>61 μA</u>

Sketch the resultant wave for Exercises 11 through 15.

Chapter 17
Sine Waves

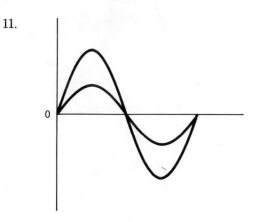

FIGURE 17–15a Sine waves for Evaluation Exercise (11)

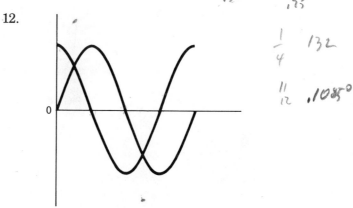

FIGURE 17–15b Sine waves for Evaluation Exercise (12)

13.

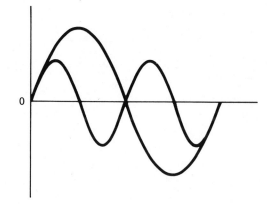

FIGURE 17–15c   Sine waves for Evaluation Exercise (13)

14.

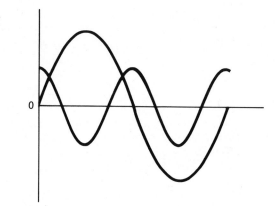

FIGURE 17–15d   Sine waves for Evaluation Exercise (14)

15.

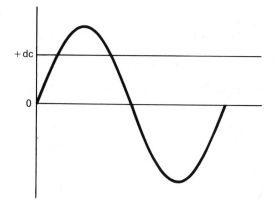

FIGURE 17–15e   Sine wave and dc voltage for Evaluation Exercise (15)

CHAPTER 18

Phasors

Objectives

After completing this chapter you will be able to:
- Convert sine wave equations to phasors
- Express phasors in both polar and rectangular notation
- Add sine waves using phasors
- Express quantities as j operators
- Express phasors in polar notation as j operators
- Simplify j operator expressions

Vectors have been defined as quantities with both magnitude and direction. But for many years no distinction was made between a phasor and a vector. Because impedance and voltage and current sine waves do not fit this definition completely, they are called phasors (the term "phasor" is formed by combining the words *phase* and *vector*). The mathematical method used to solve phasor problems is, in all respects, the same as that used to solve vector problems. Phasor problems are solved just as vector problems are solved.

18-1 SINE WAVES AND PHASORS. In Chapter 17 we saw sine waves of voltage and current written in equation form. Sine waves can also be expressed as phasors. Example A illustrates an equation and its phasor expression. The magnitude expressed in the equation is the peak value of the sine wave, and the phasor is expressed as a peak value. This statement should not be confused with the general statement made in electronics that a voltage or current value is in rms unless otherwise stated. A sine wave equation is written in peak values unless otherwise stated.

Example A

Equation	Phasor in polar form
$v = 100 \sin \omega t$ V	$100 \underline{/0°}$ V
$v = 20 \sin (\omega t + 25°)$ mV	$20 \underline{/25°}$ mV
$i = 50 \sin \omega t$ A	$50 \underline{/0°}$ A
$i = 35 \sin (\omega t - 120°)$ mA	$35 \underline{/-120°}$ mA
$v = 60 \cos \omega t$ mV	$60 \underline{/90°}$ mV

Figure 18-1 illustrates the phasors in Example A plotted within the quadrants. It can be seen that this operation is performed as though the phasors were vectors.

278

Part 2
Mathematics for DC
and AC Circuits

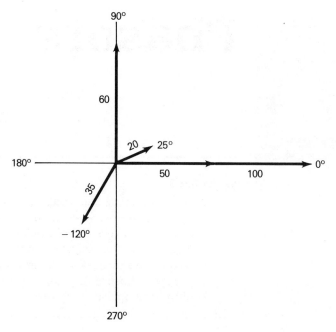

FIGURE 18-1 Plot of phasors in Example A

The equations in Example A are expressed in their polar coordinates. Like vectors, they can also be expressed in rectangular coordinates by determining the x and y values (see Figure 18-2).

$$x = \cos \theta \, V_m \qquad y = \sin \theta \, V_m$$

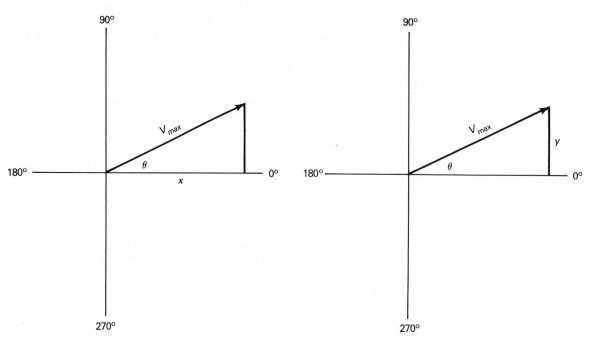

FIGURE 18-2 Determining x and y

279

Chapter 18
Phasors

	x	y
100/0° V	100	0
20/25° mV	18.1	8.45
50/0° A	50	0
35/−120° mA	−17.5	−30.3
60/90° mV	0	60

EXERCISE 18–1 Determine the polar form and the x and y value for each equation.

	Polar	x	y
1. $v = 150 \sin(\omega t - 20°)$ V	150/−20°	140.95	−51.30
2. $i = 16 \sin(\omega t + 45°)$ A	16/45°	11.31	11.31
3. $i = -30 \sin \omega t$ A	30/0	30	0
4. $v = 250 \cos \omega t$ mV	250/90°	0	250
5. $i = 75 \sin(\omega t - 145°)$ mA	75/−145°	−61.436	−43.018
6. $i = 175 \sin(\omega t + 232°)$ A	175/232°	107.74	−137.9
7. $v = -350 \cos \omega t$ mA	350/−90	0	−350
8. $i = 12 \sin(\omega t - 30°)$ A	12/−30°	10.392	−6
9. $v = 117 \sin(\omega t + 70°)$ V	117/70°	40.016	109.94
10. $v = 18 \sin(\omega t + 135°)$ V	18/135°	14.7447	12.7279

$X = \cos \Theta V_m$
$Y = \sin \Theta V_m$

18–2 ADDING SINE WAVES USING PHASORS. Chapter 17 illustrates how sine waves can be added graphically. Resultant sine waves are determined and the equations are given for the resultant sine waves. Although accurate graphing could be used to determine the resultant, a phasor solution is usually more convenient. The equations used in Example B are the sine waves shown in Figures 17–8b and 17–8c. This example will prove the equations by use of phasors.

Example B
Given the following equations for Figure 17–8b, determine the resultant. (Refer to Figures 18–3a and 18–3b.)

$$v_1 = 100 \sin \omega t \text{ V}$$
$$v_2 = -100 \cos \omega t \text{ V}$$

First express the equation in polar form; then determine the values of x and y. Follow the same procedure used to determine resultant vectors.

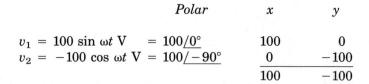

	Polar	x	y
$v_1 = 100 \sin \omega t$ V	$= 100\underline{/0°}$	100	0
$v_2 = -100 \cos \omega t$ V	$= 100\underline{/-90°}$	0	-100
		100	-100

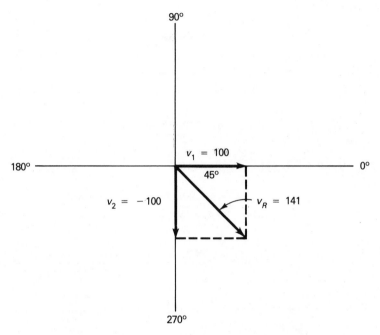

FIGURE 18–3a Graphical sketch of Figure 17-8b

Determining theta

$$\tan \theta = \frac{y}{x}$$

$$\tan \theta = \frac{-100}{100}$$

$$\theta = -45°$$

Determine the peak value of the resultant

$$\sin \theta = \frac{y}{v_R}$$

$$v_R = \frac{y}{\sin \theta}$$

$$v_R = \frac{-100}{\sin -45°}$$

$$v_R = 141 \text{ V}$$

The equation for the resultant is

$$141 \underline{/-45°} = v_R = 141 \sin (\omega t - 45°) \text{ V}$$

The solution for the sine wave equations is $v_1 = 100 \sin \omega t$ V and $v_2 = -60 \cos \omega t$ V (Figure 17–8c).

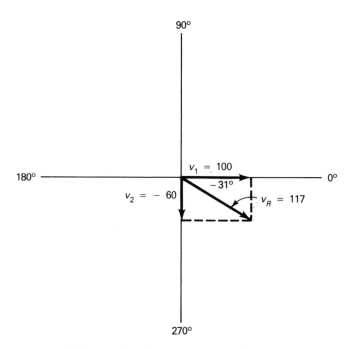

FIGURE 18–3b   Graphical sketch of Figure 17-8c

$$\begin{array}{rcccccc}
& & & & & x & y \\
v_1 = 100 \sin \omega t & = & 100\underline{/0°} & = & 100 & 0 \\
v_2 = -60 \cos \omega t & = & 60\underline{/-90°} & = & 0 & -60 \\
& & & & & 100 & -60
\end{array}$$

$$\tan \theta = \frac{-60}{100}$$
$$\theta = -31°$$

$$v_R = \frac{-60}{\sin -31°}$$
$$v_R = 117 \text{ V}$$

and
$$v_R = 117 \sin (\omega t - 31°) \text{ V}$$

EXERCISE 18–2 Draw a phasor sketch of each problem and determine the position and magnitude of the resultant. Write the resultant sine wave equation with respect to v_1 or i_1.

1. $v_1 = 50 \sin \omega t$ V, $v_2 = 30 (\sin \omega t + 40°)$ V

$v_R = \underline{75.5 \sin \omega t + 14.8°}$

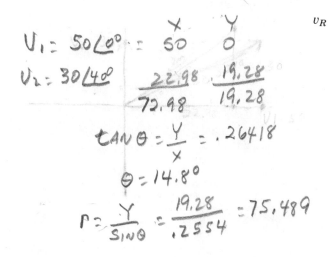

$$\begin{array}{rcccc}
& & x & y \\
V_1 = 50\underline{/0°} & = & 50 & 0 \\
V_2 = 30\underline{/40°} & & 22.98 & 19.28 \\
& & 72.98 & 19.28
\end{array}$$

$$\tan \theta = \frac{Y}{X} = .26418$$
$$\theta = 14.8°$$

$$r = \frac{Y}{\sin \theta} = \frac{19.28}{.2554} = 75.489$$

282

*Part 2
Mathematics for DC
and AC Circuits*

2. $i_1 = 250 \sin \omega t$ mA, $i_2 = 80 (\omega t - 60°)$ mA

$i_R =$ _____

3. $i_1 = 35 \sin \omega t$ A, $i_2 = 15 \cos \omega t$ A

$i_1 = 35\underline{/0}$ 35   0
$i_2 = 15\underline{/90}$ 0   15
            35   15

$\tan \theta = \frac{Y}{X} = .42857$

$\theta = 23.2°$

$r = \frac{Y}{\sin \theta} = 38.1$

$i_R = \underline{38.1 \sin \omega t + 23.2°}$

4. $v_1 = -80 \sin \omega t$ mV, $v_2 = -30 \cos \omega t$ mV

$v_R =$ _____

5. $v_1 = 350 \sin \omega t$ V, $v_2 = 220 \sin (\omega t + 120°)$ V

$V_1 = 350\underline{/0°}$ 350   0
$V_2 = 220\underline{/120°}$ -110   190.5255
              240   190.5255

$\tan \theta = \frac{Y}{X} =$

$\theta = 38.4°$

$r = \frac{Y}{\sin \theta} = 306.73$

$v_R = \underline{307 \sin \omega t + 38.4°}$

6. $i_1 = 16 \sin \omega t$ A, $i_2 = 30 \sin (\omega t - 135°)$ A

$v_R = $ _____

Chapter 18
Phasors

7. $v_1 = 8 \sin \omega t$ mV, $v_2 = 14 \sin (\omega t + 65°)$ mV, $v_3 = 6 \sin (\omega t - 35°)$ mV

$V_1 = 8\angle 0$ X   Y
$V_2 = 14\angle 65$ 8   0
$V_3 = 6\angle -35$ 5.91665   12.6883
             4.9149   −3.4414
             ───────   ───────
             18.83155   9.2469

$v_R = \underline{21 \sin \omega t + 26.2°}$

$r = \dfrac{Y}{\sin \theta}$

$r = 20.979$

$\tan \theta = \dfrac{Y}{X} = .491032$

$\theta = 26.1525°$

8. $i_1 = 40 \sin \omega t$ A, $i_2 = 40 \sin (\omega t + 120°)$ A, $i_3 = 40 \sin (\omega t - 120°)$ A

$i_R = $ _____

18–3 THE J OPERATOR. The term "*j* operator" is synonymous with analyzing ac circuits and is derived from the use of imaginary numbers. Real numbers when multiplied by themselves result in positive values.

$$(-3)(-3) = 9$$
$$(5)(5) = 25$$

Therefore the square root of such negative numbers as −9 and −25 do not exist as real numbers. Problem solving requires the use of square roots of negative numbers; to do so, the imaginary number is developed. The square root of −25 is imaginary and is simplified by breaking $\sqrt{-25}$ into two factors of $(25)(-1)$, where the square root of 25 can be extracted.

$$\sqrt{-25} = \sqrt{(25)(-1)} = 5\sqrt{-1}$$

Any negative number can be factored into $\sqrt{n(-1)}$, where the square root of n can be extracted and -1 remains under the radical. The square root of -1 is defined as imaginary and is given the symbol i.

$$\sqrt{-9} = 3\sqrt{-1} = 3i$$
$$\sqrt{-36} = 6\sqrt{-1} = 6i$$
$$\sqrt{-100} = 10\sqrt{-1} = 10i$$

In electronics, i is used as the symbol for current, and $\sqrt{-1}$ is defined as j.

$$\sqrt{-9} = 3j \text{ (or } j3 \text{ is preferred)}$$
$$\sqrt{-15} = j3.87$$
$$\sqrt{-113} = j10.6$$

EXERCISE 18-3 Express the numbers as j operators.

1. $\sqrt{-49}$ *j7*
2. $\sqrt{-16}$ *j4*
3. $\sqrt{-45}$ *j6.7082*
4. $\sqrt{-V^2}$ *jV*
5. $\sqrt{-16Z^2}$ *j4Z*
6. $\sqrt{\dfrac{-V^2}{16}}$ *jV/4*
7. $\sqrt{-32i^2}$ *j5.656i*
8. $\sqrt{-(IZ)^2}$ *jIZ*
9. $\sqrt{-\left(\dfrac{V}{I}\right)^2}$ *jV/I*
10. $\sqrt{\dfrac{-I^2}{49}}$ *jI/7*

18-4 PHASORS AND THE J OPERATOR. The j operator can be related to phasors by defining it in the four-quadrant system. By definition, multiplying a number by -1 rotates the number $180°$ on any given number line (Figures 18-4a and 18-4b). In working with the j operator, the definitions below are valid.

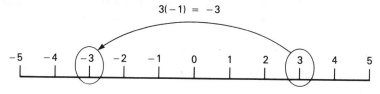

FIGURE 18-4a   Number line rotation

$$j^2 = \sqrt{-1}\sqrt{-1} = -1$$

Thus, multiplying by j^2 is the same as multiplying by -1, or multiplying by j^2 is a rotation of $180°$. Each time a number is multiplied by j, it rotates $90°$. The following definitions are valid within the four quadrants.

$j = \sqrt{-1}$ (or 90° rotation)

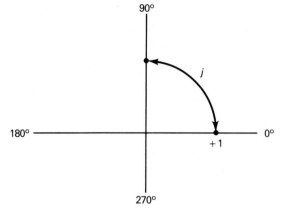

$j^2 = \sqrt{-1}\,\sqrt{-1}$
$j^2 = -1$ (or 180° rotation)

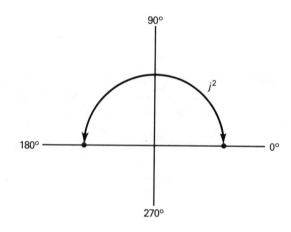

$j^3 = \sqrt{-1}\,\sqrt{-1}\,\sqrt{-1}$
$j^3 = (-1)\sqrt{-1}$
$j^3 = (-1)(j)$
$j^3 = -j$ (or $-90°$ rotation)

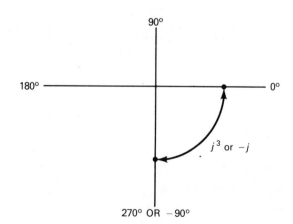

FIGURE 18–4b   *j*-operator rotation

$j^4 = j^2 j^2$
since j^2 is rotation of $180°$
j^4 = a rotation of $360°$

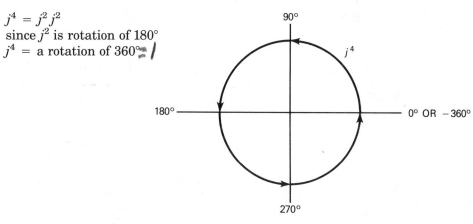

FIGURE 18–4b (continued)

Phasors with phase shifts at other than 90° are expressed using complex numbers. A complex number is one with an imaginary part and a real part. The value of x is the real part and the value of y is the imaginary part.

The phasor $10\underline{/60°}$ can also be expressed in rectangular values of x and y. Figure 18–5 illustrates this expression.

$$\begin{array}{ccc} & x & y \\ 10\underline{/60°} = & 5 & 8.66 \end{array}$$

Because y is at a 90° phase shift with x it is expressed in j operator as $j8.66$ and is the imaginary part. The real part is the value of x, or 5. It is the vector, or phasor, sum of x and y that equals the resultant.

then $\qquad 10\underline{/60°} = 5 + j8.66$

If the example just given were an equation, $v = 10 \sin(\omega t + 60°)$, it could be expressed:

$$v = 10 \sin(\omega t + 60°) \text{ V} \qquad v = 10\underline{/60°} \text{ V}$$
$$v = 5 + j8.66 \text{ V}$$

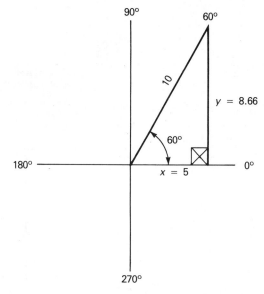

FIGURE 18–5   Phasor other than 90°

EXERCISE 18-4 Express the equations as phasors, giving answers in polar and as a j operator.

		polar	j operator
1.	$v = 50 \sin(\omega t + 30°)$ V	$50 / 30°$	$43.3 + j25$
2.	$i = 16 \cos \omega t$ mA	$16 / 0°$	$16 + j0$
3.	$i = 130 \sin(\omega t - 60°)$ A	$130 / -60°$	$65 - j112.6$
4.	$v = -30 \sin(\omega t + 130°)$ mV	$30 / 130°$	$-19.3 - j22.98$
5.	$i = 7.5 \sin(\omega t - 140°)$ A	$7.5 / -140$	$-5.75 - j4.82$ ☆
6.	$v = -150 \sin \omega t$ V	$150 / 0$	$150 + j$
7.	$v = 165 \sin(\omega t + 240°)$ V	$165 / 240°$	$-82.5 - j142.89$
8.	$i = -1250 \sin \omega t$ mA	$1250 / -0$	$1250 + 0$
9.	$v = -440 \cos \omega t$ V	$440 / -90°$	$0 - j440$
10.	$i = 300 \sin(\omega t + 180°)$ A	$300 / 180°$	$-300 + j$

18-5 SIMPLIFYING J OPERATORS. In summary, j indicates a 90° rotation.

$$j = 90° \text{ rotation}$$
$$-j = -90° \text{ rotation}$$
$$j^2 = 180°$$
$$j^3 = 270° \text{ or } -90°$$
$$j^4 = 360°$$
$$j^5 = 450° \text{ or } 90°$$
etc.

In other respects the j is handled like any other symbol in algebra. Example C illustrates some manipulations.

Example C Multiply $(j2)(j3)$

$$(j2)(j3) = j^2 6$$

since
$$j^2 = -1$$

then
$$(-1)(6) = -6$$

Multiply $(j^28)(j2)$

$$(j^28)(j2) = j^316$$

since $j^3 = -j$

then $(-j)(16) = -j16$

Divide $\dfrac{j^625}{j^25} = j^45$

since $j^4 = 1$

then $j^45 = 5$

EXERCISE 18-5 Perform the indicated operations and simplify the answers.

1. $(j^2)(j^3) = j^5$ $-j$

2. $(j)(j^2) = j^3$ $-j$

3. $(j2)(j4) = j^28$ -8

4. $(-j)(j^23) = -j^33$ $j3$

5. $(j6)(j^33) = j^418$ 18

6. $\dfrac{j^8}{j^6}$ j^2 -1

7. $\dfrac{j^58}{j^32}$ j^24 -4

8. $\dfrac{j^236}{-j4}$ $= -j9$ $-j9$

9. $\dfrac{(j3)(j4)}{j^2} = \dfrac{j^212}{j^2} = 12$ 12

10. $\dfrac{(j^26)(j4)}{(j4)(j2)} = \dfrac{j^324}{j^22} = j12$ $j12$

11. $\dfrac{(-j^3 8)(j^3)}{(j2)(j2)(j4)} = \dfrac{-j^6 8}{j^3 16} = -j^3 .5$ $j.5$

289

Chapter 18
Phasors

12. $\dfrac{(j^3)(j)(j4)}{(j)(-j)} = \dfrac{j^5 4}{-j^2} = -j^4 4$ -4

EVALUATION EXERCISE In Exercises 1 through 5 write the phasor for each equation in order form and give the x and y coordinates.

$X = \cos$
$Y = \sin$

		Polar	x	y
1.	$v = 35 \sin(\omega t + 25°)$ V	$35 \angle 25°$	31.72	14.79
2.	$i = 300 \cos \omega t$ mA	$300 \angle 90°$	0	300
3.	$v = 12 \sin(\omega t - 45°)$ mV	$12 \angle -45°$	8.485	-8.485
4.	$v = -75 \sin \omega t$ V	$75 \angle 0$	75	0
5.	$i = -6 \cos \omega t$ A	$6 \angle -90$	0	-6

In Exercises 6 through 10 determine the resultant equation for the addition of the sine waves. Draw a sketch of the phasor solution. Use v_1 or i_1 as the zero reference phasor, and express the resultant with respect to y_1 or i_1.

6. $v_1 = 40 \sin \omega t$ V, $v_2 = 25 \sin(\omega t + 30°)$ V

$V_R = 62.9 \sin \omega t + 11.46°$

$v_1 = 40 \angle 0$ X=40   Y=0
$v_2 = 25 \angle 30$ 21.65   12.5
                        61.65   12.5

$\tan \theta = \dfrac{Y}{X}$

$\theta = 11.46°$

$r = \dfrac{Y}{\sin \theta} = 62.914$

290

Part 2
Mathematics for DC
and AC Circuits

$x = \cos$
$y = \sin$

7. $i_1 = 50 \sin \omega t$ mA, $i_2 = -25 \cos \omega t$ mA

$i_1 = 50 \angle 0$ x: 50   y: 0
$i_2 = 25 \angle -90$ 0   -25
     50   -25

$\tan \theta = \frac{y}{x}$ $\theta = -26.57$

$r = \frac{y}{\sin \theta} = 55.89$

$i_R = 55.89 \sin \omega t - 26.57°$

8. $i_1 = 60 \sin \omega t$ mA, $i_2 = 35 \sin(\omega t - 25°)$ mA

$i_1 = 60 \angle 0$ x: 60   y: 0
$i_2 = 35 \angle -25$ 31.72   -14.79
     91.72   -14.79

$\tan \theta = \frac{y}{x}$
$\theta = -9.16°$

$r = \frac{y}{\sin \theta} = 92.89$

$i_R = 92.89 \sin \omega t - 9.16°$

9. $v_1 = 75 \sin \omega t$ mV, $v_2 = 55 \cos \omega t$ mV, $v_3 = -30 \cos \omega t$ mV

$V_1 = 75 \angle 0$ x: 75   y: 0
$V_2 = 55 \angle 90$ 0   55
$V_3 = 30 \angle -90$ 0   -30
     75   25

$\tan \theta = \frac{y}{x}$ $\theta = 18.43°$

$r = \frac{y}{\sin \theta} = 79.06$

$v_R = 79.06 \sin \omega t + 18.43°$

10. $v_1 = 1250 \sin \omega t$ V, $v_2 = -1200 \sin \omega t$ V

$V_1 = 1250 \angle 0$ x: 1250   y: 0
$V_2 = 1200 \angle 0$ 1200   0
     2250   0

$\tan \theta = \frac{y}{x}$ $\theta = 0°$

$r = \frac{0}{\sin \theta} = 0$

$v_R = 0 \sin \omega t$

$0 \sin \omega t$

Express Exercises 11 through 15 as *j* operators.

11. $\sqrt{-81}$ $j9$

12. $\sqrt{-P^2}$ jP

13. $\sqrt{-20X^4}$ $j4.47X^2$

14. $\sqrt{\dfrac{80Z^2}{20}}$ $62Z$

15. $\sqrt{\dfrac{-125C^2}{-25B^2}} = \sqrt{\dfrac{-25C^2}{-B^2}}$ $j\dfrac{5C}{B}$

Express the answers in Exercises 16 through 20 as phasors, giving the answers in polar form and as a *j* operator expression.

	polar	*j* operator
16. $v = 36 \sin(\omega t - 45°)$ mV	$36/-45°$	$25.46 - j25.45$
17. $i = 25 \cos \omega t$ A	$25/90°$	$25 + j$
18. $i = 80 \sin(\omega t + 60°)$ mA	$80/60°$	$40 + j69.28$
19. $v = -150 \sin \omega t$ V	$150/0°$	$150 + 0$
20. $v = 58 \sin(\omega t - 125°)$ V	$58/-125°$	$-33.27 - j47.51$

Perform the indicated operations in the remainder of the exercises and simplify the answers.

21. $(j^2)(j) = j^3$ $-j$

291

Chapter 18
Phasors

$j = j$
$j^2 = -1$
$j^3 = -j$
$j^4 = 1$

22. $(j^4 5)(j8) = j^5 40$ $j40$

23. $\dfrac{(j^3)(j^4 9)}{(j^2)(j4)} = \dfrac{j^7 9}{j^3 4} = \dfrac{j^4 9}{4}$ $\dfrac{9}{4} = 2.25$

24. $\dfrac{(j^3 5)(j)}{j^2} = \dfrac{j^4 5}{j^2} = j^2 5$ -5

25. $\dfrac{(j25)(j^3)(j)}{(j^2)(j2)} = \dfrac{j^5 25}{j^3 2} = \dfrac{j^2 25}{2}$ -12.5

CHAPTER 19

Phasor Algebra

Objectives

After completing this chapter you will be able to:
- Add and subtract complex numbers containing j operator
- Multiply complex numbers containing j operator
- Divide complex numbers using the conjugate
- Multiply phasors in polar notation
- Divide phasors in polar notation
- Convert polar notation to rectangular notation containing the j operator
- Convert rectangular notation to polar notation

Phasor algebra requires a basic understanding of algebraic principles. These principles are then applied to complex numbers which can be expressed in either polar or rectangular form. At this point the student should be familiar with the equations relating to direct-current circuits. Alternating-current circuits have different characteristics and require further study; however, the equations used to solve direct-current circuits can be used to solve alternating-current circuits, providing phasor algebra is used to find the solution.

19–1 ADDING AND SUBTRACTING COMPLEX NUMBERS. There is no difference between adding or subtracting complex numbers and adding or subtracting algebraic expressions. Keep in mind that only like things can be added or subtracted. All real parts and all imaginary parts are collected. Example A illustrates the addition and subtraction of complex numbers.

Example A
Add:

$$(+)\quad \begin{array}{r} 5 + j6 \\ 2 + j4 \\ \hline 7 + j10 \end{array} \qquad \begin{array}{r} 8 + j10 \\ 6 - j4 \\ \hline 14 + j6 \end{array}$$

Subtract, changing the sign of the subtrahend and adding:

$$(-)\quad \begin{array}{r} 6 + j12 \\ 3 - j4 \\ \end{array} = \begin{array}{r} 6 + j12 \\ -3 + j4 \\ \hline 3 + j16 \end{array}$$

293

EXERCISE 19–1 Perform the indicated operations. In Exercises 1 through 10 add the complex numbers.

1. $\begin{array}{r} 7 + j6 \\ \underline{2 + j4} \end{array}$

2. $\begin{array}{r} 10 + j12 \\ \underline{3 - j6} \end{array}$

3. $\begin{array}{r} 150 - j75 \\ \underline{16 + j12} \end{array}$

4. $\begin{array}{r} 6.5 + j17 \\ \underline{2.4 - j6.8} \end{array}$

5. $\begin{array}{r} -1.75 + j12 \\ \underline{1.4 - j2} \end{array}$

6. $\begin{array}{r} 50 - j8 \\ \underline{0 + j7} \end{array}$

7. $\begin{array}{r} 12 - j5 \\ \underline{16 + j0} \end{array}$

8. $\begin{array}{r} 3.8 - j7 \\ \underline{4.2} \end{array}$

9. $\begin{array}{r} 15 + j12 \\ 5 + j6 \\ 32 - j18 \\ \underline{-5 + j10} \end{array}$

10. $\begin{array}{r} 35 + j150 \\ 15 - j60 \\ -8 - j12 \\ \underline{26 + j132} \end{array}$

Subtract the complex numbers in Exercises 11 through 18.

11. $\begin{array}{r} 1250 + j600 \\ \underline{250 - j200} \end{array}$

12. $\begin{array}{r} 16 - j12 \\ \underline{-4 + j4} \end{array}$

13. $\begin{array}{r} 25 - j0 \\ \underline{5 + j16} \end{array}$

14. $\begin{array}{r} 0 - j12 \\ \underline{4.5 + j12} \end{array}$

15. $\begin{array}{r} 4.5 - j7 \\ \underline{4.5 + j8} \end{array}$

16. $\begin{array}{r} 42 \\ \underline{16 - j12} \end{array}$

17. $\begin{array}{r} j16 \\ \underline{-6 + j17} \end{array}$

18. $\begin{array}{r} 15 - j45 \\ \underline{6 + j17} \end{array}$

Collect like terms in Exercises 19 through 25.

19. $j6 + 4 - j2 + 3$ _____

20. $16 - j2 + j4 - 6$ _____

21. $j12 - 6 + j4 - 3$ _____

22. $16 - j3 - 8 + j2 + j$ _____

23. $25 - j7 - (12 + j4 - 5)$ _____

24. $(6 - j3) - (j3 + 16)$ _____

25. $j45 - j12 + (15 - j120)$ _____

19-2 MULTIPLICATION USING THE J OPERATOR. Multiplying complex numbers is done using the same procedure as that used to multiply binomial expressions. Since $j^2 = -1$ and $j^4 = 1$, the final answer can be simplified. Example B illustrates this procedure.

Example B
Multiply $(2 + j4)(3 + j2)$

Using the rules for binomial multiplication

$$(2 + j4)(3 + j2) = 6 + j16 + j^2 8$$

since $j^2 = -1$, substitute (-1) for j^2

$$6 + j16 + (-1)(8) = 6 + j16 - 8$$

By collecting like terms, the simplified answer is

$$-2 + j16$$

Multiply $(2 - j4)(3 + j2)$

$$(2 - j4)(3 + j2) = 6 - j8 - j^2 8$$

substitute -1

$$6 - j8 - (-1)(8)$$

and collect terms

$$6 - j8 + 8 = 14 - j8$$

The two problems in Example B illustrate that j^2 changes the sign of the number it precedes. Knowing this will eliminate the need for substituting -1 for j^2. Change the sign of any number preceded by j^2.

$$-j^2 4 = +4 \quad \text{and} \quad j^2 6 = -6$$

EXERCISE 19-2 Perform the multiplications.

1. $4(6 - j2)$

2. $j3(2 + j7)$

3. $j(6 - j3)$

4. $(2 + j7)(3 - j2)$

5. $(a + jb)(a - jb)$

6. $(3 - j6)(3 + j6)$

7. $(4 - j2)(2 - j4)$

8. $(3 + j2)^2$

9. $(6 - j2)(6 + j2)$

10. $(R - jX)^2$

11. $(8 - j3)(-4 + j6)$

12. $(12 - j10)(j10 + 12)$

19-3 DIVISION USING THE J OPERATOR. Division using the *j* operator is performed by multiplying the numerator and the denominator by the "conjugate" of the denominator. The conjugate of a binomial is an expression having the same symbols but with the sign separating the symbols opposite.

Expression	Conjugate
$a + b$	$a - b$
$3a - 5$	$3a + 5$
$5 - j6$	$5 + j6$
$2 + j4$	$2 - j4$

Example C illustrates division by conjugating.

Example C

Divide $(3 + j2)$ by $(2 - j4)$

$$\frac{3 + j2}{2 - j4}$$

$$\frac{(3 + j2)(2 + j4)}{(2 - j4)(2 + j4)} \qquad \frac{6 + j14 + j^2 8}{4 - j^2 16}$$

Change the sign of the numbers preceded by j^2, and collect terms:

$$\frac{14 + j14}{4 + 16} \qquad \frac{14 + j14}{20}$$

Divide the real and imaginary parts by the denominator of 20:

$$\frac{(14 + j14)}{20} = 0.7 + j0.7$$

EXERCISE 19-3 Perform the division.

1. $\dfrac{5}{2 + j3}$

2. $\dfrac{12 - j5}{4}$

3. $\dfrac{3 - j2}{3 - j5}$

4. $\dfrac{2 + j3}{2 - j3}$

5. $\dfrac{5 - j3}{3 + j2}$

6. $\dfrac{1}{2 + j1}$

7. $\dfrac{25}{j5}$

8. $\dfrac{6 - j16}{j6}$

9. $\dfrac{4 + j4}{4 + j4}$

10. $\dfrac{1 + j}{2 - j3}$

19–4 MULTIPLYING PHASORS IN POLAR FORM. Multiplying phasors in polar form is similar to multiplying numbers in power-of-10 notation. In power-of-10 notation significant numbers are multiplied and the exponents are collected algebraically. When multiplying phasors in polar form multiply the magnitudes and algebraically add the angles. Example D illustrates this procedure.

Example D
Multiply $2\underline{/30°}$ times $4\underline{/20°}$

Multiply the magnitudes and add the angles algebraically:

$$(2\underline{/30°})(4\underline{/20°}) = 8\underline{/50°}$$

Multiply

$$(2\underline{/-10°})(4\underline{/40°})(5\underline{/-60°}) = 40\underline{/-30°}$$

Multiply

$$(3\underline{/220°})(4\underline{/-150°}) = 12\underline{/70°}$$

EXERCISE 19-4   Perform the multiplications.

1. $(8\underline{/65°})(30\underline{/42°})$

2. $(420\underline{/-36°})(2.6\underline{/32°})$

3. $(18.5\underline{/-136°})(3.5\underline{/242°})$

4. $(0.25\underline{/45°})(0.5\underline{/120°})$

5. $(6\underline{/33°})(14\underline{/-62°})(3\underline{/-12°})$

6. $(15\underline{/120°})(6\underline{/-100°})(12\underline{/-20°})$

7. $(85\underline{/-3°})(10\underline{/-75°})(3\underline{/-62°})$

8. $(0.5\underline{/16°})(1.5\underline{/-75°})(0.02\underline{/225°})$

9. $(4.5\underline{/90°})(2.5\underline{/90°})(1.5\underline{/-180°})$

Chapter 19
Phasor Algebra

10. $(125\underline{/5°})(550\underline{/355°})$

19–5 DIVIDING PHASORS IN POLAR FORM. Division in polar form is similar to dividing units expressed in power-of-10 notation. The magnitudes are divided by conventional division and the angles are subtracted algebraically. Example E illustrates this operation.

Example E
Divide $20\underline{/30°}$ by $5\underline{/10°}$

Subtract the angles

$$30° - (10°) = 20°$$

Divide the magnitudes

$$\frac{20}{5} = 4$$

$$\frac{20\underline{/30°}}{5\underline{/10°}} = 4\underline{/20°}$$

Divide $8\underline{/25°}$ by $4\underline{/-10°}$

$$\frac{8\underline{/25°}}{4\underline{/-10°}} = 2\underline{/35°}$$

Divide $6\underline{/-150°}$ by $2\underline{/-220°}$

$$\frac{6\underline{/-150°}}{2\underline{/-220°}} = 3\underline{/70°}$$

EXERCISE 19–5 Perform the operations.

1. $\dfrac{85\underline{/45°}}{15\underline{/20°}}$

2. $\dfrac{350\underline{/135°}}{40\underline{/-35°}}$

3. $\dfrac{50/-62°}{87/-25°}$

4. $\dfrac{0.75/-240°}{0.25/120°}$

5. $\dfrac{(6/36°)(4/75°)}{10/-75°}$

6. $\dfrac{(145/125°)(35/55°)}{(16/-30°)(4/25°)}$

7. $\dfrac{(0.25/-12°)(5/12°)}{(1.25/60°)(3/-125°)}$

8. $\dfrac{(1.6/-45°)(4.2/135°)}{(1.6/45°)(3/-240°)}$

9. $\dfrac{(1250/-240°)}{(125/245°)(35/-45°)}$

10. $\dfrac{(35.6/-55°)(12.8/40°)}{(4.56/-12°)(3.14/37°)}$

301

Chapter 19
Phasor Algebra

19-6 CONVERTING POLAR AND RECTANGULAR FORMS. The operations of converting polar and rectangular forms have already been studied and will not be reviewed at this point. Example F illustrates this procedure. Polar coordinates cannot be added or subtracted, and it is necessary to convert polar to rectangular before adding or subtracting. Some calculators have this function built into their operation, thus simplifying the procedure. Consult the operation manual of a calculator for rectangular and polar conversions.

Example F
Convert $40\underline{/35°}$ to rectangular form, expressing the answer by using the j operator.

$$\begin{array}{ccc} & x & y \\ 40\underline{/35°} = & 32.8 & 22.9 \end{array}$$

where
$$x = \text{the real part}$$
$$y = \text{the imaginary part}$$

(polar form to rectangular form)

then
$$40\underline{/35°} = 32.8 + j22.9$$

Convert $6 + j8$ to polar form

$$\tan \theta = \frac{\text{imaginary part}}{\text{real part}}$$

$$\tan \theta = \frac{8}{6}$$
$$\theta = 53.1°$$

The magnitude is found by dividing the imaginary part by the sin of θ:

$$\frac{8}{\sin 53.1} = 10$$

then
$$6 + j8 = 10\underline{/53.1°}$$

EXERCISE 19-6 Convert to polar form in Exercises 1 through 8.

1. $15 + j5$

2. $85 - j6$

3. $150 + j35$

4. $-0.25 + j0.65$

5. $-1250 + j1250$ _____

303

Chapter 19
Phasor Algebra

6. $0 - j12$ _____

7. $65.3 - j24.8$ _____

8. $12.5 + j76.3$ _____

Convert from polar to rectangular coordinates in Exercises 9 through 16, using the j operator.

9. $75/\underline{35°}$ _____

10. $85/\underline{-125°}$ _____

11. $0.6/\underline{240°}$ _____

12. $420/\underline{0°}$ _____

13. $36/\underline{-220°}$ _____

14. $45/\underline{-90°}$ _____

15. $25.8/\underline{22.4°}$ _____

16. $78.6/\underline{-67.4°}$ _____

Part 2
Mathematics for DC and AC Circuits

EVALUATION EXERCISE Simplify in Exercises 1 through 5.

1. $(5 - j8) - (4 + j7)$ _____

2. $(2 + j3) + (5 - j4)$ _____

3. $(j10 - 5) - (5 + j20)$ _____

4. $(6 - j4) - (j3 + 8)$ _____

5. $(25 + j12) + (10 - j12)$ _____

Multiply by the rectangular method in Exercises 6 through 10. Show your work and leave the answers in rectangular form.

6. $(3 - j3)(2 + j5)$ _____

7. $(4 - j4)^2$ _____

8. $(2 + j3)(3 - j2)(1 - j2)$ _____

9. $(j2 - 2)(3 - j4)$ _____

10. $(3 + j3)^2$ _____

Simplify in Exercises 11 through 15. Leave your answers in polar form.

11. $\dfrac{(15\underline{/60°})(4\underline{/30°})}{j5 - 16}$ _____

Chapter 19
Phasor Algebra

12. $\dfrac{8\underline{/35°}(j4 - 4)}{105\underline{/-35°}}$ _____

13. $\dfrac{(j4 - 3)(j4 - 3)}{12.5\underline{/-45°}}$ _____

14. $\dfrac{5\underline{/18°}(5 - j5)}{(j14 + 8)(6\underline{/32°})}$ _____

15. $\dfrac{(3 + j3)^2(2 + j2)}{(30\underline{/-25°})(2\underline{/15°})}$ _____

Divide in Exercises 16 through 20 using rectangular division, leaving your answers in rectangular form.

16. $\dfrac{4 + j3}{2 + j4}$ _____

17. $\dfrac{3 - j6}{j4}$ _____

18. $\dfrac{3 - j3}{4 + j4}$

19. $\dfrac{j4 + 5}{j6 - 6}$

20. $\dfrac{(j2 + 2)(j6 - 2)}{j2 - 2}$

CHAPTER 20
AC Series Circuits

Objectives

After completing this chapter you will be able to:
- Express electronic components in polar and rectangular notation
- Solve ac series R-L circuit problems using phasor algebra
- Solve ac series R-C circuit problems using phasor algebra
- Solve ac series R-L-C circuit problems using phasor algebra
- Solve ac circuits for true power, reactive power, and apparent power
- Determine power factor
- Solve ac circuits for component Q and circuit Q

Equations used to solve dc problems can be used to solve ac problems, provided phasor algebra is used to obtain the solution. In this chapter some new concepts to series ac circuits will be introduced, such as power factor, reactance power, applied power, and circuit Q.

20-1 INDUCTIVE AND CAPACITIVE REACTANCE. The opposition to the flow of alternating current that an inductor or a capacitor gives in a circuit can be calculated as follows.

$$X_L = 2\pi f L$$

where
X_L = inductive reactance, in ohms
f = frequency, in hertz
L = inductance, in henrys

$$X_C = \frac{1}{2\pi f C}$$

where
X_C = capacitive reactance, in ohms
f = frequency, in hertz
C = capacitance, in farads

Recall from Chapter 17, "Sine Waves," that $\omega = 2\pi f$. The equations just given are also written:

$$X_L = \omega L \qquad \text{and} \qquad X_C = \frac{1}{\omega C}$$

307

20–2 COMPONENTS IN THE SERIES AC CIRCUIT.

Chapter 18 illustrates the expression of quantities in rectangular and polar form. Components in ac circuits can also be expressed in either polar or rectangular form.

A resistor is a component in which the current through the resistor and the voltage across the resistor are in phase, or where both are at 0°. The resistor is expressed as a phasor (see Figure 20–1a).

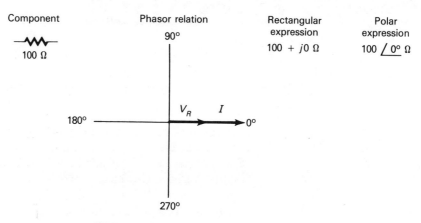

FIGURE 20–1a The resistor as a component

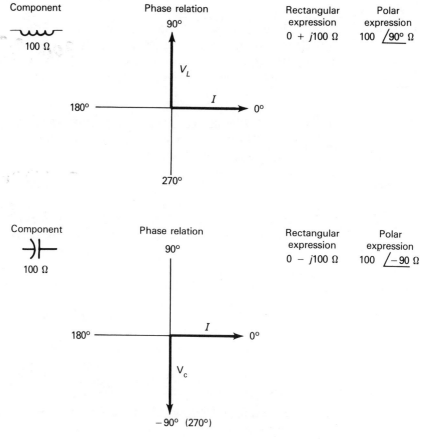

FIGURE 20–1b The inductor and capacitance as components

Because the resistor does not fall into one of the quadrants, it does not have an imaginary component and is considered a real component. In rectangular form the j operator is zero, and in the polar form the angle is at zero.

The inductor is a component in which current through the inductor lags the voltage by 90°. The capacitor is a component in which the current leads the voltage by 90°. They are expressed as shown in Figure 20–1b.

EXERCISE 20–1 Calculate the opposition each component gives in an ac circuit. Express the answers in both rectangular and polar form.

$X_L = 2\pi f L$

1. (inductor) $L = 20$ mH, $f = 20$ kHz   rect = $0 + j2512$
      polar = $2512 \angle 90°$

2. (capacitor) $C = 0.015$ µF, $f = 15$ kHz   rect = $0 - j707$
      polar = $707 \angle -90°$

3. (capacitor) $C = 20$ µF, $f = 1000$ Hz   rect = $0 - j7.96$
      polar = $7.96 \angle -90°$

4. (inductor) $L = 8$ H, $f = 60$ Hz   rect = $0 + j3016$
      polar = $3016 \angle 90°$

5. (resistor) $R = 2700$ Ω, $f = 100$ Hz   rect = _____
      polar = _____

6. (capacitor) $C = 12$ pF, $f = 10$ MHz   rect = $0 - j1326$
      polar = $1326 \angle -90°$

7. (inductor) $L = 15$ H, $f = 5$ MHz   rect = _____
      polar = _____

8. (resistor) $R = 3.3$ MΩ, $f = 18$ gHz   rect = _____
      polar = $\angle 0$

20–3 R-L SERIES CIRCUITS. The basic equations studied in dc circuits also apply to ac circuits; opposition to alternating current, however, is called impedance, with the symbol Z.

$$V = IZ$$
$$Z_T = Z_1 + Z_2 + Z_3 + \text{etc.}$$
$$V_T = V_1 + V_2 + V_3 + \text{etc.}$$
$$I_T = I_1 = I_2 = I_3 = \text{etc.}$$

The equations used to calculate consumption of power are the same, but this will be discussed as a separate topic. Examples A and B illustrate the application of phasor algebra to an R-C circuit.

Example A

Given the circuit in Figure 20–2, determine the total impedance, circuit current, and voltage across the resistance, and inductance. Impedance is

$$Z_T = Z_1 + Z_2$$

Express resistance and inductance in rectangular form for addition

$$Z_T = (100 + j0) + (0 + j300)$$

Remove the parentheses and collect

$$Z_T = 100 + j300$$

or $\quad 316\underline{/71.6°}$ (in polar form)

The circuit current is determined by Ohm's Law. In the equation for the given voltage, 17 sin ωt V, 17 is a peak value. The problem can be worked in either peak or rms. In this example the conversion is to rms. When they are not part of an equation, or when they are stated otherwise, all voltages and current are considered to be in rms values.

$$17\ V_P\ (0.707) = 12\ V_{\text{rms}}$$

$$I_T = \frac{V_T}{Z_T}$$
$$I_T = \frac{12.0\underline{/0°}}{316\underline{/71.6°}}$$

The applied voltage is at 0° before the circuit is active:

$$I_T = 38.0\underline{/-71.6°}\ \text{mA}$$

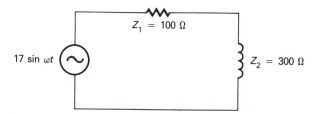

FIGURE 20–2   The R-L circuit

The voltages across Z_1 and Z_2 are determined by Ohm's Law:

$$V_1 = I_T Z_1$$
$$V_R = (0.0380\underline{/-71.6°})(100\underline{/0°})$$
$$V_R = 3.8\underline{/-71.6}\text{ V}$$

$$V_2 = I_T Z_2$$
$$V_L = 0.0380\underline{/-71.6°}(300\underline{/90°})$$
$$V_L = 11.4\underline{/18.4°}\text{ V}$$

The total circuit voltage is not the algebraic sum of V_1 and V_2, but, rather, the phasor sum of the voltages:

$$V_T = V_R + V_L$$
$$V_T = 3.80\underline{/-71.6°} + 11.4\underline{/18.4°}$$

The polar form must be converted to rectangular form for addition:

$$V_T = (1.2 - j3.6) + (10.8 + j3.6)$$
$$V_T = 12 + j0\text{ V}$$
or
$$V_T = 12\underline{/0°}\text{ V}$$

The phasor diagram illustrated in Figure 20–3 shows the phase relation that exists between the voltage and the current in this circuit. Because the current is the common element in a series circuit it is used as the reference phasor and is plotted at 0°. The voltage across the resistor, $V_R = 3.8\underline{/-71.6°}$ V, is in phase with the current, and this phasor is drawn at 0°. The voltage across the inductor, $V_L = 11.4\underline{/18.4°}$ V, is leading the current by 90° and is drawn on the 90° axis. The resultant voltage is the phasor sum of V_R and V_L. This resultant voltage is also the total voltage, $V_T = 12\underline{/0°}$ V, applied to the circuit.

The voltage across the resistor, $3.80\underline{/-71.6°}$, shows an angle of $-71.6°$. This is the angle between the voltage across the resistor, V_R, and the applied

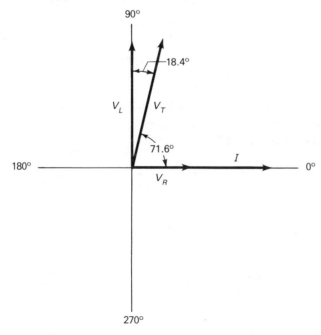

FIGURE 20–3 Phasors for series R-L circuit

312

Part 2
Mathematics for DC
and AC Circuits

voltage, V_T. These phasors indicate that the voltage across the resistor lags the total voltage by 71.6°.

The voltage across the inductor, $11.4\underline{/18.4°}$, shows an angle of 18.4°. This angle is the angle between the total voltage and the voltage across the inductor. These phasors indicate that the voltage V_L leads the voltage V_T by 18.4°.

It can also be seen in Figure 20–3 that the total applied voltage leads the current by 71.6°. In the inductive circuit the voltage leads the current, or it can be stated that the current lags the voltage. When working ac-circuit problems a phasor diagram is helpful in understanding the circuit and in checking the answer.

EXERCISE 20–2 Solve the following problems, leaving all answers in polar form (work in rms values).

1. Given the circuit in Figure 20–2, with the following changes—$V_T = 165 \sin \omega t$ V, $R = 1.2$ kΩ, $L = 1.5$ H, and the frequency equal to 100 Hz—draw the circuit diagram and label all components. Find the total impedance, circuit current, and voltage across each component. Draw and label the phasor diagram for this circuit.

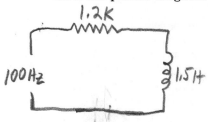

$Z_T = \underline{1.53\underline{/38.1}\ \text{K}\Omega}$

$I_T = \underline{76.5\underline{/-38.1}\ \text{mA}}$

$V_R = \underline{92\underline{/-38.1}\ \text{V}}$

$V_L = \underline{72.3\underline{/51.9}\ \text{V}}$

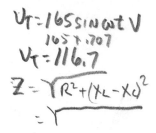

$V_T = 165 \sin \omega t$ V
$165 \times .707$
$V_T = 116.7$

$Z = \sqrt{R^2 + (X_L - X_C)^2}$
$= \sqrt{\quad}$

2. A series circuit contains an inductance and a resistance. The voltage across the inductance is $12.5\underline{/35°}$ V; across the resistance it is $8.75\underline{/-55°}$ V. What is the total voltage?

$V_T = \underline{\qquad}$

3. Use the circuit in Figure 20–2, with the following changes—$R = 22$ kΩ, $L = 12$ mH, $V_T = 75 \sin \omega t$ V, and the frequency is 1.5 MHz. Draw and label the circuit diagram; find Z_T, I_T, V_R, and V_L; and draw and label the phasor diagram.

313

Chapter 20
AC Series Circuits

$Z_T = $ _____

$I_T = $ _____

$V_R = $ _____

$V_L = $ _____

4. A resistance and an inductor are in series across a supply voltage of 9.0/0° mV and the voltage across the resistance is 4.0/62° mV. What is the voltage across the inductor? (A phasor diagram will help you find the solution.)

$V_L = $ _____

5. The current in a series R-L circuit is 3.25/−21° mA, and the resistance is 15 kΩ. If the inductance has a reactance of 5700 Ω, find the voltage across the resistance and the inductance. Write the equation for the applied voltage; draw and label the circuit and phasor diagrams.

$V_R = $ _____

$V_L = $ _____

$v = $ _____

6. In a series R-L circuit the voltage across the inductor is $82\underline{/27.3°}$ V and the inductance has a reactance of 3500 Ω. If the resistance is 1.8 kΩ, find V_R and V_T. Draw and label the circuit and phasor diagrams.

$V_R = $ _____

$V_T = $ _____

20–4 R-C SERIES CIRCUITS. The R-C circuit problems are solved in the same manner as R-L circuits. Any equation used to solve a dc circuit can be used in the ac circuit if phasor algebra is used. An R-C circuit is illustrated in Example B.

Example B

Given the circuit in Figure 20–4, find the total impedance, circuit current, and voltage across the resistor and the capacitor.

To work in rms values the peak value of 17 must be converted to rms.

$$17 \, V_P \, (0.707) = 12 \text{ V}$$

impedance
$$Z_T = Z_1 + Z_2$$
$$Z_T = (50 + j0) + (0 - j125)$$
$$Z_T = 50 - j125 \text{ Ω}$$

or
$$135\underline{/-68.2°} \text{ Ω}$$

current
$$I_T = \frac{V_T}{Z_T}$$
$$I_T = \frac{12}{135\underline{/-68.2°}}$$
$$I_T = 88.9\underline{/68.2°} \text{ mA}$$

V_R
$$V_R = I_T Z_1$$
$$V_R = (0.0889\underline{/68.2°})(50\underline{/0°})$$
$$V_R = 4.44\underline{/68.2°} \text{ V}$$

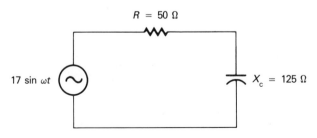

FIGURE 20–4   The R-C circuit

V_C

$V_C = I_T Z_2$
$V_C = (0.0889\underline{/68.2°})(125\underline{/-90°})$
$V_C = 11.1\underline{/-21.8°}$ V

The phasor diagram shown in Figure 20–5 gives the phase relationship of the voltage and current in the R-C circuit. In the series circuit, current is common to all components and is used as the reference phasor. The voltage phasor for resistance is drawn in phase with the current. In a capacitive circuit the current leads the voltage by 90° and the voltage V_C is drawn on the −90° axis. The resultant phasor is the total, or applied, voltage.

In this circuit the current leads the applied voltage by 68.2° and the voltage across the capacitor lags the applied voltage by 21.8°, as seen in the phasor diagram in Figure 20–5.

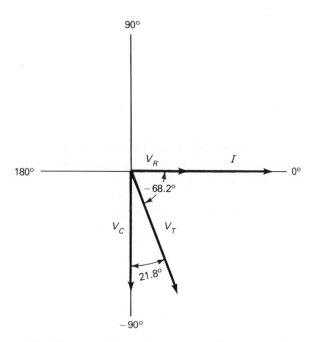

FIGURE 20–5   Phasor diagram for R-C series circuit

EXERCISE 20–3   Solve and leave your answers in polar form (work in rms values).

1. Given the circuit of Figure 20–4, with the following changes—$R = 56\ \Omega$, $C = 5\ \mu F$, $v = 311 \sin \omega t$, and frequency = 1.0 kHz—find the total impedance, circuit current, and voltage across the resistor and the voltage across the capacitor. Draw and label the circuit and phasor diagrams.

$Z_T = $ _____

$I_T = $ _____

$V_R = $ _____

$V_C = $ _____

2. The total impedance of an R-C series circuit is $563/\!-\!67°$ kΩ. If the resistance is 220 Ω, what is the reactance of the capacitor?

$X_C = $ _____

3. A series R-C circuit contains a resistance of 47 kΩ and a capacitor of 15 nF. The voltage across the resistor is $18/56.4°$ V; the frequency is 150 Hz. Find the applied voltage and the voltage across the capacitor.

$V_T = $ _____

$V_C = $ _____

4. Given the circuit in Figure 20–4, with the following changes—$R = 1.8$ kΩ, $C = 2.2$ μF, frequency = 60 Hz, and the applied voltage is 70.7 sin ωt V. Find Z_T, I_T, V_R, and V_C. Draw and label the circuit and phasor diagrams.

$Z_T = $ _____

$I_T = $ _____

$V_R = $ _____

$V_C = $ _____

5. The total voltage applied to an R-C circuit is $65/\!-\!35°$ V. Find the voltage across the capacitor and the resistor. (A phasor diagram will help you find the solution.)

$V_R = $ _____

$V_C = $ _____

6. The voltage across a capacitor in an R-C circuit is $19.7\underline{/-35°}$ V; the current is $4.5\underline{/55°}$ mA. Find the value of the resistance if the applied voltage is 24 V (use a phasor diagram).

$$R = \underline{\hspace{2cm}}$$

20–5 THE EQUIVALENT SERIES CIRCUIT. The equivalent series circuit is often used to analyze electronic circuits. The circuit contains one or two components that reflect the total impedance of the circuit. The generator "sees" only the total characteristics of the circuit, which can be resistive, capacitive, inductive, resistive, and capacitive or resistive and inductive. It is these parameters which are usually under study. Example C illustrates the equivalent series circuit.

Example C

A series circuit has three impedances: $6\underline{/30°}$, $12\underline{/-45°}$, and $15\underline{/60°}$ (Figure 20–6). Determine the impedance, circuit current, voltage across each impedance, and equivalent series circuit. The applied voltage to the circuit is $70.7 \sin \omega t$ V.

To work in rms, convert $70.7\ V_P$ to rms.

$$70.7(0.707) = 50\ V$$

Impedance

$$Z_T = Z_1 + Z_2 + Z_3$$
$$Z_T = (5.2 + j3) + (8.48 - j8.48) + (7.5 + j13)$$
$$Z_T = 21.2 + j7.52\ \Omega$$

or $22.5\underline{/19.5°}\ \Omega$

Circuit current

$$I_T = \frac{V_T}{Z_T}$$
$$I_T = \frac{(50\underline{/0°})}{22.5\underline{/19.5°}}$$
$$I_T = 2.22\underline{/-19.5°}\ A$$

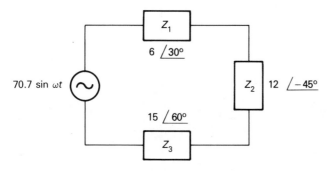

FIGURE 20–6   Three impedances in series

Voltage across each component

$$V_1 = IZ_1$$
$$V_1 = (2.22\underline{/-19.5°})(6\underline{/30°})$$
$$V_1 = 13.3\underline{/11.5°}\ V$$

$$V_2 = IZ_2$$
$$V_2 = (2.22\underline{/-19.5°})(12\underline{/-45°})$$
$$V_2 = 26.6\underline{/-64.5°}$$

$$V_3 = IZ_3$$
$$V_3 = (2.22\underline{/-19.5°})(15\underline{/60°})$$
$$V_3 = 33.3\underline{/40.5°}\ V$$

The equivalent series circuit: the total impedance of the series circuit in Example C is $22.5\underline{/19.5°}\ \Omega$. Since the circuit phase angle theta is positive 19.5°, the equivalent series circuit must contain an inductor and a resistance. If theta were 0°, the circuit would be resistive; or if theta were negative, the circuit would be capacitive and resistive. To determine the values of the resistance and the reactance, change the total impedance to its rectangular form. The real part will be the ohmic value of the resistance, and the imaginary part will be the ohmic value of the reactance. The equivalent circuit is shown in Figure 20-7.

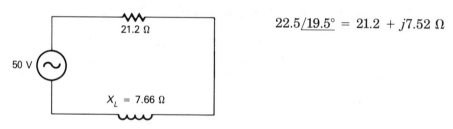

$$22.5\underline{/19.5°} = 21.2 + j7.52\ \Omega$$

FIGURE 20-7   The equivalent series circuit

EXERCISE 20-4 Solve the following series-circuit problems, using rms values.

1. Given the circuit in Figure 20-8, where $Z_1 = 22\underline{/0°}\ \Omega$, $Z_2 = 150\underline{/65°}\ \Omega$, and $Z_3 = 120\underline{/-80°}\ \Omega$, determine the impedance and circuit current. Identify the components that make up each impedance by drawing them on the schematic.

$Z_T =$ _____

$I_T =$ _____

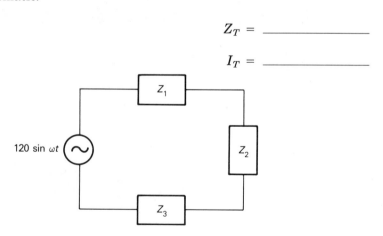

FIGURE 20-8   Impedances in series

2. In Exercise 1 determine the voltage drop across each impedance.

$V_1 = $ _____

$V_2 = $ _____

$V_3 = $ _____

Chapter 20
AC Series Circuits

3. Draw the equivalent series circuit; identify and label the ohmic value of each component in the circuit in Exercise 1.

4. Repeat Exercise 1, with impedances of $Z_1 = 22\underline{/-68°}$ kΩ, $Z_2 = 15\underline{/-73°}$ kΩ, and $Z_3 = 27\underline{/80°}$ kΩ. Draw the circuit schematic; identify and label each component before beginning any calculations.

$Z_T = $ _____

$I_T = $ _____

5. In Exercise 4 determine the voltage drop across each impedance.

$V_1 = $ _____

$V_2 = $ _____

$V_3 = $ _____

6. Draw the equivalent series circuit and identify and label the ohmic value of each component in the circuit of Exercise 4.

7. Repeat Exercise 1, with $Z_1 = 65 + j55$ Ω, $Z_2 = 150 - j125$ Ω, $Z_3 = 35 + j35$ Ω. Draw the schematic diagram; identify and label the ohmic value in each impedance before beginning any calculations.

$Z_T =$ _____

$I_T =$ _____

8. Determine the voltage across each component in Exercise 7.

$V_1 =$ _____

$V_2 =$ _____

$V_3 =$ _____

9. Draw the equivalent series circuit; identify and label the ohmic value of each component in Exercise 7.

10. The voltage measured across three impedances in series is $V_1 = 5.7\underline{/76°}$ mV, $V_2 = 8.6\underline{/-68°}$ mV, $V_3 = 3.4\underline{/0°}$ mV. Determine the applied voltage.

$V_T = $ _____

11. Determine the impedance and current in the circuit given in Figure 20-9.

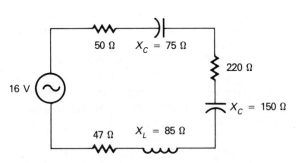

FIGURE 20-9 Circuit for problem 20-4 (11)

$Z_T = $ _____

$I_T = $ _____

12. Three impedances are in series: $Z_1 = 33 + j10$ kΩ, $Z_2 = 27\underline{/-35°}$ kΩ, and $Z_3 = 12 - j15$ kΩ. The voltage across Z_2 is $75\underline{/-18°}$ V. Find the voltage across Z_1 and Z_3.

$V_1 = $ _____

$V_3 = $ _____

20-6 POWER IN AN AC CIRCUIT. Power is consumed in the form of heat, and the component that consumes power is resistance. The watt is the basic unit that measures power in any circuit. Rms values of a sine wave, 0.707 times V, or I, are used to compute power consumption in an ac circuit. Rms values of current and voltage are equivalent to dc values and will dissipate the same amount of heat. A sine wave with a peak value of 10 V has an rms value of 7.07 V. If 10 V peak were placed across a 100 Ω resistor, it would be equivalent to placing a battery of 7.07 V dc across a 100 Ω resistor. In each case the resistor dissipates the same amount of power. Therefore the equations $\frac{V^2}{R}$ and I^2R can be used to calculate power consumption in an ac circuit, providing the current and voltages are in rms. It is not necessary to use phasor algebra to calculate power dissipated.

Resistance is the component that consumes power, but there are other characteristics of power that should be studied in ac circuits. An inductor stores energy in its magnetic field but returns this energy when the field collapses. Similarly, a capacitor stores energy and returns this energy to the circuit. In either case energy must be supplied to the circuit for the inductor and capacitor to function. This type of energy is called "reactance power" and is measured in volt amperes reactance (VAR).

The actual voltage and current requirements in an ac circuit are measured in volts-amperes (VA). The relation of the power characteristics can be seen in the R-L phasor diagram in Figure 20–10. Figure 20–10a illustrates a voltage phasor diagram for an R-L circuit; Figure 20–10b illustrates the characteristics of power in an R-L circuit.

$$P_A = IV_T$$

where P_A is the applied power measured in volts-amperes. This is the voltage and current that must be supplied to the circuit; but it is not entirely consumed. I is the circuit current and V_T is the total voltage applied to the circuit.

$$P_R = IV_L$$

where P_R is the reactance power measured in volt ampere reactance (VAR). This is the energy used when the inductor or capacitor functions. This energy is not consumed but is returned to the circuit instead. I is the circuit current and V_L is the voltage across the inductor or capacitor if it were an R-C circuit.

$$P_T = IV_R$$

where P_T is the true, or actual, power consumed by the resistance; it is measured in watts. This energy is consumed in the form of heat. I is the circuit current, and V_R is the voltage across the circuit resistance.

20-7 POWER FACTOR. In Figure 20–11 the power relation is shown as part of a right triangle. The most common measurements in a circuit are the applied voltage and current. If these values are known, the true power, P_T, can be found by using the cosine function. The cosine of theta is called the power factor (PF) of the circuit; the equation can be written:

$$P_T = (PF)(I_T V_T)$$

323

Chapter 20
AC Series Circuits

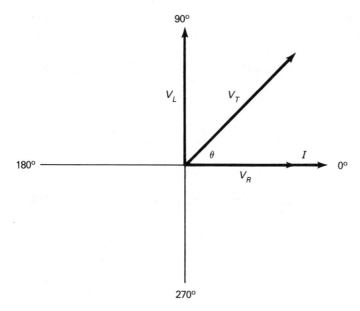

FIGURE 20–10a   Voltage phasors

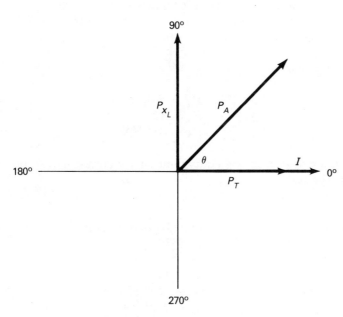

FIGURE 20–10b   Power phasors

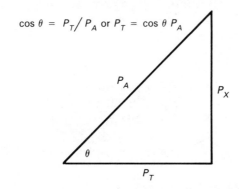

FIGURE 20–11   Using cosine of θ to find true power

Because the power factor is the cosine of theta it is always a number less than 1. For convenience, the power factor is often expressed in percentage form. When theta is a negative angle the power factor is expressed as a lagging power factor; where it is a positive angle it is expressed as a leading power factor.

$$\theta = 15° \qquad PF = 0.966 \text{ or } 96.6\% \text{ leading}$$
$$\theta = -15° \qquad PF = 0.966 \text{ or } 96.6\% \text{ lagging}$$

This equation can be used to determine true power or I^2R, or $\dfrac{V^2}{R}$ can be used. The power factor method is more useful in solving parallel-circuit problems.

Example D

Determine the applied power, the reactance power, and the true power dissipated in the circuit shown in Figure 20–6. Use the equivalent series circuit to determine the total characteristics of the circuit, and use rms values.

Applied power

$$P_A = IV_T$$
$$P_A = (2.22)(50)$$
$$P_A = 111 \text{ VA}$$

Reactance power: in the equivalent series circuit determine the voltage across the inductor.

$$V_L = IX_L$$
$$V_L = (2.22\underline{/-19.5°})(7.5\underline{/90°})$$
$$V_L = 16.6\underline{/70.5°} \text{ V}$$

$$P_R = IV_L$$
$$P_R = (2.22)(16.6)$$
$$P_R = 36.9 \text{ VAR}$$

or

$$P_R = IV_T (\sin \theta)$$
$$P_R = (2.22)(50)(\sin 19.5°)$$
$$P_R = 36.9 \text{ VAR}$$

True power

$$P_T = IV_T \cos \theta$$
$$P_T = (2.22)(50)(\cos 19.5°)$$
$$P_T = 104 \text{ W}$$

True power can also be calculated by using the resistance in the equivalent series circuit and either I^2R or $\dfrac{V^2}{R}$.

$$P_T = I^2R$$
$$P_T = (2.22)^2(21.2)$$
$$P_T = 104 \text{ W}$$

20–8 QUALITY OR Q. The inductance of a coil or inductor is measured in henrys. Another measurement associated with the inductor is the "figure of merit," or Q. The letter Q (for "quality") describes the ratio of the inductive reactance to the resistance of an inductor:

$$Q = \frac{X_L}{R}$$

An inductor is constructed so as to reflect the properties of inductance. Because it is made of wire, however, it also has the properties of resistance. In most cases this is undesirable. This property can be seen by observing the Q of the inductor. For most applications the capacitor is considered a pure component, and Q measurements are not used to reflect the quality of the capacitor.

In observing the impedance triangle it can be seen that the ratio of X_L to R is the tangent function. It can also be seen that the higher the resistance of the inductor, the lower the Q. An inductor is built to reflect the properties of inductance and should have a phase angle of 90° or as close to that angle as possible. When the phase angle of an inductor is 85° or more it is generally considered a pure component, and the resistance can be ignored. In high-frequency work Qs of several hundred are desirable (these points will be discussed when studying high frequency). Example E illustrates some low-Q and high-Q inductors.

Example E

Determine the Q of an inductor with a reactance of 1500 Ω and a resistance of 500 Ω.

$$Q = \frac{X_L}{R}$$
$$Q = \frac{1500}{500}$$
$$Q = 3$$

The phase angle can be determined by using the inverse tangent of 3.

$$\arctan 3 = 71.6°$$

Determine the Q of an inductor with a reactance of 1500 Ω and a resistance of 50 Ω.

$$Q = \frac{X_L}{R}$$
$$Q = \frac{1500}{50}$$
$$Q = 30$$

The arctan of 30 is 88.1°, which would be a much better inductor than the coil with a Q of 3 and a phase angle of 71.6°.

The Q of a circuit is often an important factor. It can be calculated by using the equivalent series circuit. It would then be the ratio of the reactance in the circuit to the resistance. The reactance of the circuit can be inductive or capacitive.

$$Q_{cir} = \frac{X_?}{R}$$

EXERCISE 20–5 Solve the series-circuit problems, using rms values.

1. Two impedances are in series: $Z_1 = 165\underline{/55°}$ Ω and $Z_2 = 250\underline{/35°}$ Ω. The applied voltage is 170 sin ωt V. Determine the power factor and wattage dissipated in the circuit.

$$PF = \underline{73.2} \%$$
$$P_T = \underline{25.8 \text{ w}}$$

2. In Exercise 1 determine the applied and reactance power.

$$P_A = \underline{\hspace{2cm}}$$
$$P_R = \underline{\hspace{2cm}}$$

3. The power factor of a circuit is 0.82. What is the circuit phase angle and the circuit Q?

$\theta =$ _____

$Q =$ _____

4. A coil with a value of 250 µH and a resistance of 0.75 Ω is in a circuit with a frequency of 100 kHz. What is the Q of the coil?

$Q =$ _____

5. The ac voltage applied to a circuit is 27.5 V and the current is $218\underline{/-72°}$ µA. What is the true power dissipated by the circuit?

$P_T =$ _____

6. A series ac circuit dissipates 120 mW and the phase angle is −34°. If the circuit current is 25 mA, what is the value of the applied voltage?

$V_A =$ _____

7. A generator rated at 19.8 sin ωt V is across an inductor with a reactance of 45 Ω and a resistance of 12 Ω. Determine the applied power, reactance power, and power dissipated.

$P_A =$ _____

$P_R =$ _____

$P_T =$ _____

328

Part 2
Mathematics for DC
and AC Circuits

8. Given three impedances in series across a voltage of 12.7 sin ωt V, where $Z_1 = 1250 - j1875$ kΩ, $Z_2 = 300 + j450$ kΩ, and $Z_3 = 2500 + j2750$ Ω, find the equivalent series circuit, the Q of each impedance, and the circuit Q. Draw the equivalent series circuit and identify and label the ohmic value of each component.

Q_{Z1} = _____

Q_{Z2} = _____

Q_{Z3} = _____

Q_{cir} = _____

9. In Exercise 8 determine the power factor and the true power dissipated in the circuit.

PF = _____ %

P_T = _____

10. What would be the true power dissipated in Exercise 8 if a fourth impedance of $56/\!-\!90°$ kΩ were added in series to the circuit?

P_T = _____

EVALUATION EXERCISE Solve the following problems. Work in rms values and leave the answers in three significant figures and in a convenient prefix.

1. A series circuit has a generator voltage of $57 \sin \omega t$ V; it is across a resistor of 50 Ω, an inductor of 1.5 H, and a capacitor of 12 μF. The frequency is 60 Hz. Find:

$$X_L = \underline{555}$$

$$X_C = \underline{221}$$

$$Z_T = \underline{394.6 \quad 81.7°}$$

$$I_T = \underline{102}$$

$$PF = \underline{14.4\ \%}$$

$$P_{true} = \underline{.68\ w}$$

2. Draw and label the ohmic value of the equivalent series circuit for Exercise 1 and determine the circuit Q.

$$Q = \underline{\hspace{3cm}}$$

3. Z_1, Z_2, and Z_3 are connected across a voltage source: $V_1 = 2.0\underline{/38°}$ V, $V_2 = 13\underline{/-65°}$ V, and $V_3 = 6\underline{/15°}$ V. What is the applied voltage?

4. A coil with a resistance of 40 Ω and a Q of 1.7 is in series with a capacitor with a reactance of 150 Ω. What is the voltage across the capacitor if the applied voltage is 300 mV?

$V_C =$ _____

5. A series circuit contains impedance of $Z_1 = 150\underline{/0°}$ kΩ, $Z_2 = 60 + j255$ kΩ, and $Z_3 = 120\underline{/-45°}$ kΩ. If the applied voltage is $75\underline{/0°}$ V, find the true power dissipated, the reactance power, and the applied power.

$P_T =$ __14.3____

$P_R =$ __55.9____

$P_A =$ __28.6____

6. In Exercise 5 determine the Q of impedances Z_2 and Z_3, the circuit Q, and the power factor.

$Q_{Z2} =$ _____

$Q_{Z3} =$ _____

$Q_{cir} =$ _____

$PF =$ _____

CHAPTER 21
AC Parallel Circuits

Objectives

After completing this chapter you will be able to:
- Solve ac parallel R-C circuit problems using phasor algebra
- Solve ac parallel R-L circuit problems using phasor algebra
- Solve ac parallel R-L-C circuit problems using phasor algebra
- Solve power dissipation in ac parallel circuits
- Determine power factor correction

Like the dc parallel circuit, the ac parallel circuit has a common voltage across each branch. The equations used to solve dc parallel circuits are also used to solve ac parallel circuits, but they must be solved by using phasor algebra. Impedance is the opposition to the flow of current in the ac circuit, and the equations reflect these impedances.

$$\frac{1}{Z_T} = \frac{1}{Z_1} + \frac{1}{Z_2} + \frac{1}{Z_3} + \text{etc.}$$
$$I_T = \frac{V_T}{Z_T}$$
$$V_T = V_1 = V_2 = V_3 = \text{etc.}$$
$$P_T = I_T V_T \cos \theta$$
$$I_T = I_1 + I_2 + I_3 + \text{etc.}$$

21-1 R-L CIRCUITS. The generator supplying the voltage to a circuit encounters only the total impedance. The current divides inversely with the branch impedance, as it does in the dc circuit. The total current is the phasor sum of the branch current. Example A illustrates an R-L circuit.

Example A

Given the circuit in Figure 21-1, find the total impedance and the total current.

Work in rms values. The applied voltage is given as a sine wave equation; the peak value should be changed to rms for convenience. When measuring voltage with an oscilloscope, peak-to-peak values are usually measured and are generally converted to rms for calculations.

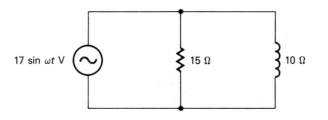

FIGURE 21-1 R-L parallel circuit

$$(17\ V_P)(0.707) = 12\ V_{rms}$$

Total impedance

$$Z_T = \frac{Z_1 Z_2}{Z_1 + Z_2}$$

$$Z_T = \frac{(15 \underline{/0°})(10 \underline{/90°})}{(15 + j0) + (0 + j10)}$$

$$Z_T = \frac{150 \underline{/90°}}{15 + j10}$$

$$Z_T = \frac{150 \underline{/90°}}{18 \underline{/33.7°}}$$

$$Z_T = 8.33 \underline{/56.3°}\ \Omega$$

Total current

$$I_T = \frac{V_T}{Z_T}$$
$$I_T = \frac{12 \underline{/0°}}{8.33 \underline{/56.3°}}$$
$$I_T = 1.44 \underline{/-56.3°}$$

21-2 THE PHASOR DIAGRAM FOR R-L CIRCUITS. Because the voltage is common to all branches of the ac circuit, it is used as the reference phasor, and the branch currents are drawn with respect to the voltage. Figure 21-2

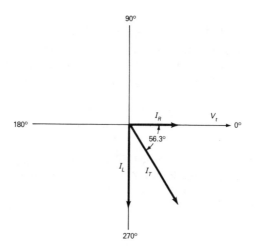

FIGURE 21-2 Phasor diagram for an R-L circuit

illustrates the phasor diagram for the circuit in Figure 21–1. In the phasor diagram it can be seen that the current through the resistance and the voltage across the resistance are in phase. The current through the inductance lags the voltage by 90°; the total current lags the applied voltage by 56.3°.

21–3 R-C CIRCUITS. The R-C circuit reacts like the R-L circuit, but the total current leads the total applied voltage. Example B illustrates the R-C circuit.

Example B

Given the circuit in Figure 21–3, find the total impedance and the total current.

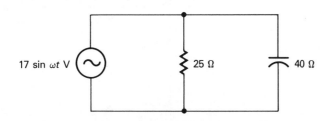

FIGURE 21–3   R-C parallel circuit

Convert to rms

$$(17\ V_P)(0.707) = 12\ V$$

Total impedance

$$Z_T = \frac{Z_1 Z_2}{Z_1 + Z_2}$$
$$Z_T = \frac{(25\underline{/0°})(40\underline{/-90°})}{(25 + j0) + (0 - j40)}$$
$$Z_T = \frac{1000\underline{/-90°}}{25 + j40}$$
$$Z_T = \frac{1000\underline{/-90°}}{47.2\underline{/-58°}}$$
$$Z_T = 21.2\underline{/-32°}\ \Omega$$

Total current

$$I_T = \frac{V_T}{Z_T}$$
$$I_T = \frac{12\underline{/0°}}{21.2\underline{/-32°}}$$
$$I_T = 566\underline{/32°}\ \text{mA}$$

21-4 THE PHASOR DIAGRAM FOR R-C CIRCUITS. Figure 21-4 is the phasor diagram for the circuit used in Example B. It can be seen that the current through the resistor is in phase with the voltage. The current through the capacitor is leading the voltage by 90°. The total current leads the total applied voltage by 32°.

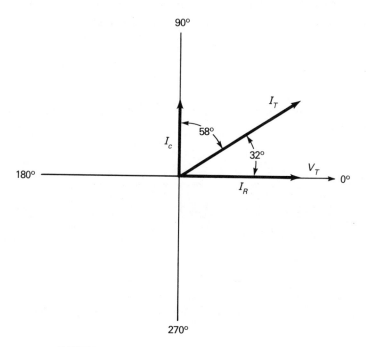

FIGURE 21-4   Phasor diagram for the R-C circuit

Example C

Given the circuit in Figure 21-5, find the total impedance and total current and draw the phasor diagram (Figure 21-6).

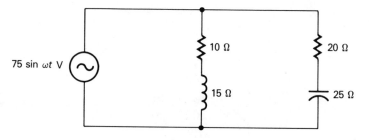

FIGURE 21-5   Circuit for Example C

Convert to rms

$$(75\ V_P)(0.707) = 53\ V$$

Total impedance

$$Z_T = \frac{Z_1 Z_2}{Z_1 + Z_2}$$

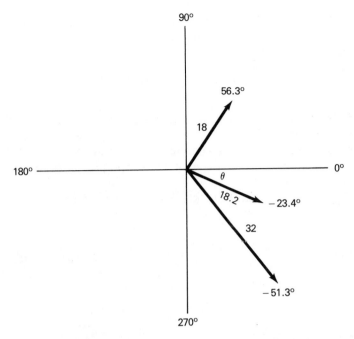

FIGURE 21-6   Phasor diagram for Example C

$$Z_T = \frac{(10 + j15)(20 - j25)}{(10 + j15) + (20 - j25)}$$

$$Z_T = \frac{(18\underline{/56.3°})(32\underline{/-51.3°})}{30 - j10}$$

$$Z_T = \frac{576\underline{/5°}}{31.6\underline{/-18.4°}}$$

$$Z_T = 18.2\underline{/23.4°} \ \Omega$$

Total current

$$I_T = \frac{V_T}{Z_T}$$
$$I_T = \frac{53\underline{/0°}}{(18.2\underline{/23.4°})}$$
$$I_T = 2.91\underline{/-23.4°} \ A$$

21-5 CURRENT RATIO IN AC PARALLEL CIRCUITS. The current-ratio method can be used to solve ac parallel circuits. These circuits are inverse proportions; however, the ratios are developed between current and impedance, and the equations given are for two parallel impedances. When more than two impedances are in parallel, it is often more convenient to use conventional theory methods to solve the problems.

$$I_T = \frac{Z_1 + Z_2}{Z_2} (I_1)$$

$$I_T = \frac{Z_1 + Z_2}{Z_1} (I_2)$$

Chapter 21
AC Parallel Circuits

$$I_1 = \frac{Z_2}{Z_1 + Z_2}(I_T)$$

$$I_2 = \frac{Z_1}{Z_1 + Z_2}(I_T)$$

EXERCISE 21–1 Each problem in this exercise is for a two-branch parallel circuit. The impedance of each branch and the applied voltage are given. Draw the circuit and identify the components in each branch. Determine the total impedance and current. Draw a sketch of the phasor diagram.

1. $Z_1 = 250\underline{/0°}$ kΩ, $Z_2 = 450\underline{/90°}$ kΩ, $V_T = 57 \sin \omega t$ V.

$Z_T =$ 218 /29.1 KΩ

$I_T =$ 185 /-29.1 mA

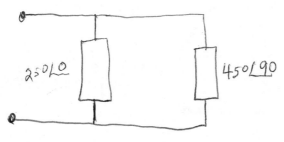

$250\underline{/0}$ $450\underline{/90}$

$250 + j0$
$0 + j450$
$\overline{250 + j450}$

$\tan \frac{450}{250} = 60.9°$

$\frac{450}{\sin 60.9} = 515 \underline{/60.9}$

$\frac{250\underline{/0} \times 450\underline{/90}}{515\underline{/60.9}}$

2. $Z_1 = 85 + j0$ Ω, $Z_2 = 150 - j500$ Ω, $V_T = 350 \sin \omega t$ mV.

$Z_T =$ _____

$I_T =$ _____

3. $Z_1 = 50\underline{/90°}$ kΩ, $Z_2 = 25\underline{/0°}$ kΩ, $V_T = 12.7 \sin \omega t$ V.

$Z_T = $ _____

$I_T = $ _____

4. $Z_1 = 12 + j18$ Ω, $Z_2 = 20 - j25$ Ω, $V_T = 6.4 \sin \theta$ mV.

$Z_T = $ _____

$I_T = $ _____

5. $Z = 50 - j75$ Ω, $Z_2 = 125\underline{/-30°}$ Ω, $V_T = 65 \sin \theta$ mV.

$Z_T = $ _____

$I_T = $ _____

6. $Z_1 = 250\underline{/45°}\ \Omega$, $Z_2 = 450\underline{/60°}\ \Omega$, $V_T = 170 \sin \omega t$ V.

$Z_T =$ _____

$I_T =$ _____

7. $Z_1 = 80 - j140$ kΩ, $Z_2 = 100 + j65$ kΩ, $V_T = 70.7 \sin \omega t$ V.

$Z_T =$ _____

$I_T =$ _____

8. $Z_1 = 1.2\underline{/25°}$ MΩ, $Z_2 = 2.6\underline{/-42°}$ MΩ, $V_T = 450 \sin \omega t$ V.

$Z_T =$ _____

$I_T =$ _____

21-6 THE EQUIVALENT SERIES CIRCUIT.
The generator supplying voltage to the circuit encounters only the total characteristics of the circuit; it does not distinguish between individual branches. Complex circuits are analyzed by determining the equivalent series circuit. That circuit will be resistive, inductive, capacitive, resistive and inductive, or resistive and capacitive. Example D shows the equivalent series circuit for the parallel circuits used in Figure 21-5.

Example D

Find the equivalent series circuit for the circuit given in Example C, where

$$Z_T = 18.2 \underline{/-13.4°} \, \Omega$$

The total impedance has a negative phase angle, $-13.4°$; it also contains a resistor and a capacitor. The ohmic values of the components can be determined by converting the polar form of the impedance to the rectangular form. The real part is the resistance value of the equivalent circuit; the imaginary part is the reactance of the capacitor.

$$Z_T = 18.2 \underline{/-13.4°} = 17.7 - j4.22 \, \Omega$$

The equivalent series circuit is shown in Figure 21-7.

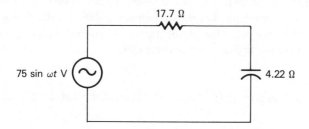

FIGURE 21-7 Equivalent series circuit of Example C

All circuits can be expressed as an equivalent series circuit containing no more than two components.

21-7 POWER DISSIPATION IN PARALLEL CIRCUITS.
Power is dissipated in the form of heat; the component that dissipates the power is the resistor. The equivalent series circuit is convenient for calculating power dissipation, because it contains the resistance total for the circuit. The equations are the same for all ac circuits.

True power

$$P_T = I_T V_T \cos \theta$$

$$P_T = I^2 R_T$$

$$P_T = \frac{V^2}{R_T}$$

Reactance power

$$P_R = I_X V_X$$

$$P_R = I_T V_T \sin \theta$$

Applied power

$$P_A = I_T V_T$$

Refer to Chapter 20 for a detailed explanation.

21-8 CALCULATIONS OF Q. The parallel circuit has a total circuit Q and a Q for each branch.

Total circuit Q

$$Q_{\text{cir}} = \frac{X_T}{R_T}$$

where X_T is the total reactance of the circuit, either inductive or capacitive, and R_T is the total circuit resistance (see Chapter 20).

21-9 PARALLEL CIRCUITS WITH MORE THAN TWO BRANCHES. Parallel circuits with more than two branches are solved in the same manner as a direct-current circuit; but the latter must be solved using phasor algebra. Example E illustrates a three-branch circuit.

Example E
Given the circuit in Figure 21-8, find the total impedance and total current.

The voltage, given in rms, need not be converted.

Total impedance

$$Z_T = \frac{1}{\dfrac{1}{Z_1} + \dfrac{1}{Z_2} + \dfrac{1}{Z_3}}$$

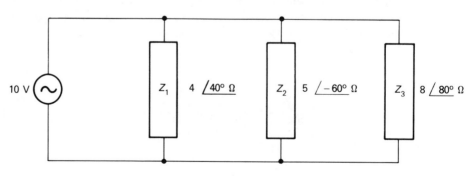

FIGURE 21-8   Three branch parallel circuit

$$Z_T = \cfrac{1}{\cfrac{1}{4/40°} + \cfrac{1}{5/-60°} + \cfrac{1}{8/80°}}$$

Take the reciprocal of each impedance

$$Z_T = \cfrac{1}{0.25/-40° + 0.2/60° + 0.125/-80°}$$

Convert to rectangular and add

$$\begin{aligned} 0.25/-40° &= 0.192 - j0.161 \\ 0.2/60° &= 0.100 + j0.173 \\ 0.125/-80° &= \underline{0.0217 - j0.123} \\ &0.314 - j0.111 \end{aligned}$$

$$Z_T = \cfrac{1}{0.314 - j0.111}$$

$$Z_T = \cfrac{1}{0.333/-19.5°}$$

$$Z_T = 3/19.5° \; \Omega$$

Total current

$$I_T = \cfrac{V_T}{Z_T}$$

$$I_T = \cfrac{10}{3/19.5°}$$

$$I_T = 3.33/-19.5° \text{ A}$$

Example F
Determine the equivalent series circuit for Figure 21–8.

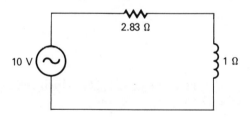

FIGURE 21–9   The equivalent series circuit

$$Z_T = 3/19.5° \; \Omega$$

In rectangular form, $Z_T = 2.83 + j1 \; \Omega$ (refer to Figure 21–9).

Example G
Find the power dissipated by the circuit.

$$P_T = I_T V_T \cos \theta$$
$$P_T = (3.33)(10) \cos 19.5°$$
$$P_T = 31.4 \text{ W}$$

Example H

Draw the phasor diagram of the circuit that is used in Figure 21–8 (refer to Figure 21–10).

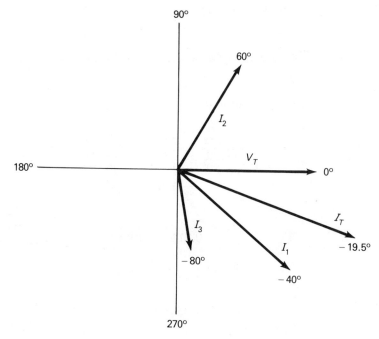

FIGURE 21–10 Phasor diagram for Figure 21-8

Example I

Determine the circuit Q and the Q for each branch. The equivalent series circuit contains an inductor of 1 Ω reactance and a resistor of 2.83 Ω.

$$Q_{cir} = \frac{X_L}{R}$$
$$Q_{cir} = \frac{1}{2.83}$$
$$Q_{cir} = 0.353$$

Branch 1 has an impedance of $4/\underline{40°}$ Ω, which is a resistor of 3.06 Ω and an inductor of 2.57 Ω.

$$Q_1 = \frac{2.57}{3.06}$$
$$Q_1 = 0.84$$

Branch 2

$$5/\underline{-60°} = 2.5 - j4.33 \text{ Ω}$$

$$Q_2 = \frac{4.33}{2.5}$$
$$Q_2 = 1.73$$

Branch 3

$$8\underline{/80°} = 1.39 + j7.88 \ \Omega$$
$$Q_3 = \frac{7.88}{1.39}$$
$$Q_3 = 5.67$$

EXERCISE 21–2 Solve the problems.

1. Find the total impedance of three impedances in parallel:
   $Z_1 = 250\underline{/-45°} \ \Omega$, $Z_2 = 175\underline{/30°} \ \Omega$, $Z_3 = 150\underline{/10°} \ \Omega$.

   $Z_T = \underline{69.2 \underline{/4.72°} \ \Omega}$

2. What is the total impedance for three impedances in parallel if branch 1 contains a resistance of 1.5 kΩ and a capacitance of 3.0 kΩ, branch 2 has an inductor of 1.2 kΩ, and branch 3 a resistor of 2.7 kΩ?

   $Z_T = $ _____

3. In Exercise 1 the current in branch 1 is $200\underline{/45°}$ mA. What is the current through branches 2 and 3?

$I_2 = $ _____

$I_3 = $ _____

4. In Exercise 1, if the current in branch 3 were $5\underline{/-10°}$ mA, what would be the total current?

$I_T = $ _____

5. The total impedance of a circuit is found to be $23.8\underline{/-82.3°}$. What is the circuit Q?

$Q = $ _____

6. Draw the phasor diagram for Exercise 3. Label each phasor and the resultant.

7. Draw and label the phasor diagram for Exercise 4.

8. Find the total impedance of the circuit in Figure 21–11.

$Z_T =$ _____

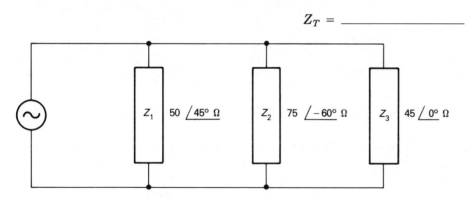

FIGURE 21–11   Circuit for Exercise 21-2 (8)

9. If the voltage applied to the circuit in Exercise 8 were 25 V, what would be the power dissipated in the circuit?

$P_T =$ _____

10. Determine each branch Q in Exercise 8.

$Q_1 =$ _____

$Q_2 =$ _____

$Q_3 =$ _____

11. The current through branch 3 in Exercise 8 is 150 mA. What is the total current?

$I_T =$ _____

12. Draw and label the equivalent series circuit for Exercise 8.

Chapter 21
AC Parallel Circuits

21–10 POWER FACTOR CORRECTION. The power factor, introduced in Chapter 20, is defined as the cosine of the circuit phase angle. The power factor relates to the position of the circuit current. If the current is leading, the power factor is leading; and if the current is lagging, the power factor is lagging. For convenience, the power factor is often given as a percent, for instance,

$$\text{cosine } 15° = 0.966 = 96.6\% \text{ leading}$$
$$\text{cosine } -15° = -0.966 = 96.6\% \text{ lagging}$$

If the parallel circuit were capacitive the circuit phase angle would be positive and in the first quadrant; the total current would be leading the voltage, and the power factor would be positive. If the circuit were inductive the current would be in the fourth quadrant, and the circuit phase angle would be negative; the total current would be lagging the voltage, and the power factor would be negative.

In applications where large inductive loads are present, such as a machine shop, the current lags the voltage. The inductive loads of motors in a machine shop use the energy supplied but do not consume all of it. The resistive portion of the load consumes power. This results in a waste of energy, since the power must be supplied to the machine shop. The wattmeter recording the use of power records only the amount of energy consumed, in the form of heat. Figure 21–12 illustrates the relationship of true power, reactance power, and applied power in an inductive circuit. The reactance power is shown as a lagging power factor in the fourth quadrant. To correct this the phase angle should be as close to zero as possible; an angle of 3° or less is usually acceptable. Parallel capacitance added to the circuit reduces the phase angle. In industry capacitance is added to the line to maintain the power factor. Companies that supply the power also correct the power factor by adding capacitance. Large capacitors can be seen on power poles. Example J shows how to make a power-factor correction.

In a pure inductive load a capacitance equal to the inductive reactance could be added across the load and achieve unity power factor. In an electric motor inductance and resistance are physically the same properties. The capacitance placed across the motor is in parallel with the inductive and resistance properties. Reactive power can be used to determine the capacitive reactance needed to achieve unity power factor.

The true power dissipated in a circuit can be determined by the equation $P_T = \dfrac{V^2}{R}$, and the reactive power can be determined by the equation $P_R = \dfrac{V^2}{X_C}$. These equations are helpful in making power factor corrections.

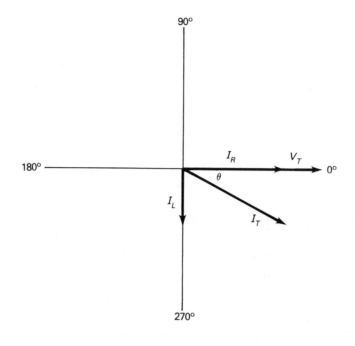

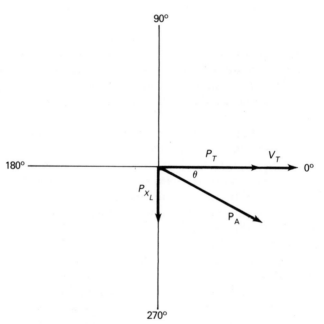

FIGURE 21–12   Power in an ac circuit

Example J

A motor is operating on a 120 V, 60 Hz line. The line current is 2.4 A and the power factor is 90% lagging. What value of capacitance must be added to obtain unity power factor?

Draw the phasor diagram (Figure 21–13).

The circuit phase angle is 25.8°

$$90\% \text{ lagging} = -0.90 \text{ and the arccos} = -25.8°$$

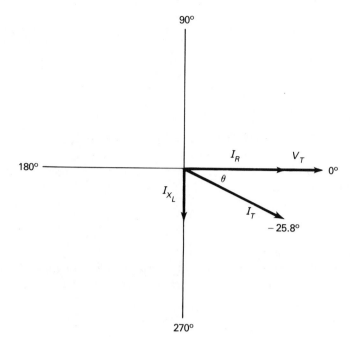

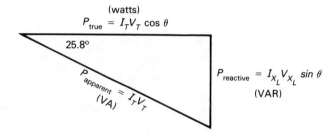

FIGURE 21–13 Phasor diagram for Example J

Calculate the reactive power

$$P_R = I_T V_T \sin \theta$$
$$P_R = (2.4)(120)(\sin 25.8°)$$
$$P_R = 125 \text{ VAR}$$

Determine X_C

$$P_R = \frac{V^2}{X_C}$$
$$X_C = \frac{V^2}{P_R}$$
$$X_C = \frac{120^2}{125}$$
$$X_C = 115 \text{ } \Omega$$

Determine the capacitance

$$X_C = \frac{1}{2\pi f C}$$

$$C = \frac{1}{2\pi f X_C}$$

$$C = \frac{1}{(6.28)(60)(115)}$$

$$C = 23.1 \ \mu F$$

Adding 23.1 µF in parallel to the line results in a zero phase shift, or unity power factor. When the power factor is kept to unity, less power must be supplied to the user, which results in less power loss on transmission to the user.

EXERCISE 21–3

1. An inductive circuit has a power factor of 87% lagging. It dissipates 8000 W. What is the reactive power, the applied power?

$P_R = $ _____

$P_A = $ _____

2. In Exercise 1 what value of capacitance would be added in parallel to achieve a power factor of 100% if the frequency is 60 Hz and the applied voltage is 120 V?

_____

3. A motor operating at 120 V, 60 Hz, and 3.6 A has a power factor of 93% lagging. What value of parallel capacitance is required to achieve unity power factor?

4. A 220 V, 60 Hz motor has a line current of 11.5 A. It is operating at a lagging power factor of 84%. What value of pure capacitance should be added in parallel to achieve a power factor of 100%?

5. A 220 V, 60 Hz motor has a load current of $5.6 / -22°$ A. What value of capacitance must be added in parallel to achieve unity power factor?

6. The power factor of a 220 V, 60 Hz motor is determined to be 89%. If the line current is 18.5 A, what value of capacitance should be added to achieve unity power factor?

EVALUATION EXERCISE

1. Find the total impedance of two impedances in parallel if $Z_1 = 45\underline{/25°}\ \Omega$ and $Z_2 = 80\underline{/-60°}\ \Omega$.

2. What is the total impedance of three parallel impedances, where $Z_1 = 12 + j10\ \Omega$, $Z_2 = 10 - j10\ \Omega$, $Z_3 = 15 + j20\ \Omega$?

3. The total impedance of a two-branch parallel circuit is $35/-30°$ kΩ. One of the impedances is $60/-45°$ kΩ. What is the other impedance?

_____

4. In a two-branch parallel circuit, one branch current is $10/-25°$ mA and the other is $5/25°$ mA. What is the total current?

_____

5. In Figure 21–14, find the total current and the power dissipated by the circuit.

$I_T = $ _____

$P_T = $ _____

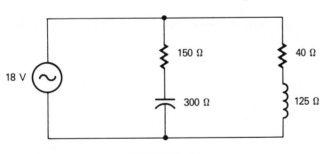

FIGURE 21–14

6. Draw the equivalent series circuit and label the value of each component in Exercise 5.

7. Determine the circuit Q and each branch Q for the circuit in Figure 21–14.

Q_{cir} = _____

Q_1 = _____

Q_2 = _____

8. The current through Z_1 in Figure 21–14 is $45\underline{/63.4°}$ mA. What is the current through Z_2?

_____

9. The power factor in a 60 Hz 220 V inductive circuit is 80% lagging. The current is 14.5 A. What ideal value of capacitance should be added in parallel to achieve unity power factor?

CHAPTER 22
AC Series-Parallel Circuits and Resonance

Objectives

After completing this chapter you will be able to:
- Solve ac series-parallel circuit problems using phasor algebra
- Observe the effects of resonance in a series circuit
- Observe the effects of resonance in a parallel circuit
- Solve resonant circuit problems

Series-parallel circuits are common in electronic devices. They are a combination of series circuits and parallel circuits. In solving problems the parallel portion of the circuit can be simplified in an equivalent series circuit and combined with other series components. This is done in the same manner as the dc series-parallel circuit, by using phasor algebra. Series and parallel resonance circuits are also studied in this chapter.

22-1 SERIES-PARALLEL CIRCUITS. Power sources have a measurable internal resistance that is in series with the load. If this resistance is great enough with respect to the load, it must be considered when the load is connected to the source. Example A illustrates a source impedance and a parallel circuit load. This combination forms a simple series-parallel circuit.

Example A
Determine the voltage delivered to the parallel load in the circuit in Figure 22-1.

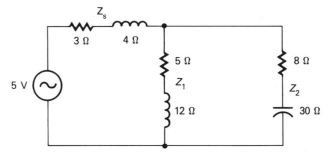

FIGURE 22-1 Series-parallel circuit for Example A

356

Total impedance

$$Z_T = Z_S + \frac{Z_1 Z_2}{Z_1 + Z_2}$$

$$Z_T = 3 + j4 + \frac{(13\underline{/67.4°})(31\underline{/-75.1°})}{(5 + j12) + (8 - j30)}$$

$$Z_T = 3 + j4 + \frac{403\underline{/7.7°}}{22.2\underline{/-54.2°}}$$

$$Z_T = 3 + j4 + 18.2\underline{/61.9°}$$

$$Z_T = 3 + j4 + 8.57 + j16.1$$

$$Z_T = 11.6 + j20.1 \ \Omega \text{ or } 23.2\underline{/60°} \ \Omega$$

Total current

$$I_T = \frac{V_T}{Z_T}$$

$$I_T = \frac{5\underline{/0°}}{23.2\underline{/60°}}$$

$$I_T = 216\underline{/-60°} \text{ mA}$$

The impedance of the parallel load was determined as part of the calculations for the total impedance. The product of the total current and the impedance of the parallel load give the voltage across the load.

Voltage across the parallel load

$$V_{par} = I_T Z_{par}$$
$$V_{par} = (0.216\underline{/-60°})(18.2\underline{/61.9°})$$
$$V_{par} = 3.93\underline{/1.9°} \text{ V}$$

The equivalent series circuit is expressed

$$Z_T = 11.6 + j20.1 \ \Omega$$

The equivalent series circuit contains a resistor of 11.6 Ω and an inductor with a reactance of 20.1 Ω.

EXERCISE 22–1

1. Determine the total impedance of the circuit in Figure 22–2.

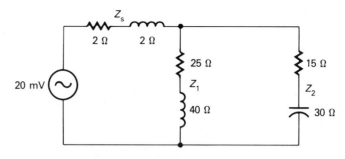

FIGURE 22–2   Circuit for Exercise 22–1 (1)

2. What is the voltage across the parallel load in the circuit given in Figure 22–2?

3. Draw and label the equivalent series circuit for the circuit in Figure 22–2. If the frequency of the voltage applied to the circuit were 60 Hz, what would be the pure value of the reactive component?

4. Determine the Q of the parallel load for the circuit given in Figure 22–2.

Chapter 22
AC Series-Parallel Circuits and Resonance

5. In the circuit given in Figure 22–2, with the 24 V applied, what is the power dissipated by the load and the power dissipated by the source impedance?

$P_L = $ _____

$P_S = $ _____

6. Determine the total impedance of the circuit given in Figure 22–3.

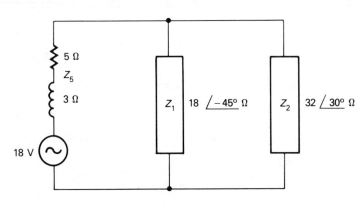

FIGURE 22–3   Circuit for Exercise 22–1 (6)

360

*Part 2
Mathematics for DC
and AC Circuits*

7. What is the voltage across the parallel load in the circuit in Figure 22–3?

   _____

8. Draw and label the equivalent series circuit for the circuit given in Figure 22–3. If the frequency of the applied voltage is 1000 Hz, determine the pure value of the reactance in the equivalent circuit.

   _____

9. Determine the power dissipated by the parallel load in the circuit in Figure 22–3 and the power dissipated by the source impedance.

   $P_L =$ _____

   $P_S =$ _____

10. Determine the total impedance of the circuit given in Figure 22–4.

11. Determine the total current in the circuit in Figure 22–4.

12. What is the voltage delivered to the load in the circuit in Figure 22–4?

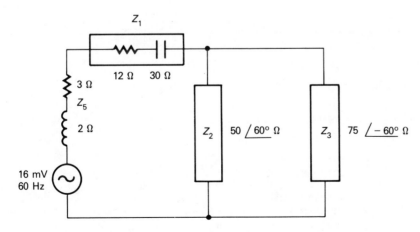

FIGURE 22–4 Circuit for Exercise 22–1 (10)

13. What is the power dissipated by the load in the circuit in Figure 22–4?

   _____

14. Draw and label the equivalent series circuit for Figure 22–4. Find the pure value of the reactive component.

   _____

15. Determine the Q of the circuit load in Figure 22–4.

   _____

22–2 SERIES RESONANCE. In a series circuit a special condition exists when the inductive reactance equals the capacitive reactance. Figure 22–5a illustrates a series circuit where $X_L = X_C$. Figure 22–5b illustrates the voltage phasor diagram for this circuit. The phasor diagram shows the voltage across the inductor and the capacitor to be equal and 180° out of phase. The voltages cancel each other, and the circuit becomes pure resistance. The current is at maximum value, limited only by the resistance. Voltage across the inductor and the capacitor is at maximum because the current is at maximum. Example B illustrates the conditions for the circuit in Figure 22–5.

Chapter 22
AC Series-Parallel
Circuits and Resonance

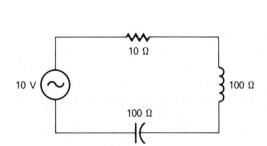

FIGURE 22–5a   Series resonant circuit

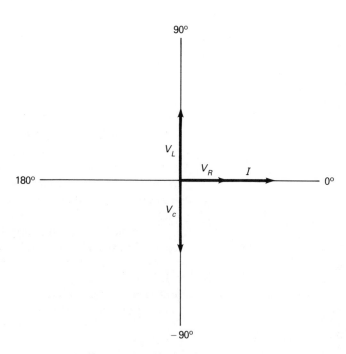

FIGURE 22–5b   Phasors for series resonance

Example B
Circuit impedance

$$\text{since } X_L = X_C$$

$$\text{then } Z_T = R_T$$

Circuit current

$$I_T = \frac{V_T}{R_T}$$
$$I_T = \frac{10 \text{ V}}{10 \text{ }\Omega}$$
$$I_T = 1 \text{ A}$$

Voltage across L and C

$$V_L = I_T X_L$$
$$V_L = (1)(100)$$
$$V_L = 100 \text{ V}$$

$$V_C = I_T X_C$$
$$V_C = (1)(100)$$
$$V_C = 100 \text{ V}$$

It can be seen that the voltages across the inductor and the capacitor can rise to high values. The applied voltage is only 10 V, and the voltage across the inductor and the capacitor is 100 V. To summarize:

$$X_L = X_C$$

$$Z_T = R_T$$

$$I_T = \text{maximum}$$

$$V_L = V_C = \text{high voltage}$$

22–3 PARALLEL RESONANCE. Resonance in a parallel circuit is defined as the condition that exists when $X_L = X_C$. There are two other conditions of resonance, however, that can be present in a parallel resonant circuit. Figure 22–6a illustrates a parallel resonant circuit, and Figure 22–6b illustrates the phasors for the circuit. When $X_L = X_C$, the phasors show that the currents are equal and that they are 180° out of phase. The current is then at minimum value; theoretically, it is zero.

A capacitor is considered a pure component. The current leads the voltage by 90°, whereas the inductor is made of wire and has some resistance. The current does not lag the voltage by 90°. The impedance of the inductive branch will be greater because of the resistance, and the current is less than the current through the capacitance branch. There are three conditions of resonance in parallel circuits that occur at slightly different frequencies. When $X_L = X_C$ line current is minimum. Unity power factor is achieved or the circuit is pure resistive.

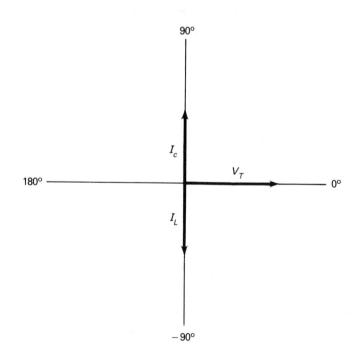

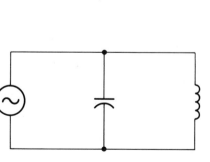

FIGURE 22–6a Parallel resonant circuit

FIGURE 22–6b Phasors for parallel resonance

In most cases parallel resonance is defined as the condition where $X_L = X_C$. This text will be concerned only with the condition where $X_L = X_C$.

The parallel resonant circuit is sometimes referred to as a tank circuit and can be described in the following manner. When a voltage is first applied to the circuit, the capacitor is a direct short and will charge as shown in Figure 22–7a. When the capacitor is fully charged it will discharge current through the inductor, creating a magnetic field about the coil (Figure 22–7b). When the capacitor is neutralized, the field about the coil will discharge, and current will flow, as indicated in Figure 22–7c. The capacitor will again charge with the opposite polarity. When fully charged, the capacitor will discharge in the opposite direction (Figure 22–7d). Current oscillates back and forth between the capacitor and the inductor. This pattern would continue indefinitely if the components were pure and the conditions perfect; because that is not possible, energy is needed to keep up the oscillations. Energy is needed only to overcome the losses due to impurities in the circuit. The generator supplying the circuit encounters a high impedance and low line current, since it need only supply the losses to the circuit.

To summarize the parallel resonant circuit,

$X_L = X_C$ (for most conditions)
Z_T is high
I_T is low

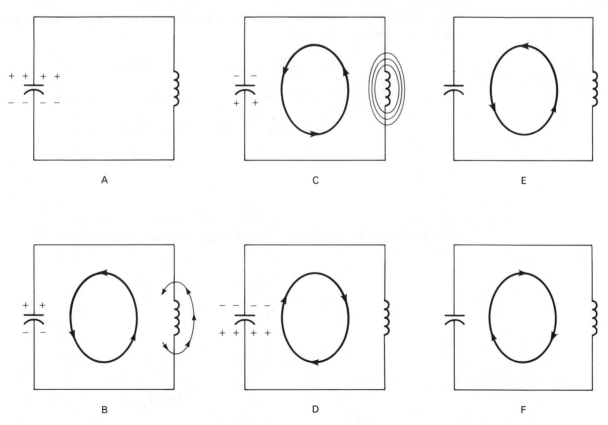

FIGURE 22–7 The tank circuit

22–4 RESONANT FREQUENCY. The frequency at which resonance occurs can be developed from the condition $X_L = X_C$.

$$X_L = X_C$$
$$2\pi fL = \frac{1}{2\pi fC}$$

Transposing f and $2\pi L$

$$f^2 = \frac{1}{4\pi^2 LC}$$

Extracting the square root of both sides

$$f = \frac{1}{2\pi\sqrt{LC}}$$

Resonant frequency is denoted by f_r or f_o

where
  f is the resonant frequency, in hertz
  L is the inductance, in henrys
  C is the capacitance, in farads

22–5 Q AND THE RESONANT CIRCUIT. The quality of merit, or Q, has already been introduced. In a resonant circuit it can be helpful in determining currents and voltages.

In the series resonant circuit the voltage across the inductor and capacitor can be found by using Q:

$$V_L = QV_T \quad \text{or} \quad V_C = QV_T$$

In the parallel resonant circuit, branch currents can be found by using Q:

$$I_L = QI_T \quad \text{or} \quad I_C = QI_T$$

The impedance in the parallel resonant circuit can be determined with Q.

$$Z_T = X_L Q$$

Following is a summary of the properties of series and parallel resonant circuits.

	Series	Parallel
f_o	$\frac{1}{2\pi\sqrt{LC}}$	$\frac{1}{2\pi\sqrt{LC}}$
Z	$Z = R$ (low)	QX_L (high)
I	max	min I_{source}
I_L	I_T	QI_T

	Series	Parallel
I_C	I_T	QI_T
V_L	QV_T	V_T
V_C	QV_T	V_T
Q	$\dfrac{X_L}{R}$	$\dfrac{X_L}{R}$
PF	1	1

EXERCISE 22-2

1. A series circuit contains an inductor of 15 mH and a capacitor of 15 nF. At what frequency will it become a resonant circuit?

   _____

2. In Exercise 1, at what frequency would the circuit become resonant if the capacitor were doubled?

   _____

3. In Exercise 1 a series resistance of 150 Ω is added and a voltage of 50 mV is applied in series. What is the voltage drop across each component?

   $V_R =$ _____

   $V_L =$ _____

   $V_C =$ _____

4. A capacitor of 12 pF, an inductor of 50 μH, and a resistor of 2.2 kΩ are connected in series across a voltage of 300 μV. What is the resonant frequency of the circuit, and what is the line current?

$f_o =$ _____

$I_L =$ _____

5. Determine the voltage drop across the inductor, capacitor, and resistor in the circuit in Exercise 4.

$V_L =$ _____

$V_C =$ _____

$V_R =$ _____

6. A 75 mH coil and a 50 nF capacitor are connected in parallel across a voltage of 65 mV. At what frequency will the circuit resonate?

_____

7. In Exercise 6 what are the line current and circuit impedance if the circuit Q is 150?

I_{line} = _____

Z_{cir} = _____

8. A parallel resonant circuit has a coil of 200 µH and a capacitor of 8 pF across a voltage of 120 mV. At what frequency will it resonate?

_____

9. The Q of the circuit in Exercise 8 is 250. What are the circuit impedance and the line current?

Z_{cir} = _____

I_{line} = _____

10. What value capacitance should be placed in parallel with a 50 mH coil to make the circuit resonate at 10 kHz?

EVALUATION EXERCISE

1. The signal source shown in Figure 22-8 has an internal resistance of 2.5 +j2 Ω. A load of two parallel branches is connected to the signal source. What is the value of the rms voltage across the load when the signal generator is set to an output of 25 V_{P-P} and 1 kHz?

2. What power is dissipated by the load in Exercise 1?

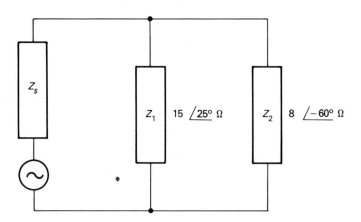

FIGURE 22-8 Circuit for Evaluation Exercise (1)

3. Draw and label the equivalent series circuit for the circuit in Exercise 1. Determine the pure value of the reactive component.

Chapter 22
AC Series-Parallel Circuits and Resonance

4. Determine the total impedance of the circuit in Figure 22–9.

5. If a voltage of 300 mV were applied to the circuit in Figure 22–9, what would be the total power dissipated by the circuit?

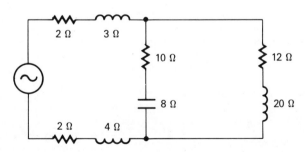

FIGURE 22–9   Circuit for Evaluation Exercise (4)

6. At what frequency will a series circuit resonate if it contains an inductor of 30 mH and a capacitor of 22 nF?

_____

7. If a resistance of 40 Ω is placed in series in the circuit in Exercise 6, and a voltage of 25 mV is applied, what is the voltage across the capacitor?

_____

8. A parallel circuit resonates at 500 kH, and the circuit has a capacitance of 10 pF. What is the value of the inductance in mH?

_____

9. Using the circuit in Figure 22–10, find the total impedance.

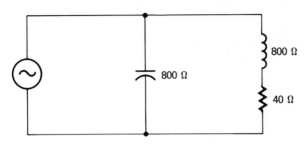

FIGURE 22–10   Circuit for Evaluation Exercise (9)

10. The impedance of a parallel resonant circuit is 180 kΩ, and the applied voltage is 15 mV. If the Q of the circuit is 400, what is the current through the inductance?

CHAPTER 23
AC Circuit Analysis

Objectives

After completing this chapter you will be able to:
- Identify delta and wye circuits
- Convert dc and ac delta-to-wye circuits
- Convert dc and ac wye-to-delta circuits
- Solve dc bridge circuits using delta-to-wye conversions
- Solve an ac bridge circuit problem using delta-to-wye conversions
- Solve ac circuit problems using Norton's Theorem

The dc bridge circuit, already studied, should be reviewed if need be. The ac bridge circuit, studied in this chapter, will be used to apply the principles of "delta-to-wye" (Δ-to-Y) conversions. Principles of Thevenin's Theorem will be applied to ac circuits. AC bridge circuits can be solved using the conventional loop or Thevenin's method; another method of simplifying the bridge circuit will become evident in the delta-to-wye conversion.

23-1 DELTA-TO-WYE CONVERSIONS. The delta (Δ), or pi (π), circuit is so named because of its physical configuration. Figure 23-1a shows the delta circuit drawn as it is seen within the bridge circuit, where the configuration resembles the symbol (Δ).

In Figure 23-1b the same circuit is drawn in the configuration of the symbol π. The wye (Y) circuit is shown in Figure 23-1c, resembling the letter Y, and Figure 23-1d shows the Y drawn in the configuration of the letter T. The term "delta" is synonymous with the term "pi," and the term "tee" is synonymous with the term "wye."

Equations are used to explain the delta-to-wye conversion, but it is not recommended that the equations be depended on to solve problems. Follow the method used in the explanation rather than depending on the equations. Figure 23-2a shows a delta circuit to be converted to the wye circuit (shown in Figure 23-2c). The first step is to sketch a Y circuit within the delta, as shown in Figure 23-2b. Pure resistive components are used for ease of explanation.

375

Chapter 23
AC Circuit Analysis

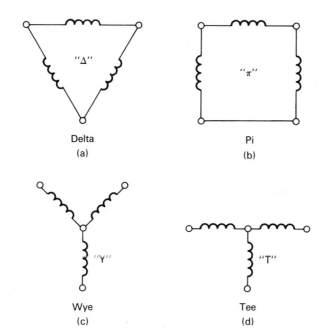

FIGURE 23-1 (a–d)   Delta and wye circuits

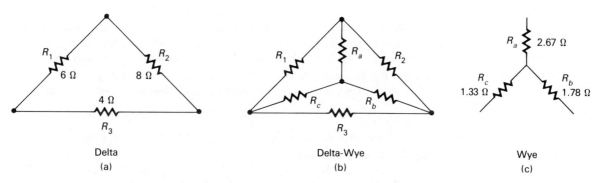

FIGURE 23-2 (a–c)   Relation between the delta to wye circuit

The following equations illustrate the circuit conversion from Δ to Y or π to T.

$$R_A = \frac{R_1 R_2}{(\Sigma R_\Delta)}$$

$$R_B = \frac{R_2 R_3}{(\Sigma R_\Delta)}$$

$$R_C = \frac{R_1 R_3}{(\Sigma R_\Delta)}$$

where $\quad \Sigma R_\Delta = R_1 + R_2 + R_3$

Determining the value of R_A

$$R_A = \frac{R_1 R_2}{(\Sigma R_\Delta)}$$

$$R_A = \frac{(6)(8)}{18}$$

$$R_A = 2.67 \; \Omega$$

Determining the value of R_B

$$R_B = \frac{R_2 R_3}{(\Sigma R_\Delta)}$$

$$R_B = \frac{(4)(8)}{18}$$

$$R_B = 1.78 \ \Omega$$

Determining the value of R_C

$$R_C = \frac{R_1 R_3}{(\Sigma R_\Delta)}$$

$$R_C = \frac{(6)(4)}{18}$$

$$R_C = 1.33 \ \Omega$$

In the procedure for determining the Y circuit, the summation of the delta is the same in each case: 18 Ω. The product of the two resistors adjacent to the resistor in the equivalent Y is used to determine the value of the resistor in the Y circuit. Equations are used to illustrate this procedure, but it is not necessary to rely on them. In practical problems the components are not labeled R_1, R_2, R_A, etc.; it is usually better to rely on the procedure than to memorize equations. Example A illustrates a delta-to-wye conversion.

Example A

Given the delta circuit in Figure 23–3a, convert the circuit to the equivalent wye circuit.

First, sketch the equivalent Y inside the delta, as shown in Figure 23–3b. The equivalent Y is labeled R_A, R_B, and R_C for convenience and is shown in Figure 23–3c.

Determine the summation of the delta

$$\Sigma R_\Delta = 25 + 50 + 100$$
$$\Sigma R_\Delta = 175 \ \Omega$$

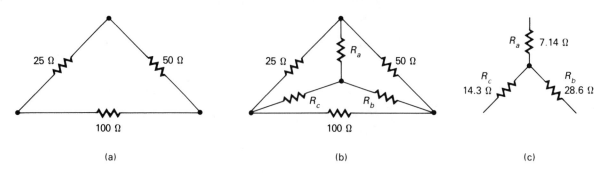

FIGURE 23–3 (a–c)   Circuits for Example A

Determine R_A. This is the product of the two resistors adjacent R_A, 25 Ω and 50 Ω, divided by the summation of the delta.

$$R_A = \frac{(25)(50)}{175}$$
$$R_A = 7.14 \, \Omega$$

Determine R_B and R_C

$$R_B = \frac{(50)(100)}{175}$$
$$R_B = 28.6 \, \Omega$$

$$R_C = \frac{(25)(100)}{175}$$
$$R_C = 14.3 \, \Omega$$

23–2 WYE-TO-DELTA CONVERSIONS.

Converting a wye circuit back to a delta circuit is seldom done when solving bridge-circuit problems but is used when working with industrial power transformers or servo mechanisms. Again, equations are used for the explanation. The procedure, rather than the equations, should be relied on when solving circuit problems.

The following equations illustrate the circuit conversion from Y to Δ or T to π. The components in Figures 23–4a–c are labeled for purposes of explanation.

First, draw the equivalent delta as shown in Figure 23–4b.

$$R_1 = \frac{\Sigma R_Y}{R_B}$$

$$R_2 = \frac{\Sigma R_Y}{R_C}$$

$$R_3 = \frac{\Sigma R_Y}{R_A}$$

where ΣR_Y is the summation of the products of the pairs of resistors in the Y circuit.

$$\Sigma R_Y = R_A R_B + R_A R_C + R_B R_C$$

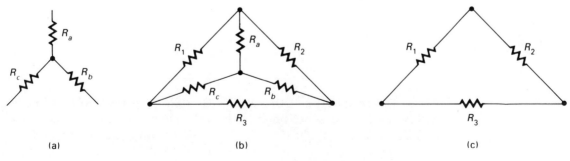

FIGURE 23–4 (a–c)   Wye to delta conversion

Note how these equations are used: the summation of the Y is the same in each case. To determine R_1, the summation is divided by the resistor opposite R_1. To determine R_2 and R_3, the summation is divided by the resistor opposite R_2 or R_3. Example B shows a conversion from Y to Δ, using this procedure. For convenience of explanation, the components in the delta are labeled R_1, R_2, and R_3.

Example B

Convert the Y circuit in Figure 23–5a to a delta. Draw the equivalent delta (Figure 23–5b).

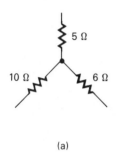 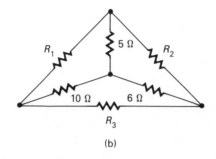 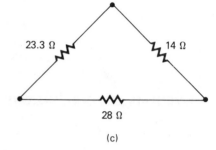

(a)   (b)   (c)

FIGURE 23–5 (a–c)   Circuit for Example B

Determine the summation of Y

$$R_Y = (5)(6) + (5)(10) + (6)(10)$$
$$R_Y = 140 \ \Omega$$

Determine R_1

$$R_1 = \frac{140}{6}$$
$$R_1 = 23.3 \ \Omega$$

(where 6 is the resistance opposite R_1 in Figure 23–5b)

Determine R_2 and R_3

$$R_2 = \frac{140}{10}$$
$$R_2 = 14 \ \Omega$$

$$R_3 = \frac{140}{5}$$
$$R_3 = 28 \ \Omega$$

23–3 DELTA-TO-WYE CONVERSIONS USING IMPEDANCE. Impedances will be used and Z substituted for R in the equations shown in sections 23–1 and 23–2. The procedure method is used to solve the circuit problem, but phasor algebra must be applied to the solution. A conversion using impedance and phasor algebra is illustrated in Example C.

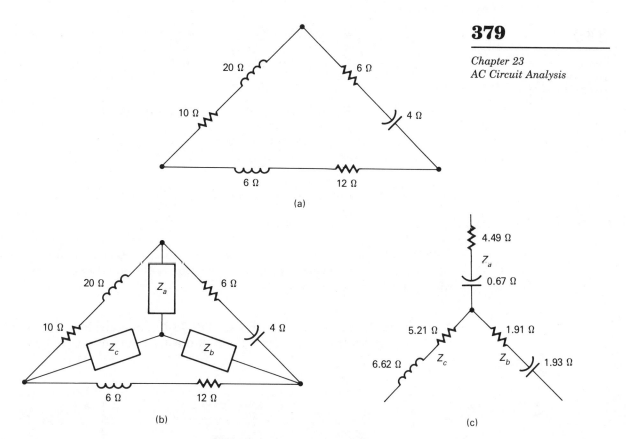

FIGURE 23-6 (a-c)   Circuit for Example C

Example C

Convert the circuit in Figure 23-6a to an equivalent Y. Sketch the equivalent Y circuit within the delta, as shown in Figure 23-6b.

Impedances are needed in both rectangular and polar form. It is more convenient first to express the impedances in both forms.

$$10 + j20 = 22.4 \underline{/63.4°} \ \Omega$$

$$6 - j4 = 7.21 \underline{/-33.7°} \ \Omega$$

$$12 + j6 = 13.4 \underline{/26.6°} \ \Omega$$

Determine the summation of the delta

$$\begin{array}{r} 10 + j20 \\ 6 - j4 \\ \underline{12 + j6} \\ 28 + j22 \end{array} \quad \text{or} \quad 35.6 \underline{/38.2°} \ \Omega$$

Determine Z_A, Z_B, and Z_C

$$Z_A = \frac{(22.4\underline{/63.4°})(7.21\underline{/-33.7°})}{35.6\underline{/38.2°}}$$

$$Z_A = 4.54\underline{/-8.5°}\ \Omega = 4.49 - j0.67\ \Omega$$

$$Z_B = \frac{(7.21\underline{/-33.7°})(13.4\underline{/26.6°})}{35.6\underline{/38.2°}}$$

$$Z_B = 2.71\underline{/-45.3°}\ \Omega = 1.91 - j1.93\ \Omega$$

$$Z_C = \frac{(22.4\underline{/63.4°})(13.4\underline{/26.6°})}{35.6\underline{/38.2°}}$$

$$Z_C = 8.43\underline{/51.8°}\ \Omega = 5.21 + j6.62\ \Omega$$

EXERCISE 23–1 The problems in this exercise should be solved using the procedure method rather than trying to apply the equations and inserting the values in the equations.

1. Convert the delta circuit in Figure 23–7a to an equivalent Y in Figure 23–7b. Label the equivalent values on the Y circuit shown.

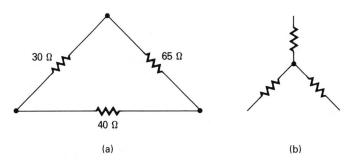

FIGURE 23–7 (a–b)   Circuit for Exercise 23–1 (1)

2. Convert the π circuit in Figure 23–8 to an equivalent T circuit. Label the values on the equivalent T circuit shown.

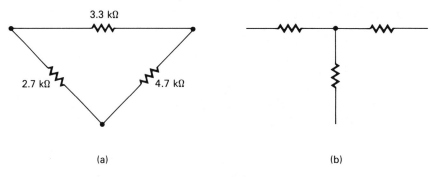

FIGURE 23–8 (a–b)   Circuit for Exercise 23–1 (2)

3. Convert the Y circuit shown in Figure 23–9 to an equivalent delta. Label the values on delta in Figure 23–9.

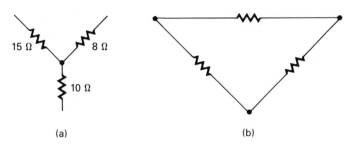

FIGURE 23–9 (a–b)   Circuit for Exercise 23–1 (3)

4. Convert the T circuit in Figure 23–10 to an equivalent π circuit and label the values on the π circuit in Figure 23–10.

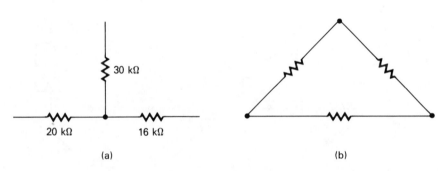

FIGURE 23–10 (a–b)   Circuit for Exercise 23–1 (4)

5. Convert the delta circuit in Figure 23–11 to an equivalent Y. Label the Y circuit given with the equivalent values.

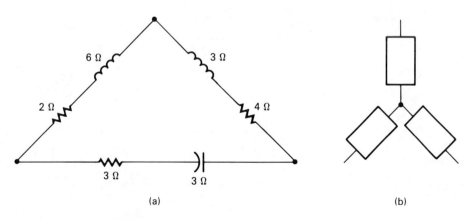

FIGURE 23–11 (a–b)   Circuit for Exercise 23–1 (5)

6. Repeat Exercise 5 for the circuit shown in Figure 23–12.

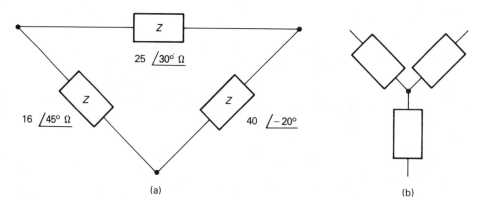

FIGURE 23–12 (a–b)   Circuit for Exercise 23–1 (6)

7. Convert the π circuit in Figure 23–13 to an equivalent T circuit. Label the equivalent values on the T circuit in Figure 23–13.

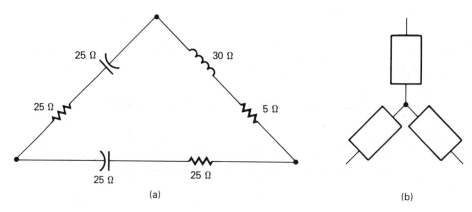

FIGURE 23–13 (a–b)   Circuit for Exercise 23–1 (7)

8. Repeat Exercise 7 for the circuit in Figure 23–14.

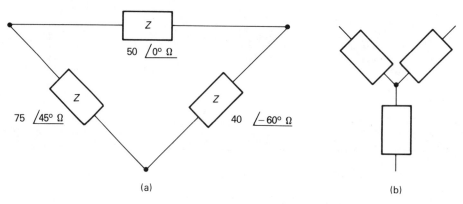

FIGURE 23–14a and b   Circuit for Exercise 23–1 (8)

23–4 ANALYZING THE BRIDGE WITH DELTA-TO-WYE CONVERSIONS.

The fundamentals of the unbalanced bridge circuit are given in Chapter 15 and should be reviewed if necessary.

The delta circuit can easily be seen in the bridge circuit. When the delta is converted to a Y circuit the bridge becomes a series parallel circuit. Figure 23–15a illustrates a bridge circuit, and Figure 23–15b illustrates the converted circuit. Resistors R_3 and R_4 are the same values as the original bridge circuit. The current through these resistors in the equivalent circuit is the same current as in the original bridge. The current and voltage drops in the bridge can be found by using the current through R_3 and R_4. Example D analyzes a resistive bridge circuit, using a delta-to-wye conversion.

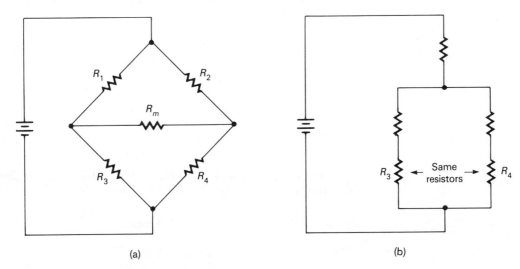

FIGURE 23–15 (a–b)   Delta and Y circuits in the bridge

Example D

Determine the current through R_m in the circuit shown in Figure 23–16a.

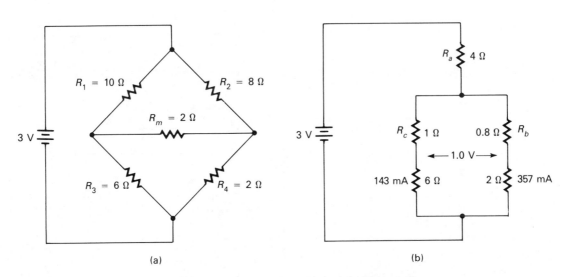

FIGURE 23–16 (a–b)   Bridge circuit for Example D

Convert the upper delta circuit to an equivalent Y and draw the equivalent circuit, labeling the components as shown in Figure 23–16b.

$$\Sigma R_\Delta = 20\ \Omega$$

$$R_A = \frac{(8)(10)}{20} = 4\ \Omega$$

$$R_B = \frac{(8)(2)}{20} = 0.8\ \Omega$$

$$R_C = \frac{(10)(2)}{20} = 1\ \Omega$$

The circuit is series-parallel. The total resistance and total current can now be calculated.

$$R_T = 6\ \Omega$$

$$I_T = 500\ \text{mA}$$

The voltage across the parallel combination can be found:

$$V_\text{par} = 1\ \text{V}$$

The current through R_3 and R_4 is determined by calculating the current through each parallel branch:

$$I_3 = 143\ \text{mA}$$

$$I_4 = 357\ \text{mA}$$

Return to the original circuit and determine the voltage across R_3 and R_4 (Figure 23–17).

The voltage across R_3 and R_4 can now be calculated, using Ohm's Law or the voltage-ratio method.

$$V_3 = 858\ \text{mV}$$

$$V_4 = 714\ \text{mV}$$

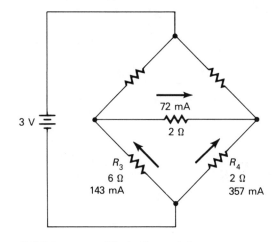

FIGURE 23–17 The bridge with known currents

The voltage across R_m is the difference between V_3 and V_4.

$$V_m = 144 \text{ mV}$$

$$I_m = 72 \text{ mA}$$

An ac bridge circuit with impedance measurements rather than resistive measurements can be analyzed in the same manner, using phasor algebra.

EXERCISE 23-2

1. Determine the current through the meter in the circuit in Figure 23-18, using the delta-to-wye conversion method.

$I_m = $ _____

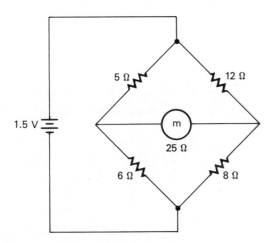

FIGURE 23-18   Circuit for Exercise 23-2 (1)

2. Repeat Exercise 1 for the circuit in Figure 23-19.

$I_m = $ _____

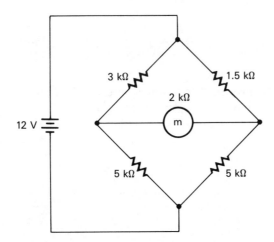

FIGURE 23-19   Circuit for Exercise 23-2 (2)

3. Determine the total impedance across the power source for the circuit in Figure 23–20.

$Z_T = $ _____

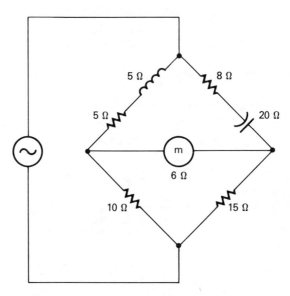

FIGURE 23–20   Circuit for Exercise 23-2 (3)

4. If the voltage applied to the circuit in Exercise 3 were 6 V, what would be the total current?

$I_T = $ _____

5. In Exercise 4, how much current would flow through the meter? Redraw the circuit before beginning the calculation.

$I_m = $ _____

23-5 REVIEWING THEVENIN'S THEOREM.

Thevenin's Theorem is given in Chapter 14 and will be reviewed only briefly in this chapter.

Basically, Thevenin's Theorem states that the current through a load can be determined if what is known about that load is the open-circuit voltage, the resistance of the load, and Thevenin's equivalent resistance.

$$I_L = \frac{V_{OPEN}}{R_L + R_{Th}}$$

Example E reviews an application of Thevenin's Theorem and the bridge circuit.

Example E

Determine the current through R_m in Figure 23–21 with the use of Thevenin's Theorem.

Remove the load R_m from the circuit, and the circuit becomes a series-parallel circuit. The load resistance is then 10 Ω. Calculate the voltage across R_1 and R_2. The difference between these two voltages is the open-circuit voltage, V_{OPEN}.

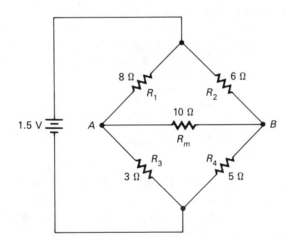

FIGURE 23–21   Circuit for Example E

$$V_1 = 1.09 \text{ V}$$
$$V_2 = 0.818 \text{ V}$$

$$V_{OPEN} = V_1 - V_2$$
$$V_{OPEN} = 1.09 - 0.818$$
$$V_{OPEN} = 272 \text{ mV}$$

Thevenin's equivalent resistance is the resistance between points A and B, with the power source a short circuit. The resistance between points A and B is the resistance calculated for the circuit in Figure 23–22.

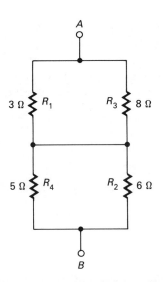

FIGURE 23-22 Circuit for Thevenin's resistance

$$R_{Th} = \frac{R_1 R_3}{R_1 + R_3} + \frac{R_2 R_4}{R_2 + R_4}$$

$$R_{Th} = 4.91 \, \Omega$$

The current through the meter is:

$$I_m = \frac{V_{\text{OPEN}}}{R_L + R_{Th}}$$

$$I_m = \frac{0.272}{10 + 4.91}$$

$$I_m = 18.3 \text{ mA}$$

23-6 SOLVING THE IMPEDANCE BRIDGE WITH THEVENIN'S THEOREM.
The same principles that are applied to the dc bridge can be applied to the impedance bridge as long as phasor algebra is used to solve the problem. Example F illustrates an impedance bridge problem.

Example F
Determine the current through the Z_L in the circuit in Figure 23-23.
Remove Z_m from the circuit and calculate the voltage across Z_1 and Z_2.
Voltage across Z_1

$$I_1 = \frac{V_T}{Z_1 + Z_3}$$

$$I_1 = \frac{12}{6 + j4 + 3 - j5}$$

$$I_1 = \frac{12}{9.06 \underline{/-6.34°}}$$

$$I_1 = 1.32 \underline{/6.34°} \text{ A}$$

$$V_1 = I_1 Z_1$$
$$V_1 = (1.32\underline{/6.34°})(7.21\underline{/33.7°})$$
$$V_1 = 9.52\underline{/40°} \text{ V}$$

Voltage across Z_2

$$I_2 = \frac{V_T}{Z_2 + Z_4}$$
$$I_2 = \frac{12}{2 + j4 + 4 - j4}$$
$$I_2 = \frac{12}{6\underline{/0°}}$$
$$I_2 = 2\underline{/0°} \text{ A}$$

$$V_2 = I_2 Z_2$$
$$V_2 = (2\underline{/0°})(4.47\underline{/63.4°})$$
$$V_2 = 8.94\underline{/63.4°} \text{ V}$$

Open-circuit voltage

$$V_O = V_1 - V_2$$
$$V_O = (9.52\underline{/40°}) - (8.94\underline{/63.4°})$$
$$V_O = (7.29 + j6.11) - (4 - 7.99)$$
$$V_O = 3.29 - j1.88 \text{ V or } 3.78\underline{/-29.7°} \text{ V}$$

Thevenin's internal impedance

$$Z_{Th} = \frac{Z_1 Z_3}{Z_1 + Z_3} + \frac{Z_2 Z_4}{Z_2 + Z_4}$$

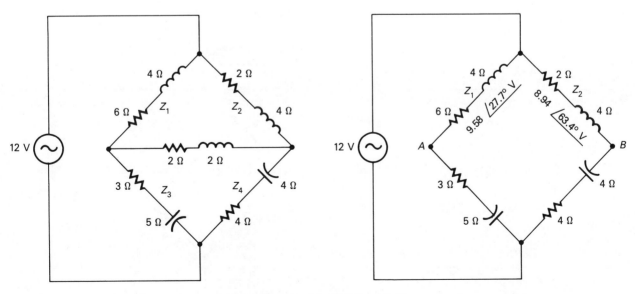

FIGURE 23-23   The impedance bridge

$$Z_{Th} = \frac{(7.21\underline{/33.7°})(5.83\underline{/-59°})}{(6 + j4) + (3 - j5)} + \frac{(4.47\underline{/63.4°})(5.66\underline{/-45°})}{(2 + j4) + (4 - j4)}$$

$$Z_{Th} = \frac{42\underline{/-25.3°}}{9.06\underline{/-6.3°}} + \frac{25.3\underline{/18.4°}}{6\underline{/0°}}$$

$$Z_{Th} = (4.64\underline{/-19°}) + (4.22\underline{/18.4°})$$
$$Z_{Th} = (4.38 - j1.51) + (4 + j1.33)$$
$$Z_{Th} = 8.38 - j0.18 \; \Omega$$

Finding the load current

$$I_m = \frac{V_O}{Z_L + Z_{Th}}$$

$$I_m = \frac{3.78\underline{/-29.7°}}{2 + j2 + 8.38 - j0.18}$$

$$I_m = \frac{3.78\underline{/-29.7°}}{10.5\underline{/9.93°}}$$

$$I_m = 360\underline{/-39.6°} \; \text{mA}$$

EXERCISE 23-3   Solve the exercises with Thevenin's Theorem.

1. Find the current through the meter in the bridge circuit given in Figure 23-24.

$$I_m = \underline{}$$

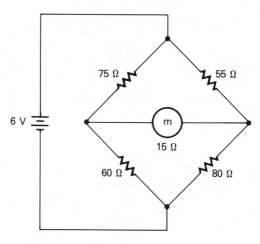

FIGURE 23-24   Circuit for Exercise 23-3 (1)

2. Find the meter current in the bridge circuit given in Figure 23–25.

$I_m = $ _____

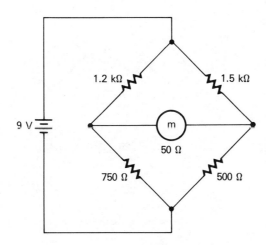

FIGURE 23–25   Circuit for Exercise 23–3 (2)

3. Determine the current through the meter in the impedance bridge shown in Figure 23–26.

$I_m = $ _____

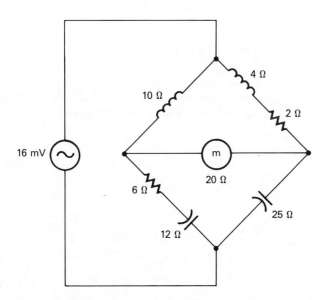

FIGURE 23–26   Circuit for Exercise 23–3 (3)

4. Find the current through the meter in the impedance bridge in Figure 23-27.

$I_m = $ _____

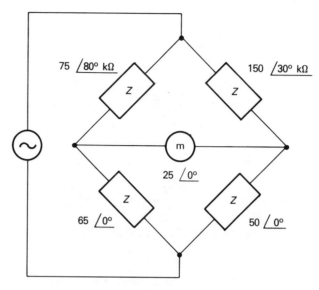

FIGURE 23-27   Circuit for Exercise 23-3 (4)

23-7 NORTON'S THEOREM.

Norton's Theorem allows a circuit to be analyzed with a constant current. Current devices such as transistors are often analyzed by applying Norton's Theorem. Developing this constant current source requires using the open-circuit voltage and Thevenin's equivalent resistance. The circuit in Figure 23-28a illustrates a Thevenin's equivalent circuit with the open-circuit voltage and Thevenin's equivalent resistance. Figure 23-28b illustrates Norton's equivalent of the same circuit, converted by using Norton's Theorem.

$$I_N = \frac{V_O}{Z_{Th}}$$

where I_N is Norton's constant current

V_O is the open-circuit voltage

Z_{Th} is Thevenin's equivalent impedance

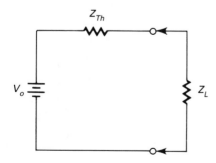

FIGURE 23-28a   Thevenin's equivalent circuit

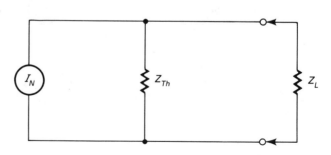

FIGURE 23-28b   Norton's equivalent circuit

Example G illustrates a circuit converted first to Thevenin's equivalent and then to Norton's.

Example G

Given the circuit in Figure 23–29, determine the load current by Norton's Theorem.

The open-circuit voltage, or the voltage that would be across points A and B, is the voltage across Z_2. The voltage from the source is divided across Z_1 and Z_2 and can be determined by Ohm's Law or voltage ratio.

$$V_2 = \left(\frac{V_T}{Z_1 + Z_2}\right)(Z_2)$$

$$V_2 = \left(\frac{10}{5 + j5 + 10 - j5}\right)(10 - j5)$$

$$V_2 = V_O = 7.47\underline{/-26.6°} \text{ V}$$

Thevenin's equivalent resistance:

$$Z_{Th} = Z_3 + \frac{Z_1 Z_2}{Z_1 + Z_2}$$

$$Z_{Th} = (8 + j10) + \frac{(7.07\underline{/45°})(11.2\underline{/-26.6°})}{(5 + j5) + (10 - j5)}$$

$$Z_{Th} = (8 + j10) + (5.01 + j1.67)$$
$$Z_{Th} = 13 + j11.7 = 17.5\underline{/42°} \text{ }\Omega$$

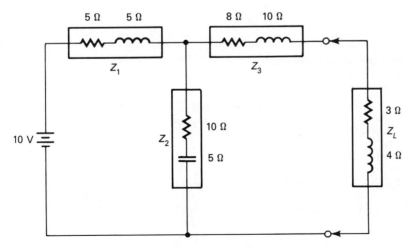

FIGURE 23–29 Circuit for Example G

Norton's constant current:

$$I_N = \frac{V_O}{Z_{Th}}$$

$$I_N = \frac{7.47\underline{/-26.6°}}{17.5\underline{/42°}}$$

$$I_N = 427\underline{/-42°} \text{ mA}$$

Norton's equivalent circuit for Example G is shown in Figure 23–30. The current through the load can be determined by parallel-circuit calculations or by the current-ratio method.

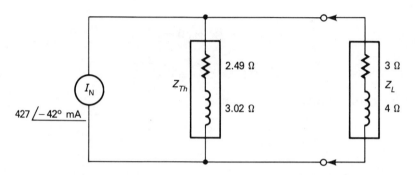

FIGURE 23–30   Norton's equivalent circuit

Calculating Load Current, I_L.

Total impedance:

$$Z_T = \frac{Z_{Th}Z_L}{Z_{Th} + Z_L}$$

$$Z_T = \frac{(17.5\underline{/42°})(5\underline{/53.1°})}{(13 + j11.7) + (3 + j4)}$$

$$Z_T = 3.91\underline{/50.5°} \text{ }\Omega$$

Voltage across the load:

$$V_L = I_N Z_T$$
$$V_L = (0.427\underline{/-42°})(3.91\underline{/50.5°})$$
$$V_L = 1.67\underline{/8.5°} \text{ V}$$

Load current:

$$I_L = \frac{V_L}{Z_L}$$
$$I_L = \frac{1.67\underline{/8.5°}}{(5\underline{/53.1°})}$$
$$I_L = 334\underline{/-44.6°} \text{ mA}$$

EXERCISE 23-4

1. Determine the Norton's equivalent circuit for the load R_L in Figure 23-31. Draw the schematic for the equivalent circuit and label the resistance and constant current values on the circuit.

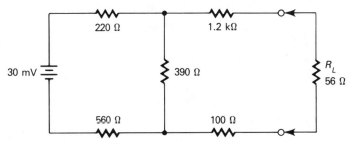

FIGURE 23-31 Circuit for Exercise 23-4 (1)

2. Determine the current through the load in Exercise 1.

$$I_L = \underline{\hspace{2cm}}$$

3. Determine the Norton's equivalent circuit for the load R_L in Figure 23–32. Draw the schematic for the equivalent circuit and label the resistance and constant current values on the drawing.

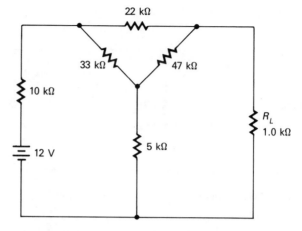

FIGURE 23–32 Circuit for Exercise 23–4 (3)

4. Determine the current through the load in Exercise 3.

$I_L =$ _____

5. Determine Norton's equivalent circuit for the circuit given in Figure 23–33. Draw the schematic for the equivalent circuit and label the resistance and constant current values on the drawing.

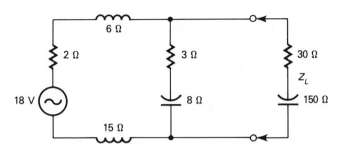

FIGURE 23–33   Circuit for Exercise 23–4 (5)

6. Determine Norton's equivalent circuit for the circuit in Figure 23–34. Draw the schematic for the equivalent circuit and label the resistance and constant current values on the drawing.

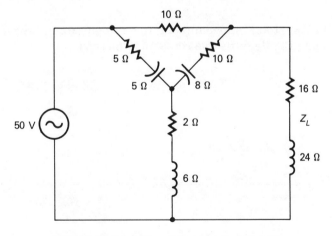

FIGURE 23–34   Circuit for Exercise 23–4 (6)

EVALUATION EXERCISE

1. Convert the delta circuit in Figure 23–35 to an equivalent Y circuit. Draw and label the component values on the Y circuit.

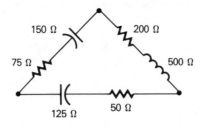

FIGURE 23–35   Circuit for Evaluation Exercise (1)

2. Determine the total impedance of the bridge circuit given in Figure 23–36, using the delta-to-wye conversion method.

$Z_T = $ _____

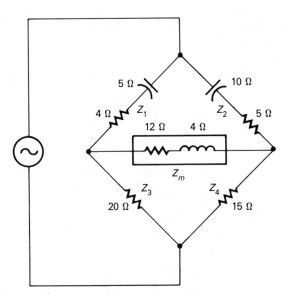

FIGURE 23–36 Circuit for Evaluation Exercise (2)

3. In Exercise 2, if voltage across $Z_3 = 16$ mV and the voltage across $Z_4 = 8.5$ mV, what is the current through Z_m?

$$I_m = \underline{}$$

4. In the circuit shown in Figure 23–37, find the current through Z_m, using Thevenin's Theorem.

$I_m = $ _____

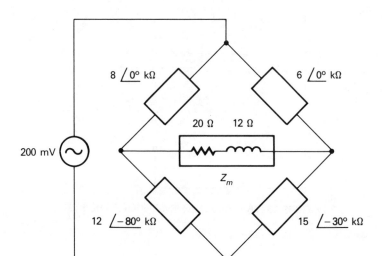

FIGURE 23–37   Circuit for Evaluation Exercise (4)

5. Using the circuit in Figure 23–38, find Norton's constant current. Draw and label the component values for Norton's constant current circuit.

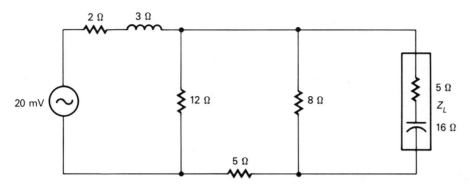

FIGURE 23–38 Circuit for Evaluation Exercise (5)

6. In the circuit in Figure 23–38, find the load current.

$$I_L = \underline{\hspace{2in}}$$

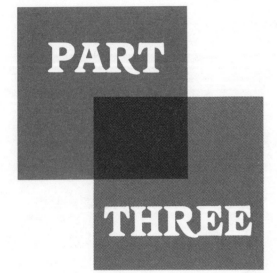

PART THREE

Computer Mathematics, Logarithms, and Quadrat Equations

CHAPTER 24
Math for the Computer

Objectives

After completing this chapter you will be able to:
- Identify numbering systems
- Convert base 10 to binary, octal and hexadecimal
- Convert binary, octal, and hexadecimal to base 10
- Convert binary to octal
- Convert octal to binary
- Convert binary to hexadecimal
- Convert hexadecimal to binary
- Convert decimal fractions to binary and octal fractions
- Convert binary and octal fractions to decimal fractions
- Add binary numbers
- Subtract binary numbers by complementation

This chapter covers the basic concepts of math related to computer operation. It is not intended to show how math is used within the computer; that should be done in a study of computer operation. Therefore, while this chapter is not a complete study of the math, it does show the relation of decimal, binary, octal, and hexadecimal numbering systems.

24–1 BASES. The base 10 numbering system is well established into our society, as it should be. The system is easily learned and is convenient for calculation and practical applications. There are other bases in common use, such as base 12; eggs and doughnuts, for instance, are packaged by the dozen, or in base 12. Another is the base 60 system, which is used in measuring time —60 seconds in a minute, 60 minutes in an hour, and so on. The convenience of having ten fingers makes it possible to use the hands as a simple computer, especially when a beginner learns the base 10 system. The drawback of this approach is that the beginner perceives that the base 10 system begins at 1 and goes to 10. The base 10 system, like all numbering systems, begins at 0, or zero. Zero is as valid a number in the system as any other. The base 10 numbering system is defined as "0 to 9." It might have been more convenient, though perhaps not too practical, if people lacked a finger, so the count would be 0 to 9.

Numbering systems are described in a given base. The count begins at zero and goes forward. Another word used to describe a base is "radix."

- *Binary* is a base, or radix, of 2, and the count is 0 to 1.
- *Octal* is a base, or radix, of 8, and the count is 0 to 7.

- *Decimal* is a base, or radix, of 10, and the count is 0 to 9.
- *Hexadecimal* is a base, or radix, of 16, and the count is 0 to 15.

The highest number that can be counted in any base is a number 1 less than the base or radix. For instance, 7 is the highest count in the base 8, and 9 is the highest number counted in base 10, and so on. Quantities counted beyond the base are counted in the next column. In the decimal system the count is 0 through 9, then 10 through 19, 20 through 29, and so forth.

In the decimal system a number such as 845 can be broken down into components, as follows.

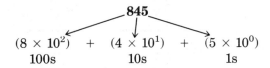

$$(8 \times 10^2) + (4 \times 10^1) + (5 \times 10^0)$$
$$\text{100s} \qquad \text{10s} \qquad \text{1s}$$

The number 845 consists of the components:

$$\begin{aligned} 8 \text{ counts of } 100 &= 800 \\ 4 \text{ counts of } 10 &= 40 \\ 5 \text{ counts of } 1 &= 5 \\ \text{total} &= 845 \end{aligned}$$

A binary number consists of 0s and 1s and can be divided into components in the same manner. A binary digit is referred to as a "bit."

$$(1 \times 2^3) + (0 \times 2^2) + (1 \times 2^1) + (1 \times 2^0)$$

$$8 \text{ counts} + 0 \text{ counts} + 2 \text{ counts} + 1 \text{ count}$$

$$8 + 0 + 2 + 1 = 11$$

Table 24–1 shows the components of the binary, octal, decimal, and hexadecimal radices and will help in counting quantities in the different radices.

TABLE 24–1

	Hundred Thousands	Ten Thousands	Thousands	Hundreds	Tens	Units
Base 10	1×10^5	1×10^4	1×10^3	1×10^2	1×10^1	1×10^0
Base 8	1×8^5	1×8^4	1×8^3	1×8^2	1×8^1	1×8^0
Base 2	1×2^5	1×2^4	1×2^3	1×2^2	1×2^1	1×2^0
Base 16	1×16^5	1×16^4	1×16^3	1×16^2	1×16^1	1×16^0

24–2 CHANGING BASE 10 TO BASE 2. To convert from base 10 to another base, the base 10 number is divided by the new base, as in Example A.

Example A

Convert 39_{10} to base 2. (Read the remainders from bottom to top as indicated by the arrow.)

```
2 | 39
2 | 19  ↑1       (39 ÷ 2 = 19, with a remainder of 1)
2 |  9   1       (19 ÷ 2 = 9, with a remainder of 1)
2 |  4   1       (9 ÷ 2 = 4, with a remainder of 1)
2 |  2   0       (4 ÷ 2 = 2, with a remainder of 0)
2 |  1   0       (2 ÷ 2 = 1, with a remainder of 0)
     0   1       (1 ÷ 2 = 0, with a remainder of 1)
```

and $\qquad 39_{10} = 100111_2$

EXERCISE 24–1 Change the decimal numbers to their binary equivalents.

1. 7 _____ 2. 15 _____

3. 19 _____ 4. 83 _____

5. 25 _____ 6. 45 _____

7. 115 _____ 8. 154 _____

24–3 CHANGING DECIMAL BASE TO OCTAL. Converting from base 10 to base 8 is done in the same manner as changing to base 2. The process is shown in Example B.

Example B

Convert 175_{10} to base 8. (Read the remainders from bottom to top as indicated by the arrow.)

```
8 | 175
8 |  21  ↑7      (175 ÷ 8 = 21, with a remainder of 7)
8 |   2   5      (21 ÷ 8 = 2, with a remainder of 5)
      0   2      (2 ÷ 8 = 0, with a remainder of 2)
```

and $\qquad 175_{10} = 257_8$

Chapter 24
Math for the Computer

EXERCISE 24-2 Change the decimal numbers to their octal equivalents.

1. 88 _____   2. 188 _____

3. 456 _____   4. 73 _____

5. 84 _____   6. 985 _____

7. 399 _____   8. 1000 _____

24-4 CHANGING DECIMAL TO HEXADECIMAL. Changing base 10 numbers to base 16 is done in the same manner as base 2 and base 8 (see Example C). Because the highest count that can be used in the decimal system is 9, and the highest count in the hexadecimal is 15, symbols are used to represent counts above 9. Table 24-2 illustrates the use of these symbols.

TABLE 24-2

decimal value	0	1	2	3	4	5	6	7	8	9	10	11	12	13	14	15
hexadecimal symbol	0	1	2	3	4	5	6	7	8	9	A	B	C	D	E	F

Example C

Convert 2525_{10} to base 16. (Read the remainders from bottom to top as indicated by the arrow.)

$$\begin{array}{r|r|r} 16 & 2525 & \\ 16 & 157 & 13 \\ 16 & 9 & 13 \\ & 0 & 9 \end{array}$$

(2525 ÷ 16 = 157, with a remainder of 13)
(157 ÷ 16 = 9, with a remainder of 13)
(9 ÷ 16 = 0, with a remainder of 9)

Substitute the symbol D for the count 13:
$$2525_{10} = 9DD_{16}$$

EXERCISE 24-3 Change the decimal numbers to their hexadecimal equivalents.

1. 88 _____
2. 175 _____
3. 355 _____
4. 4453 _____
5. 7836 _____
6. 5762 _____
7. 34,812 _____
8. 15,278 _____

24-5 CONVERTING BINARY TO DECIMAL. Table 24-1 provides a basis for converting to decimal equivalents. Example D illustrates this operation.

Example D

Convert the number 10111_2 to its decimal equivalent. (Determine the counts using Table 24-1.)

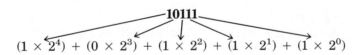

$$1 \times 2^4 = 16_{10}$$
$$0 \times 2^3 = 0_{10}$$
$$1 \times 2^2 = 4_{10}$$
$$1 \times 2^1 = 2_{10}$$
$$\underline{1 \times 2^0 = 1_{10}}$$
$$\text{total} = 23_{10}$$

and $10111_2 = 23_{10}$

EXERCISE 24-4 Change the binary numbers to their decimal equivalents.

1. 111 _____
2. 1111 _____
3. 100 _____
4. 101 _____
5. 111001 _____
6. 1011010 _____
7. 111101101 _____
8. 10110111 _____

24-6 CONVERTING OCTAL TO DECIMAL. Example E illustrates the conversion of octal to decimal equivalents. Table 24-1 can be useful in this operation.

Example E

Convert the number 132_8 to its decimal equivalent.

$$132$$
$$(1 \times 8^2) + (3 \times 8^1) + (2 \times 8^0)$$

$$1 \times 8^2 = 64_{10}$$
$$3 \times 8^1 = 24_{10}$$
$$\underline{2 \times 8^0 = 2_{10}}$$
$$\text{total} = 90_{10}$$

Thus

$$132_8 = 90_{10}$$

EXERCISE 24-5 Change the octal numbers to their decimal equivalents.

1. 77 _____

2. 63 _____

3. 126 _____

4. 147 _____

5. 377 _____

6. 442 _____

7. 527 _____

8. 1136 _____

24-7 CONVERTING HEXADECIMAL TO DECIMAL. The conversion of hexadecimal to decimal is performed in the same manner as octal and binary conversion. Example F illustrates the operation.

Example F

Convert $2A6_{16}$ to its decimal equivalent.

$$2A6$$

$$(2 \times 16^2) + (10 \times 16^1) + (6 \times 16^0)$$

$$2 \times 16^2 = 512_{10}$$

$$10 \times 16^1 = 160_{10} \text{ (where A = 10)}$$

$$\frac{6 \times 16^0 = 6_{10}}{\text{total} = 678_{10}}$$

Thus

$$2A6_{16} = 678_{10}$$

EXERCISE 24–6 Change the hexadecimal numbers to their decimal equivalents.

1. 134 _____   2. 256 _____

3. 4E3 _____   4. 16A _____

5. 2B12 _____   6. 1C62 _____

7. FAC _____   8. A2BC _____

24–8 CONVERTING BINARY TO OCTAL. A binary number such as 1101110101011 is difficult to read in ones and zeros. The number, however, can be converted to an octal number with little difficulty. Since the highest count in base 8 is 7, an octal number is expressed in binary with only three symbols (see Example G).

Example G

The number 111_2 expresses 7 in base 8, 110_2 expresses 6, and so on.

$$111 \qquad 110$$
$$4 + 2 + 1 = 7_8 \qquad 4 + 2 + 0 = 6_8$$
$$101 \qquad 100$$
$$4 + 0 + 1 = 5_8 \qquad 4 + 0 + 0 = 4_8$$
$$011 \qquad 010$$
$$0 + 2 + 1 = 3_8 \qquad 0 + 2 + 0 = 2_8$$

$$0 + 0 + 1 = 1_8$$

To "read out" a binary number in octal, first place the binary number in groups of three, starting at the octal point. Then move right to left, reading each group of three in base 8.

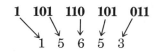

Thus $1101110101011_2 = 15653_8$

EXERCISE 24–7 Read out the numbers in their octal equivalents in Exercises 1 through 16.

1. 10110101
2. 10000011111
3. 11011010110100
4. 10011110011
5. 10011011110111
6. 11100110110
7. 11000101001101
8. 10111111010

In problems 9 through 16 convert the octal numbers to their binary equivalents.

9. 66
10. 24
11. 154
12. 444
13. 132
14. 1023
15. 1144
16. 2431

24–9 CONVERTING BINARY TO HEXADECIMAL. The procedure for converting binary to hexadecimal is the same as that used to convert binary to octal, except that the binary number is now divided into groups of four. A binary group of four numbers can express any number in hexadecimal.

Example H
Convert the number 1111_2 to hexadecimal.

$$1111$$

$$8 + 4 + 2 + 1 = 15$$

Because 15 is the highest count in base 16, any hexadecimal number can be expressed by four binary numbers.

Example I
Convert 11001001_2 to hexadecimal. Place the number in groups of four, starting at the decimal point and moving right to left.

$$1100 \qquad 1001$$

$$8 + 4 + 0 + 0 = C \qquad 8 + 0 + 0 + 1 = 9$$

This yields $\qquad 11001001_2 = C9_{16}.$

A convenient way to convert hexadecimal to octal is first convert to binary, then to octal.

EXERCISE 24-8 Read out the binary numbers in their hexadecimal equivalent in Exercises 1 through 8.

1. 10100011110101100
2. 1110010010111111
3. 1111010101011111
4. 110010100000111110
5. 111111100000000100
6. 11000011100000
7. 1111111110000
8. 1111111100000110

Convert the hexadecimal numbers in Exercises 9 through 16 to their binary equivalents.

9. 89
10. A16
11. 2D5
12. AC7D
13. C21D
14. AB6F

15. 8431

16. DF6C

24–10 CONVERTING DECIMAL FRACTIONS TO A NEW BASE. Fractions can be expressed in different bases, just as whole numbers are expressed in different bases. The process is one of multiplication (Example J).

Example J

Convert $\frac{9}{10}$ to its binary equivalent. (Read the units column from top to bottom, as indicated by the arrow.)

$$\frac{9}{10} = 0.9 \quad \text{(convert to decimal fraction)}$$

$$\begin{array}{rl}
0.9 \times 2 = 1 \,|\, .8 & \text{(multiply by 2)} \\
0.8 \times 2 = 1 \,|\, .6 & \text{(multiply the decimal fraction by the base 2} \\
0.6 \times 2 = 1 \,|\, .2 & \text{and repeat the operation)} \\
0.2 \times 2 = 0 \,|\, .4 & \\
0.4 \times 2 = 0 \,|\, .8 & \\
0.8 \times 2 = 1 \,|\, .6 & \\
0.6 \times 2 = 1 \,|\, .2 & \\
0.2 \times 2 = 0 \,|\, .4 & \\
0.4 \times 2 = 0 \downarrow .8 &
\end{array}$$

Thus

$$0.9_{10} = 0.11100110_2$$

Example K

Convert $\frac{5}{8}$ to its binary equivalent.

$$\frac{5}{8} = 0.625$$

$$\begin{array}{r}
0.625 \times 2 = 1 \,|\, .25 \\
0.25 \times 2 = 0 \,|\, .5 \\
0.5 \times 2 = 1 \downarrow .0
\end{array}$$

Thus

$$0.625_{10} = 0.101_2$$

The fraction 0.625 can be expressed with only three binary digits, whereas the fraction 0.9 is expressed using eight binary digits. The third step in Example K results in a decimal of 0, or no remainder. Unless this occurs, eight binary digits are usually required to maintain accuracy.

Example L

Fractions can be converted into other bases as well. Example L illustrates a base 8 conversion.

Convert $\frac{9}{10}$ to its octal equivalent.

$$\frac{9}{10} = 0.9$$

$$0.9 \times 8 = 7 \,|\, .2$$
$$0.2 \times 8 = 1 \,|\, .6$$
$$0.6 \times 8 = 4 \,|\, .8$$

Thus $\qquad 0.9_{10} = 0.714_8$

This result is easily checked by reading the binary number in Example J in octal readout.

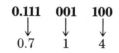

Thus $\qquad 0.111001100_2 = 0.714_8$

EXERCISE 24–9 Convert the decimal fractions to binary equivalents, then express them as octal equivalents. Carry the answer to only six binary digits or three octal digits.

	Number	Binary	Octal
1.	0.16		
2.	$\frac{9}{16}$		
3.	0.785		
4.	0.666		

5. $\dfrac{5}{6}$ _____ _____

6. 0.75 _____ _____

7. $\dfrac{3}{16}$ _____ _____

8. 0.125 _____ _____

24–11 CONVERTING BINARY AND OCTAL FRACTIONS TO DECIMAL FRACTIONS. The operation of converting binary and octal fractions to decimal fractions is done in the same manner as converting whole numbers. Now, however, the exponents are negative. Table 24–3 illustrates the negative exponent, and examples M and N illustrate such a conversion.

TABLE 24–3
1×2^{-1}, 1×2^{-2}, 1×2^{-3}, 1×2^{-4}, etc.
$X \times 8^{-1}$, $X \times 8^{-2}$, $X \times 8^{-3}$, $X \times 8^{-4}$, etc.

Example M
  Convert 0.1111_2 to its decimal equivalent.

$$0.1111$$
$$(1 \times 2^{-1}) + (1 \times 2^{-2}) + (1 \times 2^{-3}) + (1 \times 2^{-4})$$
$$0.5 \;+\; 0.25 \;+\; 0.125 \;+\; 0.0625 \;=\; 0.9375$$

Notice the pattern that develops: the first bit begins with an equivalent of 0.5; then it is divided by 2, to obtain the second bit. Then that number is divided by 2, for the third bit, and so on, yielding
$$0.1011_2 = 0.9375_{10}$$

Example N

Convert 0.6243_8 to its decimal equivalent.

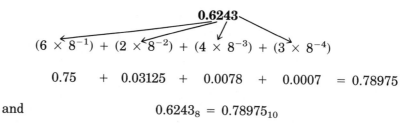

$$0.75 + 0.03125 + 0.0078 + 0.0007 = 0.78975$$

and $0.6243_8 = 0.78975_{10}$

EXERCISE 24-10 Convert the binary or octal numbers to their decimal equivalents.

1. 0.1111_2

2. 0.1001_2

3. 0.10011_2

4. 0.11101_2

5. 0.254_8

6. 0.637_8

7. 0.156_8

8. 0.346_8

24–12 BINARY ADDITION. When adding quantities in the decimal system some basic concepts or rules must be followed. The binary-system rules for adding binary numbers are:

$$1 + 1 = 0 \quad \text{(and carry 1 to the next column)}$$

$$1 + 0 = 1$$

$$0 + 0 = 0$$

Example O
Add the binary numbers 10 and 11.

```
carry 1 ──→ ①
           1 0       (first column: 0 + 1 = 1)
           1 1       (second column: 1 + 1 = 0 and carry 1)
         ─────
         1 0 1       (carry 1 to the third column, where
                      1 + 0 + 1)
```

Check your answer:

$$\begin{aligned} 10 &= 2_{10} \\ 11 &= 3_{10} \\ \hline 101 &= 5_{10} \end{aligned}$$

Example P
Add the binary numbers 1101 and 111.

```
    ①①①①        (carried numbers are in circles)
    1 1 0 1      (first column: 1 + 1 = 0 and carry 1)
        1 1 1    (second column: 1 + 0 = 1 and 1 + 1 = 0; carry 1)
  ─────────
  1 0 1 0 0      (third column: 1 + 1 = 0; carry 1; 0 + 1 = 1)
                 (fourth column: 1 + 1 = 0 and carry 1)
                 (fifth column: 1 + 0 = 1)
```

Check your answer:

$$\begin{aligned} 1101 &= 13_{10} \\ 111 &= 7_{10} \\ \hline 10100 &= 20_{10} \end{aligned}$$

EXERCISE 24–11 Add the binary numbers and check the answer in decimal calculation.

1. $\begin{aligned} 1\,0\,1\,1 \\ \underline{1\,0\,0\,1} \end{aligned}$

    $1011 = \underline{}_{10}$
    $1001 = \underline{}_{10}$
    $\underline{}_{10}$

418

*Part 3
Computer Mathematics,
Logarithms, and
Quadratic Equations*

2. $\begin{array}{r} 10111 \\ \underline{11111} \end{array}$

    $10111 = \underline{\hspace{1cm}}_{10}$
    $11111 = \underline{\hspace{1cm}}_{10}$
    $ \underline{\hspace{1cm}}_{10}$

3. $\begin{array}{r} 100011 \\ \underline{111} \end{array}$

    $100011 = \underline{\hspace{1cm}}_{10}$
    $111 = \underline{\hspace{1cm}}_{10}$
    $ \underline{\hspace{1cm}}_{10}$

4. $\begin{array}{r} 1111110 \\ \underline{1110101} \end{array}$

    $1111110 = \underline{\hspace{1cm}}_{10}$
    $1110101 = \underline{\hspace{1cm}}_{10}$
    $ \underline{\hspace{1cm}}_{10}$

5. $\begin{array}{r} 1010 \\ 1110 \\ \underline{1001} \end{array}$

    $1010 = \underline{\hspace{1cm}}_{10}$
    $1110 = \underline{\hspace{1cm}}_{10}$
    $1001 = \underline{\hspace{1cm}}_{10}$
    $ \underline{\hspace{1cm}}_{10}$

6. $\begin{array}{r} 111001 \\ 10011 \\ \underline{111} \end{array}$

    $111001 = \underline{\hspace{1cm}}_{10}$
    $10011 = \underline{\hspace{1cm}}_{10}$
    $111 = \underline{\hspace{1cm}}_{10}$
    $ \underline{\hspace{1cm}}_{10}$

7. $\begin{array}{r} 1111001.000 \\ \underline{1000111.001} \end{array}$

    $1111001.000 = \underline{\hspace{1cm}}_{10}$
    $1000111.001 = \underline{\hspace{1cm}}_{10}$
    $ \underline{\hspace{1cm}}_{10}$

8. $\begin{array}{r} 10100110.01 \\ \underline{01011001.10} \end{array}$

    $10100110.01 = \underline{\hspace{1cm}}_{10}$
    $01011001.10 = \underline{\hspace{1cm}}_{10}$
    $ = \underline{\hspace{1cm}}_{10}$

9. $\begin{array}{r} 11110 \\ 1000 \\ \underline{101} \end{array}$

    $11110 = \underline{\hspace{1cm}}_{10}$
    $1000 = \underline{\hspace{1cm}}_{10}$
    $101 = \underline{\hspace{1cm}}_{10}$
    $ \underline{\hspace{1cm}}_{10}$

10. $\begin{array}{r} 111 \\ 101 \\ 110 \\ \underline{011} \end{array}$

    $111 = \underline{\hspace{1cm}}_{10}$
    $101 = \underline{\hspace{1cm}}_{10}$
    $110 = \underline{\hspace{1cm}}_{10}$
    $011 = \underline{\hspace{1cm}}_{10}$
    $ \underline{\hspace{1cm}}_{10}$

11. $\begin{array}{r} 111.100 \\ \underline{101.101} \end{array}$

    $111.100 = \underline{\hspace{1cm}}_{10}$
    $101.101 = \underline{\hspace{1cm}}_{10}$
    $ \underline{\hspace{1cm}}_{10}$

12. $\begin{array}{r}10001.010\\1111.110\\\hline\end{array}$ $\begin{array}{r}10001.010 = \underline{\hspace{2cm}}_{10}\\1111.110 = \underline{\hspace{2cm}}_{10}\\\underline{\hspace{2cm}}_{10}\end{array}$

24–13 BINARY SUBTRACTION BY COMPLEMENTS.

The basic rules for subtracting binary numbers are much like those applied to decimal subtraction.

$$0 - 0 = 0;\ 1 - 0 = 1;\ 1 - 1 = 0;\ 0 - 1 = 1;\ \text{and borrow } 1$$

Although these rules are basic, the borrow-1 aspect of the operation is sometimes confusing. Computers do not use this operation in functioning; they use the operation of subtraction by complements instead. This operation is done by addition, which makes the computer an adding machine. Multiplication and division are also performed by addition. Example Q illustrates the addition process of subtraction by complements.

The complement of a binary number can be obtained by changing all 1s to 0s and all 0s to 1s. The complement of 101 is 010, the complement of 11001 is 00110, and so on.

Example Q

Subtract 1001 from 1011. Determine the complement of the subtrahend 1001 = 0110. Do not change the minuend.

```
 1 0 1 1  (minuend)          (add by conventional method)
 0 1 1 0  (subtrahend)
⓵0 0 0 1                     (move 1 on the far left, as shown, and re-add)
    →1
 0 0 1 0  (difference)
```

Check your answer in base 10

$$\begin{array}{r}1011 = 11_{10}\\1001 = -9_{10}\\\hline 0010 = 2\end{array}$$

Subtract

$$\begin{array}{r}11100011 = 227\\1010011 = -83\\\hline 144\end{array} \qquad \begin{array}{r}11100011\\10101100\\\hline ⓵10001111\\\to 1\\\hline 10010000\end{array}$$

EXERCISE 24–12 Subtract by using complements. Check the answers in base 10 subtraction.

1. $\begin{array}{r}1100\\101\\\hline\end{array}$ $\begin{array}{r}1100 = \underline{\hspace{2cm}}_{10}\\101 = \underline{\hspace{2cm}}_{10}\\\underline{\hspace{2cm}}_{10}\end{array}$

2. 11111   11111 = _____₁₀
    10101   10101 = _____₁₀
               = _____₁₀

3. 10111010   10111010 = _____₁₀
    10001101   10001101 = _____₁₀
               _____₁₀

4. 11100001   11100001 = _____₁₀
     1001111   1001111 = _____₁₀
               _____₁₀

5. 11110001   11110001 = _____₁₀
    10000111   10000111 = _____₁₀
               _____₁₀

6. 111001   111001 = _____₁₀
    100110   100110 = _____₁₀
               _____₁₀

7. 11001   11001 = _____₁₀
     1111   1111 = _____₁₀
               _____₁₀

8. 100011   100011 = _____₁₀
     11001   11001 = _____₁₀
               _____₁₀

9. 101111   101111 = _____₁₀
    100011   100011 = _____₁₀
               _____₁₀

10. 11000001   11000001 = _____₁₀
    10001110   10001110 = _____₁₀
               _____₁₀

EVALUATION EXERCISE Solve the exercises as indicated. In Exercises 1 through 6 change the decimal numbers to binary equivalents.

1. 37 _____   2. 132 _____

3. 56 _____   4. 178 _____

5. 1426 _____   6. 1224 _____

Change the decimal number to octal equivalents in Exercises 7 through 12.

7. 98 _____   8. 288 _____

9. 1224 _____   10. 876 _____

11. 2564 _____   12. 765 _____

Change the decimal numbers to hexadecimal equivalents in Exercises 13 through 18.

13. 455 _____   14. 8764 _____

15. 22,346 _____   16. 6784 _____

17. 1242 _____   18. 44,256 _____

Change the binary numbers to their decimal equivalents in Exercise 19 through 24.

19. 11101 _____   20. 101001 _____

21. 10000101 _____   22. 11110010011 _____

23. 10000110 _____   24. 10001101011 _____

Change the octal numbers to their decimal equivalents in Exercises 25 through 30.

25. 56 _____   26. 44 _____

27. 136 _____   28. 1264 _____

29. 662 _____   30. 1456 _____

Change the hexadecimal numbers to their decimal equivalents in Exercises 31 through 36.

31. A19 _____

32. 563 _____

33. 5B16 _____

34. AFBC _____

35. 16DE _____

36. AABC _____

Add the binary numbers and give the answers in hexadecimal readout (Exercises 37 through 42).

37. 101010001
    110110
    _____
    _____

38. 100010
    111101
    _____
    _____

39. 111101
    100100
    111110
    _____
    _____

40. 11110000010
    10001111011
    _____
    _____

41. 100111
       111
      1010
    _____
    _____

42. 1100011
      10011
        101
    _____
    _____

Chapter 24
Math for the Computer

Subtract the binary numbers by complementation; leave your answers in octal readout (Exercises 43 through 46).

43. 111100101
    10110010
    ―――――――   ―――――――

44. 1000011100011
    1101101101
    ―――――――――   ―――――――

45. 1000111011101
    101110111011
    ―――――――――   ―――――――

46. 111100101100110
    100101000110101
    ―――――――――――   ―――――――

Change the fractions to binary equivalents (Exercises 47 through 52). The answers should be in no more than four bits.

47. $\dfrac{3}{5}$ ―――――――   48. 0.15   ―――――――

49. 0.575   ―――――――   50. $\dfrac{5}{10}$ ―――――――

51. 0.165   ―――――――   52. 0.675   ―――――――

Change the fractions to octal equivalents (Exercises 53 through 58); leave your answers in three significant figures.

53. $\dfrac{4}{5}$ _____ 54. 0.25 _____

55. 0.125 _____ 56. 0.75 _____

57. 0.65 _____ 58. 0.225 _____

Change the binary or octal numbers in Exercises 59 through 67 to their decimal equivalents.

59. 0.111_2 _____ 60. 0.11011_2 _____

61. 0.1101_2 _____ 62. 0.10011_2 _____

63. 0.1101_2 _____ 64. 0.222_8 _____

65. 0.165_8 _____ 66. 0.675_8 _____

67. 0.175_8 _____

CHAPTER 25
Principles of Boolean Algebra

Objectives

After completing this chapter you will be able to:
- Identify the AND, OR and NOT functions and their symbols
- Prepare truth tables for AND, OR and NOT functions
- Prepare truth tables from logic statements
- Identify equivalent AND/OR gates
- Simplify logic expressions using Boolean algebra

In the mid-nineteenth century George Boole devised a mathematical system for analyzing symbolic logic. In this system all variables have the value "1" or "0," which makes the system ideal for designing a digital computer. This chapter introduces the principles of Boolean algebra; applications, however, will be studied more in depth in later studies of digital electronics.

25–1 LOGIC STATEMENTS. Symbolic logic can be defined as a type of formal logic that uses mathematical symbols to relate to science. The reasoning of simple logic statements is usually obvious. Refer to Figure 25–1 and reason the statement: "The meter will indicate a deflection when the switch is closed." The converse of this statement is: "The meter will not indicate a deflection when the switch is open." Anyone familiar with a simple electric circuit can understand the logic of those statements.

Boolean algebra deals with only two states, described as "1" or "0." Figures 25–2a and 2b illustrate the previous logic statements in the "1" and "0" states. The meter has been calibrated to indicate "0" or "1."

"The meter will indicate a deflection when the switch is closed." This statement can be reworded to read: The meter will indicate "1" when the switch is closed.

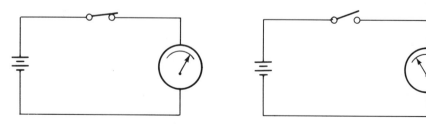

FIGURE 25–1   Simple electronic circuit

427

Chapter 25
Principles of
Boolean Algebra

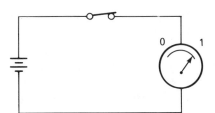

FIGURE 25–2a   The switch is closed

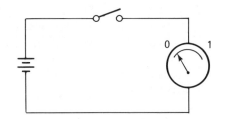

FIGURE 25–2b   The switch is open

"The meter does not indicate a deflection with the switch open" can be reworded thus: The meter indicates "0" when the switch is open.

In Figures 25–2a and 2b the "1" and "0" states are defined as an output to the meter and no output to the meter. These conditions are further defined as true or false, where "1" (Figure 25–2a) is the true state and 0 (Figure 25–2b) is the false state. The outputs can also be considered a high or a low output, where "1" is high and 0 is low. In either case, because there are only two states, both are valid Boolean relationships.

25–2 THE AND FUNCTION. The "and" function is illustrated in Figure 25–3a, where two possibilities must exist to produce a "1" condition. Both switch A and switch B must be closed before an output is indicated on the meter. The Boolean notation for the "and" function is indicated by the symbol AB. In conventional algebra, AB indicates the multiplication of A and B. In Boolean algebra this is read: "A and B" and is not the multiplication of A and B. If three switches are in series, A, B, and C, all three switches must be closed before a 1 output is indicated. This is indicated as ABC and is read "A and B and C." More than three switches in series are read "A and B and C and etc." Opening any switch changes the logic state to 0. The mathematical logic symbol for an "and" gate is shown in Figure 25–3b.

Figure 25–3c illustrates a truth table for an "and" gate that is algebraically noted as AB = X. There is an output of 1 at X only when both A and B have an input of 1. Figures 25–4a–c illustrate the equation ABC = X. There is an output at X only when there is an output at A and B and C.

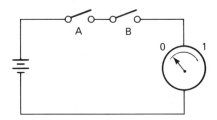

FIGURE 25–3a   Series switches

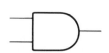

FIGURE 25–3b   Logic symbol

AB = X		
A	B	X
0	0	0
0	1	0
1	0	0
1	1	1

FIGURE 25–3c   Truth table

FIGURE 25–4a   Series switches

FIGURE 25–4b   Logic symbol

ABC = X			
A	B	C	X
0	0	0	0
0	0	1	0
0	1	0	0
0	1	1	0
1	0	0	0
1	0	1	0
1	1	0	0
1	1	1	1

FIGURE 25–4c   Truth table

25-3 THE "OR" FUNCTION. The "or" function can be symbolized by switches in parallel. Figure 25-5 illustrates the expression A + B = X. There is an output at X if there is an input at A or B. The plus sign (+) is the symbol for "or." The expression A + B = X is read: "A or B equals X."

The expression A + B + C = X is illustrated in Figure 25-6. It is read: "A or B or C equals X." There will be an output at X if there is an input at A, B, or C.

The "or" gate can be classified as either an inclusive "or" gate or an exclusive "or" gate. Figures 25-5 and 25-6 illustrate the inclusive "or"; Figure 25-7 illustrates the exclusive "or." The exclusive "or" can be symbolized by a two-position switch that can be in only one position at a time. There is an input of "1" at either A or B, but there is no input at both A and B at the same time. The point with the "0" input is designated by putting a bar over the symbol that has the "0" input. It can be written $\overline{A}B + A\overline{B}$ and is read: "A not and B or A and B not." The symbol denoting the exclusive "or" is $\oplus$ and is written $A \oplus B = X$. This can be summarized as $A \oplus B = \overline{A}B + A\overline{B}$. The logic symbol and truth table are shown in Figure 25-7.

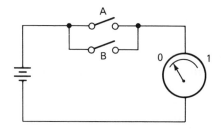

FIGURE 25-5a Parallel switches

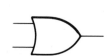

FIGURE 25-5b Logic symbol

A + B = X		
A	B	X
0	0	0
0	1	1
1	0	1
1	1	1

FIGURE 25-5c Truth table

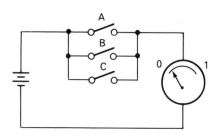

FIGURE 25-6a Parallel switches

FIGURE 25-6b Logic symbol

A + B + C = X			
A	B	C	X
0	0	0	0
0	0	1	1
0	1	0	1
0	1	1	1
1	0	0	1
1	0	1	1
1	1	0	1
1	1	1	1

FIGURE 25-6c Truth table

FIGURE 25-7a Logic symbol

A + B = X		
A	B	X
0	0	0
0	1	1
1	0	1
1	1	0

FIGURE 25-7b Truth table

25-4 THE "NOT" FUNCTION. The "not" circuit performs an inverting function and has a single input and output. The output is the complement of the input. If "1" is on the input, "0" will be on the output. If A were the input, the output would be $\overline{A}$ (not A indicated by the bar). An input of "0" places a

FIGURE 25–8a   Input = 1 and output = 0

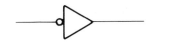
FIGURE 25–8b   Input = 0 and output = 1

Chapter 25
Principles of Boolean Algebra

FIGURE 25–9   Double inversion $\overline{\overline{A}} = A$

"1" on the output. A circle placed on either the input or the output indicates inversion at that point. Figure 25–8a shows the logic symbol for the "not" circuit with a circle on the output. The circle indicates that the output is "0" and the input is "1." Figure 25–8b shows the circle on the input; this indicates a "0" on the input and a "1" on the output.

Double inversion is done by placing two "not" circuits in series, as shown in Figure 25–9. Double inversion is indicated by a double bar.

Circles are used on the input and output of logic schematic symbols to indicate logic inversion. "Not and" or "nand" gates, as well as "not or" or "nor" gates are symbolized by circles on the output. Caution should be taken to ensure that Boolean expressions are pronounced correctly; there is a difference between the statement "A not or B not" and "not A or B."

$$A \text{ not or } B \text{ not is written } \overline{A} + \overline{B}$$
$$\text{not } A + B \text{ is written } \overline{A + B}$$

Example A

Determine the state of "X" as either "1" or "0" for the given inputs (Figure 25–10a–c).

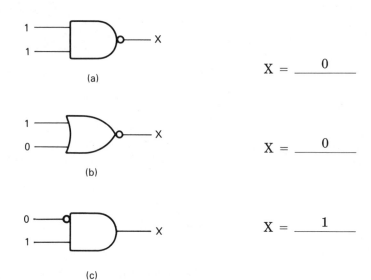

FIGURE 25–10 (a–c)   Gates for Example A

Figure 25–11 summarizes the schematic symbols and their Boolean expressions used for logic.

AB (and gate)   read "A and B"

$\overline{A}B$ read "A not and B"

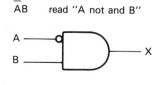

A + B (inclusive or)   read "A or B"

$\overline{A}\,\overline{B}$ read "A not and B not"

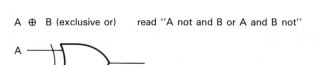

A ⊕ B (exclusive or)   read "A not and B or A and B not"

$\overline{A} + B$ read "A not or B"

A (not circuit or inverter)   read "A inverted"

$A + \overline{B}$ read "A or B not"

$\overline{AB}$ (not and – nand gate)   read "not A and B"

$\overline{A} + \overline{B}$ read "not A or B not"

$\overline{A + B}$ (not or – nor gate)   read "not A or B"

$A\overline{B}$ read "A and B not"

$\overline{A \oplus B}$ (exclusive nor)   read "A not and B or A and B not"

FIGURE 25–11   Schematic symbols used in logic circuits

EXERCISE 25-1 Determine the state of X as either 1 or 0 for the given input conditions (Figure 25-12a-j).

1.

   FIGURE 25-12a

   X = _____

2.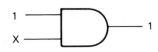

   FIGURE 25-12b

   X = _____

3.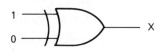

   FIGURE 25-12c

   X = _____

4.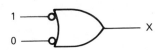

   FIGURE 25-12d

   X = _____

5. 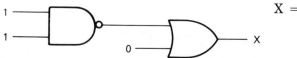 X = _____

FIGURE 25–12e

6. 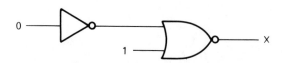 X = _____

FIGURE 25–12f

7. 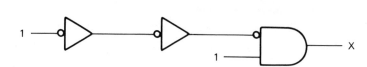 X = _____

FIGURE 25–12g

8. 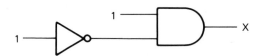 X = _____

FIGURE 25–12h

9. X = _____

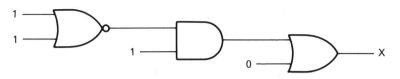

FIGURE 25–12i

10. X = _____

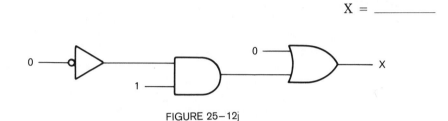

FIGURE 25–12j

25–5 TRUTH TABLES. Truth tables can be helpful in analyzing the output state of logic circuits. There are other methods, such as Karnaugh mapping, that may prove easier for complex logic design; but those methods are not considered here. The preparation of truth tables is limited to a few basic examples.

Truth tables for an "and" gate are illustrated in Figures 25–3 and 25–4; "or" gate truth tables are illustrated in Figures 25–4 and 25–5. The 1 and 0 inputs are shown, and the output is given for all possible input conditions of 1 and 0. Figure 25–13 illustrates the combination of an "and" circuit and an "or" circuit.

To make a simple truth table:

- Determine and list horizontally all possible inputs. Here, A, B, and C are the input possibilities.
- Determine the number of vertical rows by raising 2^x, where x is the number of input possibilities.

$$2^3 = 8$$

- List the input possibilities in binary order by counting from 000 to 111 (0 to 7).

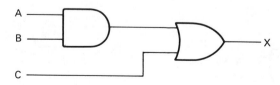

FIGURE 25–13   Circuit for $AB + C = X$

input possibilities	A	B	C	AB + C = X
1	0	0	0	0
2	0	0	1	1
3	0	1	0	0
4	0	1	1	1
5	1	0	0	0
6	1	0	1	1
7	1	1	0	1
8	1	1	1	1

Example B
Prepare a truth table for the function "A and B not = X."

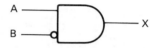

FIGURE 25–14   Schematic for $A\overline{B} = X$

The two input possibilities are A and B, and the four input possibilities are $(2^2 = 4)$.

A	B	$X = A\overline{B}$
0	0	0
0	1	0
1	0	1
1	1	1

Example C
Prepare a truth table for the function "A or B not = X."

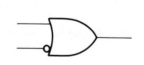

FIGURE 25–15 $A + \overline{B} = X$

A	B	$X = A + \overline{B}$
0	0	1
0	1	0
1	0	1
1	1	1

Example D
Prepare a truth table for the function "not A or B = X" ("*nor*" gate).

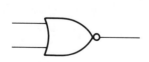

FIGURE 25–16 $\overline{A + B} = X$

A	B	$X = \overline{A + B}$
0	0	1
0	1	0
1	0	0
1	1	0

Example E
Prepare a truth table for "A not and B or A and B not."

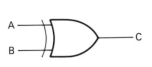

FIGURE 25–17 $A \oplus B = X$

A	B	$X = A \oplus B$
0	0	0
0	1	1
1	0	1
1	1	0

EXERCISE 25–2   Prepare truth tables for the logic circuits.

1. A not and B = X

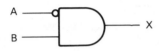

FIGURE 25–18a

2. A not or B = X

FIGURE 25–18b

3. A not and B not = X

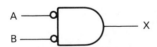

FIGURE 25–18c

4. A and B or A and C = X

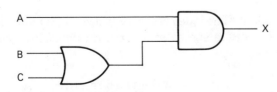

FIGURE 25–18d

5. ABC = X

FIGURE 25-18e

6. A and B or C and D

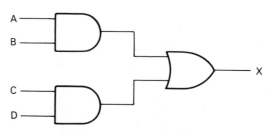

FIGURE 25-18f

25-6 LOGIC STATEMENTS. Logic statements exist in one of two states, "true" (represented by logic "1") or "false" (represented by logic "0"). These statements lend themselves to Boolean applications. The statement "Lynn owns a video recorder, and it is a VHS recorder" can be considered an "and" function. It is a compound sentence connected by the conjunction *and*. The first part of the statement is A and the second part B.

$$\text{Lynn owns a video recorder} = A$$
$$\text{It is a VHS recorder} = B$$

When the statements are combined—A and B—both parts must be true, or the statement is false. A truth table helps verify the truth statement.

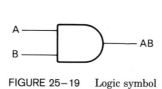

FIGURE 25-19   Logic symbol

A	B	AB
F	F	F
F	T	F
T	F	F
T	T	T

Statements connected by the conjunction "or" can be of two types: inclusive (either one or both can be true); or exclusive (either one but not both can be true).

The statement "The video recorder is owned by Lynn or Dan" is an inclusive or (A + B), and is shown in the truth table. The recorder is owned by either or both.

$$\text{Lynn owns the video recorder} = A$$
$$\text{Dan owns the video recorder} = B$$

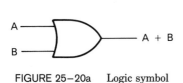

FIGURE 25–20a   Logic symbol

A	B	A + B
F	F	F
F	T	T
T	F	T
T	T	T

437

Chapter 25
Principles of
Boolean Algebra

The statement "Either Lynn or Dan owns the video recorder" is an example of the exclusive or ($A \oplus B = \overline{A}B$ or $A\overline{B}$) statement, A not and B or A and B not.

$$\text{Lynn owns the recorder} = A$$

or

$$\text{Dan owns the recorder} = B$$

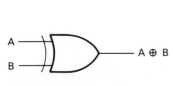

FIGURE 25–20b   Logic symbol

A	B	A ⊕ B
F	F	F
F	T	T
T	F	T
T	T	F

The statement "The video recorder can play VHS tapes but not Betamax tapes" is an example of a "not" function: A and B not.

$$\text{The recorder can play VHS tapes} = A$$
$$\text{The recorder cannot play Betamax tapes} = B$$

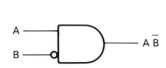

FIGURE 25–21   Logic symbol

A	B	A$\overline{B}$
F	F	F
F	T	F
T	F	T
T	T	F

EXERCISE 25–3   Prepare truth tables and draw the logic symbol for each statement. Use gates with only two input variables.

1. Keith owns a backhoe and a dump truck.

2. Scott operates the backhoe or he operates the dump truck, but not both.

3. Gerry drives the dump truck or Marte drives the dump truck.

4. It will rain and the wind will blow, or the sun will shine.

5. Either Marc will win the race or Gerry will win the race.

6. The pilot, copilot, and engineer are all needed to fly the airplane.

7. Orville, Jesse, Kelly, or Jill will ride the bus to school.

25–7 EQUIVALENT FUNCTIONS. Logic functions that have the same truth table are known as "equivalent functions." The "and" and "or" gates shown in Figure 25–22 have the same truth table; they are equivalent logic functions.

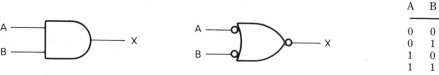

A	B	X
0	0	0
0	1	0
1	0	0
1	1	1

FIGURE 25–22   Equivalent "and" "or" functions

Table 25-1 illustrates equivalent "and" and "or" functions with two variables.

TABLE 25-1

"and"	"or"	Two variables
A AND B → X	A NOR B (inputs inverted) → X	A B X 0 0 0 0 1 0 1 0 0 1 1 1
Ā AND B → X	A OR B̄ (B inverted) → X	A B X 0 0 0 0 1 1 1 0 0 1 1 0
A AND B̄ → X	Ā OR B → X	A B X 0 0 0 0 1 0 1 0 1 1 1 0
Ā AND B̄ → X	A OR B → X (inverted output)	A B X 0 0 1 0 1 0 1 0 0 1 1 0
A NAND B → X	A OR B → X	A B X 0 0 0 0 1 1 1 0 1 1 1 1
A AND B̄ (NAND variant) → X	Ā OR B → X	A B X 0 0 1 0 1 1 1 0 0 1 1 1
Ā AND B → X	A OR B̄ → X	A B X 0 0 1 0 1 0 1 0 1 1 1 1
A AND B → X (inverted)	Ā OR B̄ → X	A B X 0 0 1 0 1 1 1 0 1 1 1 0

25-8 BOOLEAN OPERATIONS. In Boolean algebra no coefficient is greater than 1. Multiplication is used to represent the AND function; a plus sign (+) is used to represent the OR function.

$$AB \text{ is A and B} \qquad A + B \text{ is A or B}$$

Axioms. Certain relations, (or truths or axioms) in Boolean algebra are obvious and require no explanation. These axioms are illustrated in Figures 25-23 through 25-28, with their logic symbol and the binary input and outputs.

"and" function

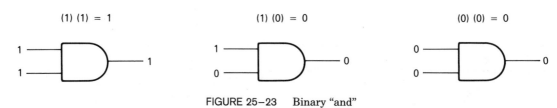

FIGURE 25-23 Binary "and"

"or" function

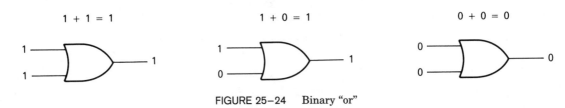

FIGURE 25-24 Binary "or"

"inverted" function

FIGURE 25-25 Binary "inverted"

Binary and Boolean relations can be shown with logic symbols, but they must be read using "and" for multiplication and "or" for the + symbol.

"and" function

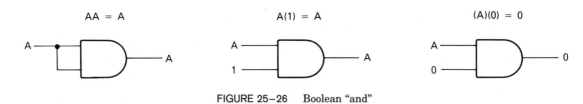

FIGURE 25-26 Boolean "and"

"or" function

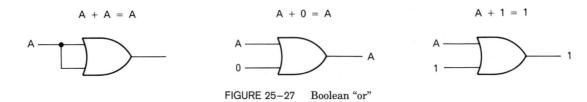

FIGURE 25-27 Boolean "or"

"not" function

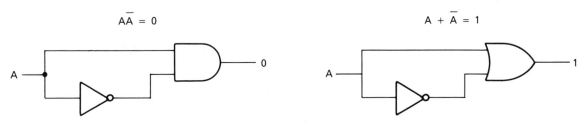

FIGURE 25-28   Boolean "inverted"

Commutative, Associative, and Distributive Laws.   Three basic laws of algebra—commutative, associative, and distributive—apply to Boolean relations as well as to conventional algebra. The Boolean relationships are:

1. *Commutative:* Output X is the same regardless of which input is called A and which is called B.

FIGURE 25-29a   Commutative "and"   FIGURE 25-29b   Commutative "or"

2. *Associative:* As in conventional algebra, the signs of grouping can be removed and the equality will be valid.

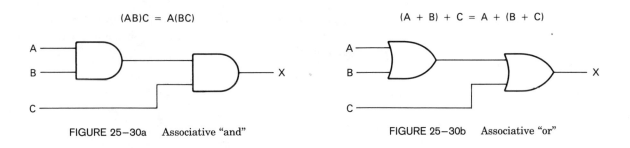

FIGURE 25-30a   Associative "and"   FIGURE 25-30b   Associative "or"

3. *Distributive:* By conventional factoring, A can be factored out of the left member, with the result, $A(B + C)$.

$$AB + AC = A(B + C)$$

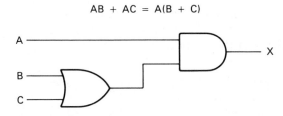

FIGURE 25-31a   Distributive "and – or" combination

441

$$(A + B)(A + C) = A + BC$$

Expand the left member by multiplication. In Boolean, $(A)(A) = A$.

$$A + AB + AC + BC = A + BC$$

Factor out A:

$$A(1 + B + C) + BC = A + BC$$

In Boolean, $(1 + B + C)$ is an "or" function and is equal to 1. Substituting 1 for $(1 + B + C)$ gives $A + BC = A + BC$.

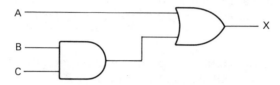

FIGURE 25–31b   Distributive "and or" combination

In binary, the complement of 1 is 0, and the complement of 0 is 1. In Boolean expressions the complement is the inverted function. In Boolean, $\overline{A}$ is the complement of A, and A is the complement of $\overline{A}$.

DeMorgan's Theorem: Demorgan's Theorem states two rules which allow an *"and"* gate to be converted to an *"or"* gate and an "or" gate to an "and" gate. Table 25–1 illustrates equivalent logic gates.

Rule 1: $AB = \overline{\overline{A} + \overline{B}}$

> The complement of an "and" gate can be obtained by changing the "and" to an "or" gate and taking the complement of the input variables and the output.

FIGURE 25–32   An "and" gate to an "or" gate

Rule 2: $A + B = \overline{\overline{A}\overline{B}}$

> The complement of an "or" gate can be obtained by changing the "or" to an "and" and taking the complements of the input variables and the output.

FIGURE 25–33   An "or" gate to an "and" gate

Double Negatives. The "not function, or complement, is designated by placing a bar, or "vinculum," over the expression, as in $\overline{A + B}$. **DeMorgan's Theorem** states that this "or" gate can be changed to an "and" gate by splitting the vinculum, with the result: $\overline{A}\overline{B}$. Vinculums can be added to expressions to help simplify them. In the expression $A\overline{(B + C)}$, variable A is not under the vinculum. Two vinculums over an A do not change the meaning of the expression ($\overline{\overline{A}} = A$). With two vinculums over A, the expression can be written:

$$\overline{\overline{A}}\overline{(B + C)}$$

Using DeMorgan's Theorem and the vinculum, the entire expression can be grouped.

$$\overline{\overline{A} + (B + C)} \quad \text{or} \quad \overline{\overline{A} + B + C}$$

The expression $\overline{A(B + \overline{CD})}$ can be simplified by splitting the vinculum (DeMorgan's Theorem) and removing the double negatives:

$$\overline{A(B + \overline{CD})} = \overline{\overline{A}} + \overline{(B\overline{CD})}$$

The double negatives are removed, and the expression is simplified:

$$A + \overline{B}CD$$

Double negatives, or vinculums, can be removed or added to expressions, thus:

$$\overline{\overline{A}} = A$$
$$\overline{\overline{A + B}} = A + B$$
$$\overline{\overline{AB}} = AB$$
$$\overline{\overline{A} + \overline{BC}} = A + BC$$

Table 25–2 summarizes the important relationships needed to use Boolean algebra.

443

Chapter 25
Principles of
Boolean Algebra

TABLE 25–2

Binary
(1)(1) = 1
(1)(0) = 0
(0)(0) = 0
1 + 1 = 1
1 + 0 = 1
0 + 0 = 0
$\overline{1} = 0$
$\overline{0} = 1$

Boolean
AA = A $\overline{\overline{A}} = A$
A(1) = A $\overline{\overline{A + B}} = A + B$
A(0) = 0 $A + \overline{A} = 1$
A + A = A   AB + 1 = 1
A + 0 = A   A + B + etc. = 1
A + 1 = 1
$A\overline{A} = 0$
A + 1 = 1

Commutative
AB = BA
A + B = B + A

Associative
(AB)C = A(BC)
(A + B) + C = A + (B + C)

Distributive
AB + AC = A(B + C)
(A + B)(A + C) = A + BC

DeMorgan's Theorem
$\overline{AB} = \overline{A} + \overline{B}$ $A + \overline{A}B = A + B$
$\overline{A + B} = \overline{A}\overline{B}$

Example F

Draw the logic circuit for the Boolean expression $\overline{\overline{A} + B} = X$. The plus sign indicates that the logic is an "or" function; $A + B$ indicates two input variables. The bar over $\overline{A}$ indicates that the A input is inverted. The long bar over $\overline{A} + B$ indicates that the output is inverted (Figure 25–34).

FIGURE 25–34   Logic circuit for $\overline{\overline{A} + B}$

Draw the logic circuit for the Boolean expression $\overline{\overline{A} + BC} = X$. It is a combination "and" and "or" gate with three variable inputs. Input A is inverted, and the output is inverted (Figure 25–35).

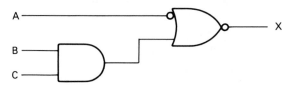

FIGURE 25–35   Logic circuit for $A + BC$

EXERCISE 25–4   Draw the logic circuits for each Boolean expression.

1. $A(B + C)$

**Chapter 25
Principles of
Boolean Algebra**

2. $\overline{A + \overline{B}}$

3. $\overline{AB}$

4. $A + \overline{B}\overline{C}$

5. $\overline{A} + \overline{B}$

6. $\overline{\overline{A} + \overline{B}}$

7. $\overline{AB + \overline{C}}$

8. $\overline{A + \overline{B} + C}$

25-9 SIMPLIFYING LOGIC EXPRESSIONS.
Symbolic logic can be simplified by using the Boolean relationships given in Table 25–2. Simplification of the logic reduces the number of gates that are required. A Boolean expression such as A + AB can be simplified to A. Figures 25–36a–e illustrate reduction of the number of logic gates.

$$A + AB \qquad \text{(factor out A)}$$
$$A(1 + B) \qquad \text{(From Table 25–2, } 1 + B = 1)$$
$$A(1) = A$$

and

$$A + AB = A$$

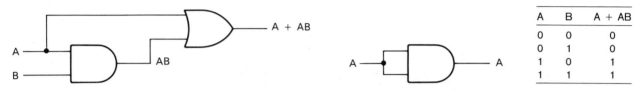

FIGURE 25–36a   Circuit for $A + AB$ simplified to A

$$A(A + B) \qquad \text{(expand by removing the parentheses)}$$
$$AA + AB \qquad \text{(from Table 25–3: } AA = A)$$
$$A + AB \qquad \text{(and } A + AB = A)$$

and

$$A(A + B) = A$$

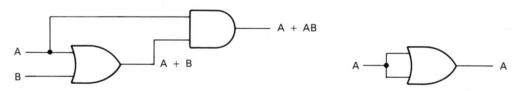

FIGURE 25–36b   Circuit for $A(A + B)$ simplified to A

$$A(\overline{A} + B) \qquad \text{(expand)}$$
$$A\overline{A} + AB \qquad \text{(from Table 25–2: } A\overline{A} = 0).$$
$$0 + AB$$

and

$$A(\overline{A} + B) = AB$$

FIGURE 25–36c   Circuit for $A(\overline{A} + B)$ simplified to AB

$C(AB + A) + B(A + \overline{B})$	(expand)
$CAB + CA + AB + B\overline{B}$	($B\overline{B} = 0$)
$AC(B + 1) + AB$	(factor out AC and $B + 1 = 1$)
$AC + AB$	(factor out A)
$A(B + C)$	

and
$$C(AB + A) + B(A + \overline{B}) = A(B + C)$$

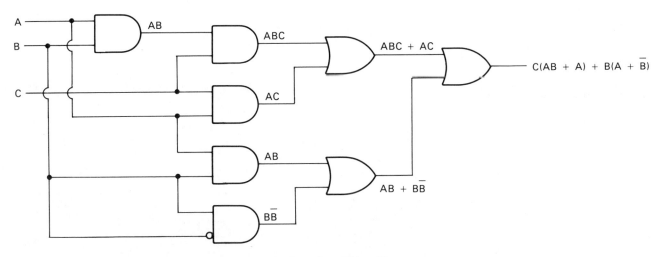

FIGURE 25–36d   Circuit for $C(AB + A) + B(A + B)$

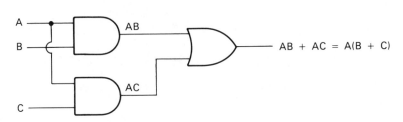

FIGURE 25–36e   Simplified to $A(B + C)$

A	B	C	A(B + C)
0	0	0	0
0	0	1	0
0	1	0	0
0	1	1	0
1	0	0	0
1	0	1	1
1	1	0	1
1	1	1	1

EXERCISE 25–5 Simplify the Boolean expressions in Exercises 1 through 10 (draw only the simplified circuit).

1. $A + (B + 0) + 0$

2. $EF\overline{E} + CC$ _____

3. $A + \overline{A} + BC\overline{C}$ _____

4. $XY + (Z + \overline{Z})$ _____

5. $(\overline{\overline{D}} + C\overline{C})E$ _____

6. $(G\overline{G} + \overline{H}H)F$ _____

7. $A + BC + D\overline{D} + C$ _____

8. $AB\overline{A}(C + D + \overline{E})$

9. $DE\overline{F} + DEG + DE + HDE$

10. $A + B\overline{A}$

Simplify the expressions in Exercises 11 through 22, but do not draw the simplified circuit.

11. $A(\overline{A} + B)$

12. $(A + B)(A + C)$

13. $DEF(D + \overline{H})(F + \overline{H})$

14. $(EF)G + (E + F)G$

15. $\overline{A}B(A + B)$

16. $\overline{(\overline{A + B})}(A + \overline{C})$

17. $XY + \overline{X}Y$

18. $AB + \overline{A}C + ABC + \overline{A}BC$

19. $(\overline{A} + \overline{B})(\overline{A} + C)$

20. $\overline{C}(C + D) + ED + E$

21. $\overline{(A + \overline{A})(\overline{BC} + \overline{C} + \overline{B})}$

22. $\overline{(\overline{EF} + G)(\overline{E} + \overline{F})\overline{G}}$

EVALUATION EXERCISE Determine the state of X as either 1 or 0 (Exercises 1 through 5).

*Chapter 25
Principles of
Boolean Algebra*

1. _____

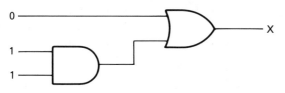

FIGURE 25–37a

2. _____

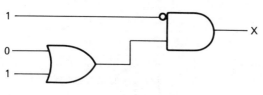

FIGURE 25–37b

3. _____

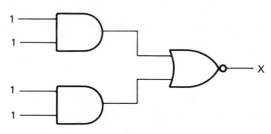

FIGURE 25–37c

4.

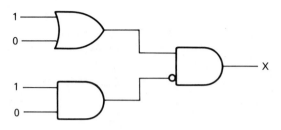

FIGURE 25–37d

5.

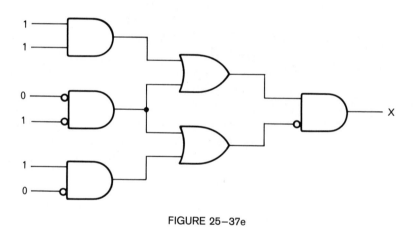

FIGURE 25–37e

Prepare a truth table for each statement in Exercises 6 through 8.

6. The swimmer will win or lose the race.

7. The shotput is round and heavy.

8. The airplane will fly on Saturday, and it will not fly on Sunday.

Prepare a truth table for the logic circuits in Exercises 9 and 10.

9.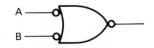

FIGURE 25-38

10.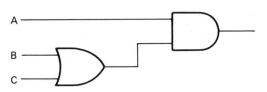

FIGURE 25-39

Draw the logic circuit for each Boolean expression in Exercises 11 through 14. Each gate may have no more than two input variables.

11. $\overline{A}\overline{B}$

12. $A(\overline{B} + \overline{C})$

13. $\overline{A} + BC$

14. $\overline{A} + \overline{B} + \overline{C}$

Simplify the Boolean expressions in Exercises 15 through 20.

15. $C + DE\overline{D} + A$ _____

16. $ABC + AB\overline{C}$ _____

17. $AB(B + D)$ _____

18. EF + F(E + G)

19. $\overline{AB} + \overline{A + B}$

20. AC + BC + AD + BD

455

Chapter 25
Principles of
Boolean Algebra

CHAPTER 26

Logarithms

Objectives

After completing this chapter you will be able to:
- identify the operation of a logarithm
- identify the log of a number
- find the log of a number using the scientific notation method
- find the antilog using the scientific notation method
- determine the natural log of a number
- solve logarithm equations

Logarithms, or "logs," were long a widely used tool in mathematics to perform the operations of multiplying and dividing. With the advent of the electronic calculator, however, logs have not been used as extensively as in the past; in some areas of math their study has been eliminated. In electronics, logs continue to play a vital role and thus cannot be eliminated.

In an R-C circuit the current decays at a certain rate and the voltage across the capacitor rises at a certain rate, as illustrated in Figure 26–1. The shape of this curve is a natural function of the circuit; it is an exponential function. The time it takes this curve to develop can be controlled by the value of R and C, but the shape of the curve cannot be changed. Logarithmic curves are an essential part of the study of electronics. In this chapter we will study the base 10 log and the natural log (ln).

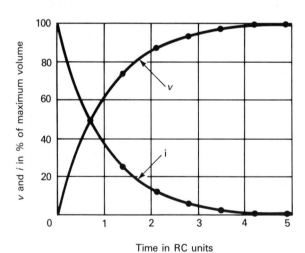

FIGURE 26–1 Time in RC units

26–1 THE OPERATION OF A LOG. In Chapter 5, power-of-10 units were introduced and their use was explained. Adding exponents is an example of the use of logs in multiplication.

Multiply 100(1000), using exponents:

$$100 \times 1000 = 100{,}000$$
$$10^2 \times 10^3 = 10^5$$

The exponent 2 is added to the exponent 3 to give the result, 10^5. The exponents of 10 are the logarithms of the numbers 100 and 1000.

> the log of 100, in the base 10, is 2
> the log of 1000, in the base 10, is 3
> the log of 100,000, in the base 10, is 5

The log, in base 10, of numbers that are even multiples of 10 are easily determined. The log of a number such as 263 can be determined by the use of a calculator or log tables. Example B illustrates how 263 fits into the system of logs.

Place the number 263 in scientific notation:

$$2.63 \times 10^2$$

the exponent 2 indicates that the number 263 is a number between 100 and 1000. This exponent is referred to as the "characteristic."

The curve in Figure 26-1 is not a linear curve and the distance that 263 lies between 100 and 1000 is not linear. Log tables or a calculator can be used to determine this distance. Enter the number 2.63 in the calculator and depress the "base 10 log" key. The number displayed on the calculator should be 0.4200. This number is referred to as the "mantissa"; it indicates how far the number 263 is between 100 and 1000. Therefore, the log, in base 10, of 263 is 2.4200.

Of course, the log of 263 can be determined on the calculator by entering 263 and depressing the "log" key (refer to the logarithmic instructions supplied with the calculator). Since the calculator can determine the log of a number, it might seem unnecessary to study this method of determining logs. Knowing the characteristic and mantissa of a log will help in studying the application of logs, especially concerning decibels.

26-2 THE LOG OF A NUMBER. The logarithm of a number is its exponent in a given base. The log of 100 in base 10 is 2.

$$10^2 = 100$$

The log of 263 in base 10 is 2.4200.

$$10^{2.4200} = 263$$

This operation should be tried on the calculator, using either the "x^y" key, the "y^x," or the "10^x" key. Consult the manual to the calculator to determine the calculator's programming. First, 10 raised to the second power equals 100, and 10 raised to the power 2.4200 equals 263.

EXERCISE 26-1 The logs given in this exercise are all base 10 numbers. Prove that the logs are correct by raising 10 to the given log, or power. Use the "x^y," "y^x," or "10^x" function on the calculator. If a statement is not true, enter NT in the answer column along with the correct answer.

1. The log of 100,000 is 5.
2. The log of 500 is 2.699.
3. The log of 285 is 4.455.
4. The log of 1285 is 3.109.
5. The log of 34.6 is 1.539.
6. The log of 0.125 is 0.9030.
7. The log of 0.000012 is -4.921.
8. The log of 10,000 is 4.
9. The log of 27.5 is 3.440.
10. The log of 0.000479 is -3.32.

26–3 FINDING THE LOGARITHM OF A NUMBER. The quickest way to find the log of a number is to enter the number on a calculator and depress the "log" key. The operation of calculators differs, so its instruction manual should be referred to for this operation (using the scientific notation method is another way). Understanding the operation of finding the log of a number will help you solve application problems.

The principles involved in the scientific notation method are discussed in Section 26–2. Example C illustrates this method.

Example A
Find the log of 58700.

5.87×10^4 (place 58700 in scientific notation)

4   (the exponent 4 becomes the characteristic)

5.87   (enter 5.87 in the calculator and determine the mantissa)

The log of 58700 is 4.769, where 4 is the characteristic and 0.769 is the mantissa.

The logarithm of a number less than 1 can be determined in one of two ways. The easier way is to enter the number in the calculator and depress the "log" key. Both the characteristic and the mantissa are negative. When using the scientific notation method the characteristic is negative and the mantissa is positive.

Example B
Determine the log of 0.0025 directly, using the calculator.

$$\log \text{ of } 0.0025 = -2.602$$

Example C
Determine the log of 0.0025, using the scientific notation method.

2.5×10^{-3}

-3 (the characteristic)

0.3979 (the mantissa of 2.5 taken from the calculator; but it is a positive value)

To avoid confusion between the two methods, the negative sign is placed above the characteristic, $\bar{3}$, indicating that only the characteristic is negative. Then the log of 0.0025 can be expressed -2.6020, taken directly from the calculator; or as $\bar{3}.3979$, using the scientific notation method.

Be cautious when finding the log of numbers below 1. The relation between these numbers can be seen by collecting the negative characteristic and the positive mantissa.

$$-3 + 0.3979 = -2.6020$$

Either system may be encountered when using text material. The calculator, however, has made more desirable the system using the negative characteristic and the negative mantissa.

NOTE: Logs of negative numbers are not defined in the real number system.

EXERCISE 26–2 Perform the following exercise by first placing each number in scientific notation and then determining the logarithm using the method given in Example C. Check each answer by taking the log directly from the calculator. Give the mantissa in four significant figures.

	Scientific notation	Characteristic	Mantissa	Log
1. 1246				
2. 8543				
3. 1.78				
4. 0.0162				
5. 2.25×10^4				
6. 477×10^7				
7. 0.166				
8. 15,700				

9. 128×10^3 _____ _____ _____ _____

10. 90×10^4 _____ _____ _____ _____

11. 0.9×10^7 _____ _____ _____ _____

12. 0.0066 _____ _____ _____ _____

26–4 ANTILOGARITHMS. Finding the antilog is also referred to as finding the inverse log. Some calculators use an inverse function; others use the "10^x" key.

Returning a logarithm to scientific notation is the reverse of the procedure discussed in Section 26–4 and is performed in Section 26–5.

Find the antilog of 3.659 (positive characteristic and mantissa).

10^3 (the characteristic 3 becomes the exponent of 10)

4.56 (the significant figures 4.56 are found by taking the inverse log of 0.659)

The antilog of 3.659 is 4.56×10^3.

Find the antilog of $\overline{3}.3979$ (negative characteristic and positive mantissa).

10^{-3} (the characteristic -3 becomes the exponent of 10)

2.50 (the significant figures 2.50 are found by taking the inverse log of 0.3979)

The antilog of $\overline{3}.3979$ is 2.5×10^{-3}.

Find the antilog of -2.602 (negative characteristic and negative mantissa). Enter -2.602 in the calculator and find the inverse log, using the "inv" and "log" keys or the 10^x function. The antilog of -2.602 is 0.0025.

The antilog of quantities having both the characteristic and the mantissa positive or negative can be taken directly from the calculator by using the "inv" and "log" keys or the 10^x function. When the characteristic is negative and the mantissa positive, the antilog can be determined using the scientific notation method or collecting the negative characteristic and positive mantissa and using the inverse function on the calculator.

Find the antilog of $\overline{3}.3937$.

$$-3 + 0.3937 = -2.6020$$

The inverse log of $-2.6020 = 0.0025$.

EXERCISE 26–3 Determine the antilogs, using the scientific notation method. Check each answer with a calculator, following the procedure described in the manual for your calculator.

		Antilog by SN *method*	*Antilog by* *calculation method*
1.	1.125	_____	_____
2.	6.258	_____	_____
3.	3̄.44	_____	_____
4.	3.000	_____	_____
5.	3̄.000	_____	_____
6.	12.65	_____	_____
7.	8.888	_____	_____
8.	1.294	_____	_____
9.	0.477	_____	_____
10.	5.00	_____	_____
11.	8̄.167	_____	_____
12.	12.66	_____	_____

26–5 NATURAL LOGS. Two widely used systems of logarithms are the base 10, or common log, and the natural log. The base of the natural log system is 2.7182818, etc. John Napier, sometimes called the inventor of logarithms, published natural log tables in 1614. A few years later Henry Briggs converted the tables to the base 10, or common system of logs.

Epsilon, or e, is used to represent the base of natural logs. Because the number 2.7182818 is an irrational number, a symbol is used to represent it, as in a circle pi is used to represent 3.14. When written log e or ln, the log is a natural one. When written log without a subscript, it is a common, or base 10, log. All other bases contain the base in the subscript: $\log_2$ is base 2, $\log_3$ is base 3, and so on.

A common log can be converted to a natural log by multiplying by 2.30259, or a natural log can be converted to a common log by dividing by 2.30259. Most scientific calculators provide for natural logs and common logs. The key generally used to find natural logs is "ln." If neither function is provided, conversions must be used.

Part 3
Computer Mathematics,
Logarithms, and
Quadratic Equations

EXERCISE 26–4 Find the common and natural logs of the quantities given in Exercises 1 through 6. Give the answers in four significant figures.

		Common log	ln
1.	6285	_____	_____
2.	100	_____	_____
3.	4.5×10^4	_____	_____
4.	2.7182818	_____	_____
5.	1614	_____	_____
6.	0.00628	_____	_____

Find the antilog of the natural logarithms in Exercises 7 through 12.

7. 1.837 _____
8. 50 _____
9. 1.00 _____
10. −2.934 _____
11. −1.000 _____
12. 0.00457 _____

26–6 LOGARITHMS AND EQUATIONS. In algebra, a logarithm is known as an "exponent." The expression $2^3 = 8$ gives the relationship of the number 8, the base 2, and the exponent 3. This expression can be written in exponential and logarithmic forms.

$$2^3 = 8 \quad \text{(exponential form)}$$

$$\log_2 8 = 3 \quad \text{(logarithmic form)}$$

The logarithmic form shows that 3 is the log of 8 in the base 2. A general equation can be derived from this relationship:

$$N = b^x$$

where N is the number, b is the base, and x, the exponent, is the log.

It is necessary to be able to convert from exponential to logarithmic form, and that is best described by following the relationship described above.

Exponential form	Log form
$4^2 = 16$	$\log_4 16 = 2$
$X^4 = V$	$\log_x V = 4$
$10^3 = 1000$	$\log_{10} 1000 = 3$
$e^{2.718} = 15.15$	$\log_e 15.15 = 2.718$

EXERCISE 26-5 Convert the exponential form to logarithmic form, or the logarithmic form to exponential form.

1. $2.5 = 32$ _____

2. $10^{1/2} = 3.16$ _____

3. $X^Y = R$ _____

4. $10^{2.178} = 151$ _____

5. $e^x = y$ _____

6. $e^{1/2} = 1.65$ _____

7. $\log_3 27 = 3$ _____

8. $\log_{10} 1000 = 3$ _____

9. $\log_m Y = p$ _____

10. $\log_{10} 3 = 0.4771$ _____

11. $\log_e 8 = 2.079$ _____

12. $\log_6 36 = 2$ _____

26-7 EXPRESSING LOGARITHMS IN DIFFERENT FORMS. Quantities can be multiplied and divided by adding or subtracting their logarithms. A multiplication can be expressed as an addition and a division as a subtraction.

Example D

$\log ab$ can be expressed: $\log a + \log b$

$\log X^4$ can be expressed: $4 \log X$

$\log I^2 R$ can be expressed: $2 \log I + \log R$

$\log \dfrac{a}{b}$ can be expressed: $\log a - \log b$

By reversing this procedure,

$$\log P + \log Q = \log PQ$$

$$3 \log A = \log A^3$$

$$2 \log b + \log p = \log b^2 p$$

$$\log R - \log K = \log \dfrac{R}{K}$$

EXERCISE 26-6 Express the logarithm form as a sum, difference, or product in Exercises 1 through 8.

1. $\log 5X$ _____

2. $\log X^4$ _____

3. $\log IR$ _____

4. $\log X^2 Y^3$ _____

5. $\log \dfrac{V}{R}$ _____

6. $\log \dfrac{V_1^2}{V_2^2}$ _____

7. $\log (ab)^{1/2}$ _____

8. $\log \dfrac{125}{a}$ _____

Reverse the procedure in Exercises 9 through 15 and express each log form as a single log expression.

9. $\log p + \log v$ _____

10. $5 \log V$ _____

11. $\log 2.71 + \log a$ _____

12. $2 \log I - \log 3$ _____

13. $\log P_2 - \log P_1$ _____

14. $2 \log V - \log R$ _____

15. $3 \log X + 2 \log Y - \log 6$ _____

26–8 LOGARITHMIC EQUATIONS. Logarithms are a natural part of some equations and are solved like any algebraic equation. Follow the basic rule when solving equations: what is done to one side of an equation must be done to the other.

Example E

Solve the equation $\log x + 2 \log x = 8$.

$3 \log x = 8$ (collect x)

$\log x = \dfrac{8}{3}$ (divide by 3)

$\log x = 2.67$

Take the antilog of both sides of the equation. The antilog of $\log x$ is x, and the antilog of 2.67 is 464. Therefore, $x = 464$.

Example F

$\log 3X = 1.2$

$\log 3 + \log X = 1.2$ (rewrite $\log 3X$)

$\log X = 1.2 - \log 3$ (transpose $\log 3$)

$\log X = 1.2 - 0.477$ (find the log of 3)

$\log X = 0.723$ (find the antilog of both sides of the equation)

$X = 5.28$

Example G

$$\log \frac{P}{2.5} = 2 \quad \text{(take the antilog of both sides of the equation)}$$

$$\frac{P}{2.5} = 100 \quad \text{(solve for } P\text{)}$$

$$P = 250$$

Example H

$$36 = 4^{2x}$$

$$\log 36 = 2X \log 4 \quad \text{(take the log of both sides)}$$

$$\log \frac{36}{2 \log 4} = X \quad \text{(divide by 2 log 4)}$$

$$\frac{1.56}{1.2} = X \quad \text{(find log of 36 and log 4)}$$

$$X = 1.3$$

Example I

$$\log 5X + 3 \log X = 16$$

$$\log 5 + \log X + 3 \log X = 16 \quad \text{(rewrite log 5}X\text{)}$$

$$\log 5 + 4 \log X = 16 \quad \text{(collect log } X \text{ and } 3 \log X\text{)}$$

$$4 \log X = 16 - \log 5 \quad \text{(find log 5 and collect)}$$

$$\log X = \frac{15.3}{4} \quad \text{(divide by 4)}$$

$$\log X = 3.83 \quad \text{(take the antilog of both sides)}$$

$$X = 6687$$

EXERCISE 26-7 Solve the equations.

1. $\log X + 1.5 \log X = 3.4$

Chapter 26
Logarithms

2. $\log 3X + 32 \log X = 40$ _____

3. $\log \dfrac{P}{3.4} = 1.8$ _____

4. $95 = 4^{2x}$ _____

5. $\log X^3 - 6 = \log X$ _____

6. $X = \log_2 550$ _____

7. $X^{6.2} = 65$ _____

8. $\log \dfrac{18}{P} = 5$ _____

9. $X = \log_7 1640$ _____

10. $X = \log_e 212$ _____

468

Part 3
*Computer Mathematics,
Logarithms, and
Quadratic Equations*

EVALUATION EXERCISE Find the common log and the natural log in Exercises 1 through 8. Give your answers in four significant figures.

		Common log	Natural log
1.	876		
2.	2.54×10^3		
3.	0.582		
4.	3.55×10^{-4}		
5.	12,850		
6.	0.00125		
7.	0.5×10^3		
8.	7575		

Find the antilog of the base 10 logs in Exercises 9 through 18. Give the answers in three significant figures.

9. 2.575 _____   10. 0.1250 _____

11. 6.000 _____   12. 2.438 _____

13. 0.0055 _____ 14. 10.5000 _____

15. 7.7500 _____ 16. 3.7552 _____

17. 0.4771 _____ 18. 1.8574 _____

Change the logarithmic form to the exponential or the exponential form to the logarithmic form in Exercises 19 through 26.

19. $8^2 = 64$ _____

20. $b^x = y$ _____

21. $10^{2.5} = 316$ _____

22. $25^{1/2} = 5$ _____

23. $\log_2 8 = 3$ _____

24. $\log_e 32 = 3.466$ _____

25. $\log 10{,}000 = 4$ _____

26. $\log_x B = A$ _____

Express the logarithmic form as a sum, difference, or product in Exercises 27 through 33.

27. $\log 2.5X$ _____

28. $\log X^5$ _____

29. $\log IV$ _____

30. $\log \dfrac{V^2}{R}$ _____

31. $\log (ab)^{0.5}$ _____

32. $\log P^2 \dfrac{R}{V^3}$ _____

33. $\log \dfrac{X^3 Y}{p^2}$ _____

Express the forms in Exercises 34 through 40 as single log expressions.

34. $\log R + \log H$ _____

35. $3 \log X$ _____

36. $2 \log Y - \log X$ _____

37. $\log V_1 - \log V_2$ _____

38. $4 \log I - 3 \log R + \log X$ _____

39. $\log 8 - (2 \log a + \log b)$ _____

40. $4(\log P - \log 3)$ _____

Solve the log equations in Exercises 41 through 48.

41. $\log Y + 2.5 \log Y = 75$ _____

42. $\log 2X + 12 \log X = 50$ _____

Chapter 26
Logarithms

43. $\log \dfrac{V}{2.5} = 5$

44. $80 = 3^{4x}$

45. $X = \log_3 150$

46. $\log \dfrac{25}{P} = 8$

47. $X = \log_e 250$

48. $R = \log_5 1510$

CHAPTER 27
Applications of Logarithms

Objectives

After completing this chapter you will be able to:
- Solve two-wire air insulated transmission line problems
- Solve coaxial cable transmission line problems
- Solve R-C transient circuit problems
- Solve R-L transient circuit problems
- Identify the decibel unit and its relationship to sound
- Determine dB gain in amplifiers
- Determine dB attenuation in electronic circuits and transmission lines
- Identify dB reference levels
- Solve dB and dBm circuit problems

There is a natural order in which the universe functions, and many of these functions are logarithmic. Circuits containing capacitance and inductance, the way in which sound is heard, and many other functions are part of this natural order. Logarithmic equations are used in solving problems relating to these functions. Both natural and common logs are studied and used in the equations in this chapter.

27–1 TRANSMISSION LINE. Two common types of transmission lines are the open wire (Figure 27–1) and coaxial cable (Figure 27–2). An important characteristic to keep in mind about a transmission line is its impedance. The impedance between the conductors of transmission lines is the characteristic impedance. The characteristic impedance is independent of the length. The equation for determining the characteristic impedance (Z_o) of the two-wire open line air-insulated is

$$Z_o = 276 \log \frac{d}{r}, \text{ where 276 is a constant for air insulation}$$

d is the distance between the centers of the conductors

r is the radius of the conductor

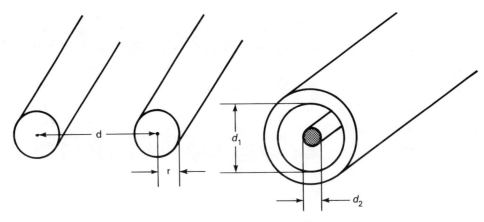

FIGURE 27-1 Two-wire, air-insulated, transmission line

FIGURE 27-2 Coaxial line

NOTE: d and r are expressed as a ratio and must be in like units.

The equation for determining the characteristic impedance of the coaxial cable, air-insulated is

$$Z_o = 138 \log \frac{d_1}{d_2}, \text{ where 138 is a constant for air insulation}$$

d_1 is the inside diameter of the outer conductor

d_2 is the diameter of the inner conductor

d_1 and d_2 must be in like units

Example A

Determine the characteristic impedance of a two-wire air-insulated transmission line. The distance between the centers is 4 cm and the wire size is no. 10.

$$Z_o = 276 \log \frac{d}{r}$$

$$Z_o = 276 \log \frac{4 \text{ cm}}{0.129 \text{ cm}} \quad \text{(refer to the Appendix for the radius of no. 10 wire)}$$

$$Z_o = 276 \log 31$$

$$Z_o = 276(1.49)$$

$$Z_o = 412 \text{ }\Omega$$

Example B

What is the center-to-center distance of a two-wire air-insulated transmission line made of no. 8 wire and having a characteristic impedance of 350 Ω?

$$Z_o = 276 \log \frac{d}{r}$$

$$350 = 276 \log \frac{d}{1.63 \text{ mm}} \quad \text{(divide both sides by 276)}$$

$$1.27 = \log \frac{d}{1.63} \quad \text{(take antilog of both sides)}$$

$$18.5 = \frac{d}{1.63} \quad \text{(solve for } d\text{)}$$

$$d = 30.2 \text{ mm}$$

EXERCISE 27-1 Solve the equations, using the formulas given in Section 27-1.

1. What is the characteristic impedance of a two-wire air-insulated transmission line with a center-to-center distance of 3.5 cm that is made of no. 20 copper wire?

2. If the center-to-center distance in Exercise 1 were changed to 2.25 inches, what would be its characteristic impedance?

474

*Part 3
Computer Mathematics,
Logarithms, and
Quadratic Equations*

3. What is the characteristic impedance of a coaxial cable with air insulation if the center conductor is of no. 24 wire and the inside diameter of the outer conductor is 20 mm?

4. What would the impedance be in Exercise 3 if the inner conductor were changed to no. 18 wire?

5. What size wire would be needed to construct a two-wire air-insulated transmission line with a center-to-center distance of 5.5 cm and a characteristic impedance of 450 Ω?

6. An 80 Ω coaxial cable with air insulation has an inner conductor made of no. 28 wire. What is the inside diameter of the outer conductor?

7. No. 12 wire is available for constructing a 350 Ω two-wire transmission line with air insulation. What will be the center-to-center distance between the wires, in inches?

8. An air-insulated coaxial cable is designed to have an impedance of 85 Ω, and the center conductor is no. 24 wire. What is the inside diameter of the outer conductor, in millimeters?

_____

9. What size wire is needed to construct a two-wire air-insulated transmission line with an impedance of 600 Ω and a center-to-center spacing of 2.75 inches?

_____

10. A 60 Ω coaxial transmission line is to be designed with the diameter of the outer conductor $\frac{5}{8}$ inch. What size wire should be used for the inner conductor?

_____

27–2 R-C CIRCUITS. The time constant in an R-C circuit is defined as the product of R and C and is represented by the Greek symbol tau (τ). It is the time required for the current to decay from maximum to approximately 37% of its maximum value and the time required for the voltage across the capacitor to rise to approximately 63% of the applied voltage. To determine currents or voltages at a more precise time, other equations are available, such as the time constant equation: $\tau = RC$.

What is the time constant of an R-C circuit containing a 150 kΩ resistor and a 22 μF capacitor?

$$\tau = RC$$

$$\tau = (1.5 \times 10^5)(2.2 \times 10^{-5})$$

$$\tau = 3.3 \text{ seconds}$$

The time constant is dependent on the value of the components, not the value of the applied voltage. The maximum current is dependent on the applied voltage and the value of the resistance; it is an Ohm's Law function. At the instant the circuit is complete, the capacitor appears as a dead short, and the current is limited only by the resistance in the circuit. The current begins at its maximum value, 2 mA, and decays to approximately 37% of maximum; or, at the end of one time constant, 3.3 seconds, the current is 0.74 mA.

At the end of one more time constant the current decays approximately another 37% of the remaining value, to a value of 0.27 mA. The current decays approximately 37% of the remaining value at the end of each time constant. In theory, the current will never reach zero, but for practical applications the current is considered to be at 0 at the end of five time constants.

Transient Response—R-C Circuits   When a dc voltage is applied to an R-C or an R-L circuit the action of the current and voltage is referred to as a "transient response." In working with electronic circuits it is sometimes necessary to determine the transient current, or voltage, at a specific time. The equation below can be used in an R-C circuit to determine the current at any time. If the current is known for a given time the voltage across the resistor and capacitor can easily be determined.

$$i = \frac{V_A}{R}(e^{-t/\tau})$$

where

i is the transient value of the current at a specific time, in amperes

V_A is the applied voltage, in volts

R is the total circuit resistance, in ohms

e is the base of the natural log epsilon: 2.72

t is the transient time or the time the current is desired, in seconds

τ is the time constant of the circuit, in seconds

Example C

Using the circuit given in Figure 27–3 determine the current at the end of 10 seconds and the voltage across the resistor and the capacitor.

$$i = \frac{V_A}{R}(e^{-t/\tau})$$

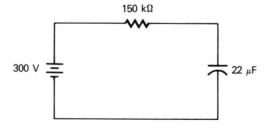

FIGURE 27–3   The R-C circuit

$$i = \frac{300}{150 \times 10^3}(2.72^{-10/3.3})$$

$$i = 2 \times 10^{-3}(2.72^{-3.03}) \quad \text{(use the ``}y^x\text{'' key to raise 2.72)}$$

$$i = 2 \times 10^{-3}(0.0488)$$

$$i = 97.6 \text{ μA}$$

To determine the voltage across the resistor at 10 seconds, use Ohm's Law:

$$V_R = IR$$

$$V_R = (97.6 \times 10^{-6})(1.5 \times 10^5)$$

$$V_R = 14.6 \text{ V}$$

Determine the voltage across the capacitor at 10 seconds:

$$V_C = V_A - V_R$$
$$V_C = 300 \text{ V} - 14.6$$
$$V_C = 285.4 \text{ V}$$

Example D

Determine transient time, using the circuit in Figure 27–3. Determine the time required for the current to decay to 0.75 mA.

$$i = \frac{V_A}{R}(e^{-t/\tau})$$

$$0.00075 = \frac{300}{150 \times 10^{-3}}(e^{-t/3.3})$$

$$0.00075 = 2 \times 10^3 (2.72^{-0.303t})$$

$$0.375 = 2.72^{-0.303t} \quad \text{(take the ln of both sides of the equation and transpose the exponent } -0.303t)$$

$$\ln 0.375 = -0.303t \ln 2.72 \quad \text{(since the ln of } e \text{ is 1, ln 2.72 drops out of the equation)}$$

$$-0.981 = -0.303t$$

$$t = 3.24 \text{ seconds}$$

This equation can be solved by using base 10 logs, but the advantage of using natural logs should be apparent.

27–3 R-L CIRCUITS. The time constant of an R-L circuit is the ratio of inductance to resistance: $\tau = \dfrac{L}{R}$. It is the time required for the current to rise to approximately 63% of maximum value and the voltage across the inductance to decay to approximately 37% of the applied voltage. The transient current can be determined by the following equation.

Transient Response—R-L Circuits

$$i = \frac{V_A}{R}(1 - e^{-t/\tau})$$

where i is the current, in amperes

V_A is the applied voltage, in volts

R is the total circuit resistance, in ohms

e is the natural log base epsilon, 2.72

t is the transient time, in seconds

τ is the circuit time constant $\left(\dfrac{L}{R}\right)$, in seconds

EXERCISE 27-2

1. A series R-C circuit with an applied voltage of 150 Vdc contains a resistor of 10 kΩ and a capacitor of 33 μF. What is the current at the end of 125 msec?

2. If the voltage in Exercise 1 were doubled, what would be the current at the end of 25 msec?

3. In Exercise 1, what is the voltage across the resistor and the capacitor at the end of 125 msec?

$V_C = $ _____

$V_R = $ _____

4. A 220 Ω resistor and a 200 μF capacitor are in series across a 9 Vdc source. What is the voltage across the resistor at the end of 15 msec?

_____

5. In Exercise 4, at what time will the current reach 35 mA?

_____

6. The circuit in Exercise 4 is to be redesigned by changing the value of the capacitor. The current must reach a value of 25 mA at the end of 22 msec. What should be the new value of the capacitor? (Use the transient formula to find the time constant.)

_____

7. An R-L circuit contains an inductance of 15 mH and a resistance of 75 Ω. What is the current at the end of 65 μS when 24 Vdc is applied to the circuit?

_____

Chapter 27
Applications
of Logarithms

8. In Exercise 7, what is the voltage across the resistor and the inductor at the end of 65 μS?

   $V_R =$ _____

   $V_L =$ _____

9. In Exercise 7, what is the voltage across the inductance if the inductance is doubled?

   _____

10. A relay coil has an inductance of 4.5 H and a resistance of 300 Ω. What is the current through the coil at the end of 8 msec when it is placed across 13.2 Vdc?

    _____

11. In the circuit in Exercise 10, in what length of time will the current reach 36 mA?

    _____

12. Referring to Exercise 10, in what length of time would the current through the choke reach 8 mA if a 270 Ω resistor were placed in series with the choke?

27–4 THE DECIBEL. The decibel unit may be one of the most confusing and misused units in electronics. This is probably due to lack of understanding of what it is that the decibel unit measures. Basically, it is a unit that measures the loudness of sound. The unit measures the ratio between two quantities at a logarithmic rate. Because this is a ratio measurement it must have a zero reference. When working with acoustics the zero reference is 10^{-16} watts per square centimeter; in electronics it is usually 1 milliwatt, though other references can be used.

The equation that describes this logarithmic ratio is $\text{bel} = \log \frac{P_2}{P_1}$, where

bel is a unit that measures a change in sound (named in honor of Alexander Graham Bell)

P_2 is a measured power output

P_1 is a reference level or input power

This equation is modified in the field of electronics to reflect the decibel unit only because it is a more convenient unit. Because there are 10 deci units in a basic unit, "bel" is multiplied by 10, which changes it to a decibel. In an equation, if the left side is multiplied by 10, the right side must also be multiplied by 10, and the equation is then modified to:

$$\text{decibels (dB)} = 10 \log \frac{P_2}{P_1}$$

The equation is used in this form in numerous texts and applications. The unit dB should not be used to designate a quantity; in applications of Ohm's Law, it is not acceptable to use "ohms $= \frac{V}{I}$." In this text the equation used is:

$$N = 10 \log \frac{P_2}{P_1}$$

where $\qquad N =$ number of decibels

Most human ears can detect changes in sound intensity of approximately 1 decibel, which represents a change of about 25% in the ratio. In the equations below, a change from 4 to 5 represents a 25% change in the ratio and is approximately 1 dB.

$$N = 10 \log \frac{P_2}{P_1}$$

$$N = 10 \log \frac{5}{4}$$

$$N = 10 \log 1.25$$

$$N = 10 \ (0.0969) = 0.969 \ (\text{approx. 1 db})$$

27–5 VOLTAGE AND CURRENT GAINS. The gain of a device such as an amplifier can be measured in wattage, voltage, or current. The voltage, or current, gain can be shown in the equations

$$A_V = \frac{A_2}{A_1} \qquad A_i = \frac{I_2}{I_1}$$

where A is the voltage, or current, gain or the ratio of input to output voltage or current

Current Gain When input and output impedances are equal, they can be canceled out of the equation.

Example E

$$N = 10 \log \frac{(I_2)^2 (R)}{(I_1)^2 (R)} \quad \text{(if equal, the } R\text{s can be canceled)}$$

$$N = 10 \log \left(\frac{I_2}{I_1}\right)^2$$

Multiplying the log of a quantity by 2 is the same operation as squaring the quantity; the equation becomes:

$$N = 20 \log \frac{I_2}{I_1}$$

The same relation can be shown for the voltage ratio. When impedances are matched, the following equation results.

$$N = 20 \log \frac{V_2}{V_1}$$

EXERCISE 27–3

1. Determine the decibel gain for power ratios of 2, 4, 8, 16, and 32.

Chapter 27
Applications of Logarithms

2. Determine the decibel gain for power ratios of 10, 20, 30, 40, 50, and 60.

3. Determine the decibel gain for current or voltage ratios that equal 5, 10, 15, 20, and 25.

4. An amplifier has an input impedance of 800 Ω and an output impedance of 5500 Ω. When 0.15 V is applied across the input, a voltage of 320 V is across the output. What power gain, in decibels, does this represent?

5. What is the power gain, in decibels, of an amplifier with 20 mV across an input impedance of 600 Ω and 30 V across an output impedance 75 Ω?

6. A voltage amplifier has an input impedance and output impedance of 1000 Ω. The voltage measured on the input is 16 mV and the voltage at the output is 36 V. What is the decibel gain?

7. The input voltage to a 50 Ω transmission line is 80 V; the voltage measured at the end of the line is 68 V. What is the loss of the line, measured in decibels?

8. An attenuator switch in an audio oscillator reduces the output voltage in 5 dB steps. The output voltage at 0 dB is 12 V. What is the output of the oscillator when the attenuator switch is set at -5 dB, -10 dB, -15 dB, -25 dB?

9. The gain of an amplifier is 27.5 dB. If the input power is 10 mW, what is the output power?

10. An amplifier has a normal output of 50 W. What power output corresponds to a reduction of 23 dB?

11. An important measurement made in quality amplifiers is the signal-to-noise ratio. Undesirable noise is generated by components and circuitry and is generally measured in decibels. If the noise power of an 80 W amplifier is measured and found to be 22 μW, what is this measurement in decibels?

12. A 600 Ω transmission line has a loss of 7 dB, and the input voltage is measured at 62 V. What is the output voltage at the end of the line?

13. If an amplifier has a gain of 43 dB and a power output of 35 W, what power is considered zero dB?

14. An attenuation pad with matched input and output impedances has an input voltage of 600 mV and an output of 420 mV. What is the attenuation, in decibels?

27-6 REFERENCE LEVELS. The most common reference level used in electronics is the dBm unit. Several other references, however, may be encountered. Regardless of what reference level is used, it is considered "zero dB." Without a reference level the decibel unit is of little value. The dBm unit uses 1 mW as its reference. The "m" in dBm implies that 1 mW is the reference. If it were written dBW, 1 W would be the reference, dBV would indicate that 1 V is the reference, or 0 dB. Six hundred ohms is a standard terminating impedance in such industries as radio and telephone. Although these industries prefer to use the dBm unit as 1 mW of power in a 600 Ω load, the dBm was established without regard to impedance. The important thing to remember is that a reference level must be established; when written as "dBm," 1 mW is the established reference.

Example F

The gain of an amplifier is 20 dBm. What is its power output?

$$N_{dBm} = 10 \log \frac{P_2}{P_1}$$

where P_1 is the reference level

$$20_{dBm} = 10 \log \frac{P_2}{1 \text{ mW}} \quad \text{(divide by 10)}$$

$$2.0 = \log \frac{P_2}{1} \quad \text{(take the antilog of both sides of the equation)}$$

$$100 = \frac{P_2}{1}$$

$$P_2 = 100 \text{ mW}$$

It can be seen in Example F that the calculations are simplified when 1 mW is used as the reference, or 0 dB. To calculate the power output using the dBm unit, divide the number of dBm by 10 and take the antilog. Remember: the answer is in milliwatts.

EXERCISE 27-4

1. Determine the power output for the decibel values given.

    10 dBm = _____

    20 dBm = _____

    25 dBm = _____

    37.5 dBm = _____

    62.5 dBm = _____

    90 dBm = _____

2. The power output of a transmission line is -12.5 dBm. What is the power at the input to the line?

3. An amplifier with a gain of 35 dBm is designed to deliver 3 W to a 4 Ω load. What input voltage is necessary to produce full output if the input resistance is 500 Ω?

4. What is the power output of the amplifier in Exercise 3 if the input voltage is 1.2 V?

5. The output power at the end of a 50 Ω coaxial cable is measured at 65 μW. What is the dBm loss in the cable?

6. An amplifier has an output impedance at 45 Ω and a gain of 42 dBm. What is the voltage measured across the load?

7. An attenuator pad has a loss of −7.5 dBm. What is the voltage at the output if the output impedance is 75 Ω?

8. The voltage measured across a load at the output of an amplifier is 8.2 V. If the amplifier has a gain of 33 dBm, what is the resistance of the load?

27–7 APPLYING DECIBELS TO OVERALL GAIN.
The decibel unit is a convenient unit for calculating the overall gain of a system. Amplifiers have a positive decibel rating, whereas volume controls, record players, microphones, tape decks, and so forth have a negative dB rating. Example G illustrates a typical audio system rated in decibels.

Example G
An audio system consists of a record player with a rating of -45 dBm, a preamplifier with a rating of 25 dB, a volume control rated at -8 dB, and an amplifier rated at 60 dB. What is the power output of the system?

record player	preamp	volume control	amplifier
-45 dBm	25 dB	-8 dB	60 dB

Adding the dB ratings, the overall gain is 32 dBm. Knowing that the reference is 1 milliwatt (dBm), the power output now can be calculated.

$$N_{dB} = 10 \log \frac{P_2}{P_1}$$
$$32 \text{ dBm} = 10 \log \frac{P_2}{1 \text{ mW}}$$
$$3.2 = \log \frac{P_2}{1}$$
$$1585 = \frac{P_2}{1}$$
$$P_2 = 1585 \text{ mW}$$

EXERCISE 27–5 Solve the problems, leaving the answers in a convenient prefix.

1. An audio system contains a tape deck rated at -20 dBm, a preamplifier at 35 dB, an equalizer at -20 dB, a volume control at -12 dB, and a final amplifier at 65 dB. What is the power output of the system?

2. If the volume control in the system in Exercise 1 were changed to −9 dB, what would be the power output of the system?

3. If the system described in Exercise 1 is increased to 175 W, how many additional decibels are needed?

4. The reference level for a microphone rated at −62 dB in an audio amplifying system is 12.5 mW. The system contains a preamp of 85 dB, an equalizer rated at −35 dB, a volume control at −8.5 dB, a speaker-matching device at −2.5 dB, and a final amplifier at 58 dB. What is the wattage output of the system?

5. The microphone in the system in Exercise 4 is changed to −56 dB, using the same reference level. What is the new power output of the system?

6. In the system used in Exercise 4 the final amplifier is changed to increase the output to 100 W. What is the increase in the decibel rating of the amplifier?

7. An audio system has a record player rated at -45 dBm, a preamplifier at 95 dB, a volume control at 0 dB to -20 dB, an equalizing network at -42 dB, and a final amplifier at 55 dB. At what setting must the volume control be set to produce an output of 50 W?

8. In the system in Exercise 7, what volume-control setting, in decibels, is needed to obtain an output of 65 W?

9. If the equalizing network in the system used in Exercise 7 were changed to -44 dB, and the volume were set at -15 dB, what would be the power output?

10. In the system in Exercise 7, the preamp is changed to 75 dB, and the volume is set at -5 dB. What is the system's power output?

EVALUATION EXERCISE

1. What is the characteristic impedance of a two-wire air-insulated transmission line with a center-to-center distance of 2.75 cm and that is constructed of no. 22 copper wire?

   Chapter 27
   Applications
   of Logarithms

   _____

2. An air-insulated coaxial cable with a characteristic impedance of 75 Ω and a center conductor of no. 20 wire has a conductor whose inside diameter is what length (in millimeters)?

   _____

3. A two-wire air-insulated transmission line is made of no. 10 wire with a characteristic impedance of 400 Ω. What is the center-to-center distance between the conductors, in inches?

   _____

4. A series R-C circuit with an applied voltage of 50 Vdc, contains a resistor of 4.7 kΩ and a capacitor of 47 nF. What is the current at the end of 85 μS?

   _____

Part 3
Computer Mathematics, Logarithms, and Quadratic Equations

5. A series R-L circuit contains an inductance of 2.25 H and a resistance of 250 Ω. When placed across a voltage of 24 Vdc, in what length of time will the current reach 60% of maximum?

   _____

6. A series R-C circuit contains a 470 Ω resistor and a 150 μF capacitor. When placed across a 6 Vdc source, in what length of time will the current reach 9.5 mA?

   _____

7. A 0 to 120 dB attenuator is available for reducing a voltage from a source. It reduces the voltage in 10 dB steps and when set at -10 the voltage output is 15 mV. What are the voltage outputs for steps of -20 dB, -30 dB, -40 dB, -50 dB, -60 dB, and -70 dB?

   -20 dB = _____

   -30 dB = _____

   -40 dB = _____

   -50 dB = _____

   -60 dB = _____

   -70 dB = _____

8. An amplifier has an input impedance of 600 Ω and an output impedance of 4500 Ω. When a signal of 0.20 V is placed across the input, the output voltage is 275 V. What power gain, in decibels, does this represent?

   _____

9. The signal-to-noise ratio of an amplifier is 70 dB and its output rating is 90 W. What is the noise power?

10. If an amplifier has a gain of 55 dB and the power output is 40 W, what power is considered 0 dB?

11. What is the dBm rating of an amplifier with an output of 75 W?

12. The output of a transmission line is -7.5 dBm. What is the power output, in watts?

13. An amplifier has an output impedance of 8 Ω and a gain of 60 dBm. What voltage is measured across the load?

14. An audio amplifying system contains a microphone rated at −80 dBm, a preamplifier at 65 dB, a matching unit at −15 dB, a volume control set at −12 dB, an equalizer rated at −35 dB, and a final amplifier rated at 100 dB. What is the system's power output?

15. An audio system has a record player rated at −38 dBm, a preamplifier at 50 dB, an equalizer at −25 dB, and a volume control set at −12 dB. What is the dB rating of the final amplifier needed to produce an output of 45 W?

CHAPTER 28
Quadratic Equations

Objectives

After completing this chapter you will be able to:
- Solve quadratic equations by factoring
- Solve quadratic equations by taking the square root of both sides
- Solve quadratics by completing the square
- Solve quadratics using the quadratic formula
- Solve resistive circuits that require quadratic solutions

The solution to some problems in electronics involves the use of second-degree equations. A second-degree equation is one in which the unknown occurs to the second power; such an equation is called a "quadratic equation." In general terms, a quadratic equation is written:

$$ax^2 + bx + c = 0$$

where a and b are coefficients, c is a constant, and x is the unknown.

In this chapter we will study quadratic solutions by factoring, completing the square and the quadratic formula. In describing solutions to quadratics the equation above will be referred to as the standard form; ax^2 will be called the first term, bx the second term, and c the third term.

28–1 PURE QUADRATICS. A pure quadratic is the simplest form of quadratic. Equations of this type have been solved in previous chapters. A pure quadratic contains only the first and third terms and is in the form:

$$x^2 - 25 = 0$$

This equation is solved by transposing 25 and extracting the square root of both sides of the equation.

$$\sqrt{x^2} = \sqrt{25} \qquad x = 5 \text{ or } x = -5$$

Both 5 and -5 are valid solutions to the equation. Negative numbers were not considered important in ancient times, so negative answers were discarded. In electronics the negative solution is as valid as the positive solution. Voltages are expressed in both positive and negative values. A negative value of resistance is not considered a valid solution. Example A illustrates other solutions to the pure quadratic.

Example A

$$(x+3)^2 = 49 \qquad \sqrt{(x+3)^2} = \sqrt{49}$$

Extract the square root of both sides:

$$x + 3 = 7 \qquad \text{or} \qquad x + 3 = -7$$
$$x = 4 \qquad\qquad\qquad x = -10$$

$$4a^2 = 256$$

Divide both sides of the equation by 4 and extract the square root of both sides:

$$a^2 = \sqrt{64}$$

$$a = +8 \text{ or } -8$$

EXERCISE 28-1 Solve the equations for the unknown.

1. $x^2 + 12 = 112$ $x = $ _____

2. $5I^2 = 1.25$ $I = $ _____

3. $2.5R^2 - 7.5 = 15$ $R = $ _____

4. $(i + 12)^2 = 60$ $i = $ _____

5. $(P - 6)^2 + 3 = 28$ $P = $ _____

6. $244 + x^2 = 2x^2$ $x = $ _____

7. $(a + b)^2 = c$ $a = $ _____

8. $\dfrac{I^2}{4} = 25$ $I = $ _____

9. $P - I^2R = 0$ $I = $ _____

10. $4(a + 6)^2 = 75$ $a = $ _____

11. $125 - R^2 = 35$ $R = $ _____

12. $2B^2 - 18 = 4B^2 - 60$ $B = $ _____

28–2 SOLUTION BY FACTORING. Equations that contain all three terms and are easy to factor can be solved by factoring. The perfect trinomial square is a special case of factoring.

$$x^2 + 10x + 100 = 25$$

Factor and extract the square root of both sides:

$$\sqrt{(x + 10)^2} = \sqrt{25}$$

$x + 10 = 5$ or $x + 10 = -5$

$x = -10 + 5$ $x = -10 - 5$
$x = -5$ $x = -15$

then

$$x = -5 \text{ or } x = -15$$

When one side of an equation is not a perfect trinomial square, the equation is placed in standard form and made equal to zero:

$$a^2 + 5a + 6 = 0$$

$$(a + 3)(a + 2) = 0$$

The equation states that $(a + 3)$ times $(a + 2)$ must equal zero. If $(a + 2)$ is made equal to zero, the equation remains true.

$$(a + 3)(0) = 0$$

Solving for a: $a + 2 = 0$
$a = -2$

The equation remains true if $(a + 3)$ is made equal to zero:

$$(0)(a + 2) = 0$$

Solving for a: $a + 3 = 0$
$a = -3$

In this equation, $a = -3$ or $a = -2$. The solutions to other quadratic equations are illustrated in examples B, C, and D.

Example B

$$4b^2 - 16b + 15 = 0$$
$$(2b - 3)(2b - 5) = 0$$

497

Chapter 28
Quadratic Equations

$$2b - 3 = 0 \qquad 2b - 5 = 0$$
$$2b = 3 \qquad 2b = 5$$
$$b = 1.5 \qquad b = 2.5$$

$$b = 1.5 \text{ or } 2.5$$

Example C

$$6a^2 + 10ab - 4b^2 = 0$$
$$(3a - b)(2a + 4b) = 0$$

$$3a - b = 0 \qquad 2a + 4b = 0$$
$$3a = b \qquad 2a = -4b$$
$$a = \frac{b}{3} \qquad a = -2b$$

$$a = \frac{b}{3} \quad \text{or} \quad -2b$$

Example D

$$16c^2 - 40c + 25 = 64$$

$$\sqrt{(4c - 5)^2} = \sqrt{64}$$

$$4c - 5 = 8 \text{ or } -8$$

$$4c = 13 \quad \text{or} \quad 4c = -3$$
$$c = 3.25 \quad \text{or} \quad c = -0.75$$

EXERCISE 28–2 Solve the quadratics by factoring.

1. $x^2 - 5x + 6 = 0$ \hfill $x =$ _____

2. $5b^2 - b - 4 = 2 - 14b$ \hfill $b =$ _____

3. $3I^2 - 5I - 2 = -6I$ \hfill $I =$ _____

4. $4R^2 - 24R + 36 = 0$ \hfill $R =$ _____

5. $6a^2 - ab - 2b^2 = 0$ \hfill $a =$ _____

6. $-8P^2 = 8P^2 + 24P + 9$ $P = $ _____

7. $3R^2 = 2(4 - 5R)$ $R = $ _____

8. $36 = -4(V^2 + 6V)$ $V = $ _____

9. $8a^2 = 2(9 - 6a - 4a^2)$ $a = $ _____

10. $6c^2 = bc + 2b^2$ $c = $ _____

28–3 SOLUTION BY COMPLETING THE SQUARE. The factoring method can be used for equations that are easily factorable; completing the square, however, can be used on any quadratic.

$$x^2 + 8x + 4 = 0$$

Transpose the third term to the opposite side of the equation:

$$x^2 + 8x = -4$$

Add the square of one-half the coefficient of x to both sides of equation. This makes the left member of the equation a perfect trinomial square.

$$x^2 + 8x + 16 = -4 + 16$$

Factor the left member of the equation and collect the right member:

$$\sqrt{(x + 4)^2} = \sqrt{12}$$

Extract the square root of both sides of the equation:

$$x + 4 = 3.46 \quad \text{or} \quad -3.46$$

$$x = 3.46 - 4 \quad \text{or} \quad x = -3.46 - 4$$
$$x = -0.54 \quad \text{or} \quad x = -7.46$$

Examples E, F, and G illustrate equations solved by completing the squares.

Example E

$$a^2 - 4a + 3 = 0$$
$$a^2 - 4a = -3$$
$$a^2 - 4a + 4 = 4 - 3$$

$$\sqrt{(a-2)^2} = \sqrt{1}$$
$$a - 2 = 1 \quad \text{or} \quad -1$$
$$a = 3 \quad \text{or} \quad a = 1$$

Example F
$$3I^2 = 9I$$

In this example the coefficient of the squared term is not 1; the equation must be divided by the coefficient 3:

$$I^2 = 3I$$
$$I^2 - 3I = 0$$
$$I^2 - 3I + 2.25 = 2.25$$
$$\sqrt{(I - 1.5)^2} = \sqrt{2.25}$$
$$I - 1.5 = 1.5 \text{ or } -1.5$$
$$I = 3 \text{ or } 0$$

Example G
$$3x^2 - 2x - 5 = 0$$
$$3x^2 - 2x = 5$$

Divide by 3 and work with decimals in practical applications.

$$x^2 - (0.667)x + 0.111 = 1.67 + 0.111$$
$$\sqrt{(x - 0.333)^2} = \sqrt{1.78}$$
$$x - 0.333 = 1.33 \text{ or } -1.33$$
$$x = 1.66 \text{ or } -1$$

EXERCISE 28-3 Solve by completing the square.

1. $a^2 + 2a - 8 = 0$ \qquad $a =$ _____

2. $I^2 - 16I - 5 = 5$ \qquad $I =$ _____

3. $R^2 + 16 = 8R$ $R = $ _____

4. $6 + 4b = b^2$ $b = $ _____

5. $2x^2 + 8x - 9 = 0$ $x = $ _____

6. $3V^2 - 18V = 15$ $V = $ _____

7. $2x(x - 4) = 8$ $x = $ _____

8. $6(a + 6) = 3a^2$ $a = $ _____

9. $3R^2 - 5R + 10 = 25$ $R = $ _____

10. $V - 8 = 4V^2 + 8$ $V = $ _____

28–4 THE QUADRATIC FORMULA. The standard equation given in the introduction to this chapter, $ax^2 + bx + c = 0$, can be solved for the value of x by using the method of completing the square. The result of this operation is an equation called the "quadratic formula." It is not necessary to know how to derive the equation but, rather, how to use it. The formula derived is:

Chapter 28
Quadratic Equations

$$x = \frac{-b \pm \sqrt{b^2 - 4ac}}{2a}$$

where a is the coefficient of the first term in the standard equation

b is the coefficient of the second term in the standard equation

c is the third term, or constant, in the standard equation

Examples H and I illustrate the use of the quadratic formula in solving quadratic equations.

Example H
Solve for x: $x^2 + 8x + 4 = 0$

$$x = \frac{-b \pm \sqrt{b^2 - 4ac}}{2a}$$

Substitute the coefficients a, b, and c in the given equation:

$$a = 1$$
$$b = 8$$
$$c = 4$$

$$x = \frac{-8 \pm \sqrt{8^2 - 4(1)(4)}}{2(1)}$$

$$x = \frac{-8 \pm \sqrt{48}}{2}$$

The positive and negative roots are used to find x:

$$x = \frac{-8 + 6.93}{2} \quad \text{or} \quad x = \frac{-8 - 6.93}{2}$$

$$x = -0.535 \quad \text{or} \quad x = -7.46$$

Example I
Solve for I: $2I^2 - 3I - 4 = 0$.

$$a = 2$$
$$b = -3$$
$$c = -4$$

$$x = \frac{-b \pm \sqrt{b^2 - 4ac}}{2a}$$

$$x = \frac{-(-3) \pm \sqrt{(-3)^2 - 4(2)(-4)}}{2(2)}$$

$$x = \frac{3 \pm \sqrt{41}}{4}$$

$$x = \frac{3 + 6.4}{4} \quad \text{or} \quad x = \frac{3 - 6.4}{4}$$

$$x = 2.35 \quad \text{or} \quad x = -0.85$$

EXERCISE 28-4 Solve the equations, using the quadratic formula.

1. $a^2 + 7a + 12 = 0$ $a = $ _____

2. $2B^2 - 3B - 4 = 0$ $B = $ _____

3. $4V^2 = 11 - 7V$ $V = $ _____

4. $I = 1 - 2I^2$ $I = $ _____

5. $a = 4a^2$ $a = $ _____

6. $9c^2 + 6c = 10$ $c =$ _____

28–5 APPLICATIONS OF QUADRATIC EQUATIONS TO ELECTRONIC CIRCUITS.
Any of the three solutions given can be used to solve quadratic problems. Most practical problems result in equations not easily factored or that may be cumbersome to work by completing the square. The quadratic formula is usually the most practical method for solving these problems. Solving some problems in electronics involves a quadratic equation.

Example J

In Figure 28–1 the wattage dissipated by R_2 is 4 W. What is the total current?

The power dissipated by R_2 can be found by:

$$P_2 = IV_2$$

The value of V_2 is not known but is equal to:

$$V_2 = V_T - V_1 \quad \text{(where } V_1 = 5I\text{)}$$
$$V_2 = 10 - 5I$$

then

$$P_2 = 4 \text{ W and } V_2 = (10 - 5I)$$

Substituting in the original equation

$$P_2 = IV_2$$
$$4 = I(10 - 5I)$$
$$4 = 10I - 5I^2$$

and

$$5I^2 - 10I + 4 = 0$$

Although this problem can be solved by completing the squares, the quadratic equation was chosen for finding the solution

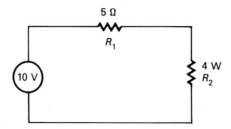

FIGURE 28–1   Circuit for Example J

where $a = 5$, $b = -10$, and $c = 4$

$$x = \frac{-b \pm \sqrt{b^2 - 4ac}}{2a}$$

$$x = \frac{-(-10) \pm \sqrt{(-10)^2 - 4(5)(4)}}{2(5)}$$

$$x = \frac{10 \pm 4.47}{10}$$

$$x = 0.553 \text{ or } x = 1.45$$

Here, the current can be either 553 mA or 1.45 A. Both values satisfy the conditions or the 4 W dissipated in R_2, with 10 V applied to the circuit.

Example K

Determine the total current delivered by the source in the circuit given in Figure 28–2.

$$I_T = I_2 + I_3$$

where

$$I_3 = 2 \text{ A}$$
$$I_2 = \frac{P_2}{V_{PAR}}$$

and

$$V_{PAR} = 12 - V_1$$
$$V_{PAR} = 12 - 0.5I_T$$

$$I_2 = \frac{30}{(12 - 0.5I_T)}$$
$$I_T = I_2 + I_3$$
$$I_T = \frac{30}{12 - 0.5I_T} + 2$$

Clear the denominator by multiplying by $12 - 0.5I_T$:

$$I_T(12 - 0.5I_T) = 30 + 2(12 - 0.5I_T)$$

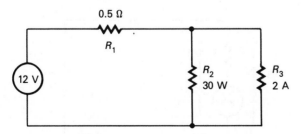

FIGURE 28–2   Circuit for Example K

$$12I_T - 0.5I_T{}^2 = 30 + 24 - I_T$$

$$0.5I_T{}^2 - 13I_T + 54 = 0$$

where $a = 0.5$, $b = -13$, and $c = 54$,

$$x = \frac{-b \pm \sqrt{b^2 - 4ac}}{2a}$$

$$x = \frac{-(-13) \pm \sqrt{(-13)^2 - 4(0.5)(54)}}{2(0.5)}$$

$$x = \frac{13 \pm 7.8}{1}$$

$$x = 5.20 \text{ A} \quad \text{or} \quad x = 20.8 \text{ A}$$

Both values of current satisfy the circuit conditions where R_2 dissipates 30 W and I_3 is 2 A.

Example L

A 9 V source has an internal resistance of 0.3 Ω and is delivering 2 W to a load. What voltage should be measured across the load?

Draw and label the known values on the circuit diagram in Figure 28–3.

$$V_{\text{LOAD}} = V_{\text{SOURCE}} - V_{\text{INT}}$$

$$V_L = 9 - V_I$$

where
$$V_I = I_T R_I \qquad P_L = I_T V_L$$
$$V_I = 0.3I \quad \text{and} \quad I = \frac{2}{V_L}$$

and

$$V_L = 9 - (0.3)\left(\frac{2}{V_L}\right)$$

Clearing the common denominator gives

$$V_L{}^2 = 9V_L - 0.6$$

$$V_L{}^2 - 9V_L + 0.6 = 0$$

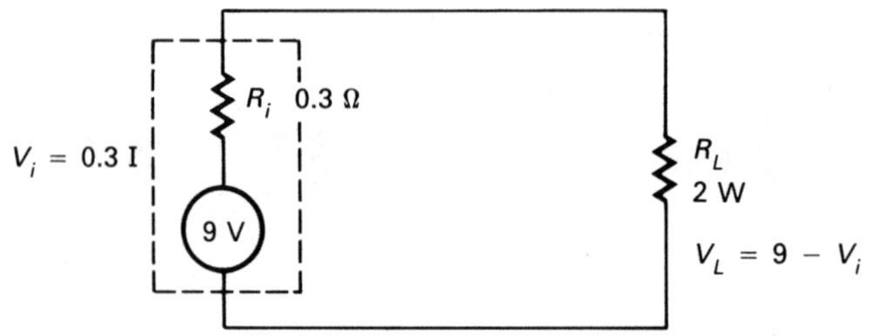

FIGURE 28–3 Circuit for Example L

where $a = 1$, $b = -9$, and $c = 0.6$,

$$x = \frac{-b \pm \sqrt{b^2 - 4ac}}{2a}$$

$$x = \frac{-(-9) \pm \sqrt{(-9)^2 - 4(1)(0.6)}}{2(1)}$$

$x = 8.94$ or $x = 0.065$

$V_L = 8.94$ V or 0.065 V

Both voltages satisfy the conditions of the circuit. If more information were known about the characteristic of the load, a practical voltage could be determined. The value of the load with 0.065 V would be 0.002 Ω and the value with 8.94 V about 40 Ω.

EXERCISE 28-5 Solve the problems, using the quadratic formula. Draw and label each circuit diagram before doing any calculations.

1. A 50 V power supply has an internal resistance of 1.5 Ω. A load connected to the supply dissipates 150 W. What values of load current will satisfy the circuit conditions?

   $I = $ _____

   $I = $ _____

2. A series-parallel circuit with 16 V applied has a series resistance of 5 Ω and two parallel resistors. The current through one parallel resistor is 250 mA; the other resistor dissipates 2 W. What values of total current will satisfy the circuit conditions?

   $I = $ _____

   $I = $ _____

3. Three resistors are in series across a 6 V source. One resistor has a value of 2.2 Ω, another dissipates 1.5 W, and the third has a voltage drop across it of 3 V. What two values of voltage can be measured across the 2.2 Ω resistor?

V = _____

V = _____

4. An 18 V power source has an internal resistance of 6 Ω, and a load connected across the source dissipates 12 W. What voltages can we expect to be measured across the load?

V = _____

V = _____

5. Two resistances, R_1 and R_2, have a combined resistance of 80 Ω. They are connected in series across a 12 V source. A meter with a resistance of 20 Ω is placed across R_1, and the current through the meter is 200 mA. What is the value of R_2?

R_2 = _____

6. Two resistors are in series across a power source; their combined resistance is 300 Ω. The power being dissipated by one resistor is 3 W; the voltage across the other is 8 V. What circuit currents will satisfy these conditions?

$I = $ _____

$I = $ _____

EVALUATION EXERCISE Solve the equations in Exercises 1 through 4, using the factoring method.

1. $a^2 + 8a + 16 = 0$ $a = $ _____

2. $2V^2 + 4V - 6 = 0$ $V = $ _____

3. $6R^2 + 12R - 18 = 0$ $R = $ _____

4. $6c^2 - 23c + 20 = 0$ $c = $ _____

Solve Exercises 5 through 8 by completing the squares.

5. $x^2 = 2x$ $x = $ _____

6. $3a^2 + 6a - 15 = 0$ $a = $ _____

7. $I^2 = 4(I + 2)$ $\qquad$ $I = $ _____

8. $V = V(V - 5)$ $\qquad$ $V = $ _____

Solve Exercises 9 through 12 by using the quadratic formula.

9. $2a^2 - 4a - 5 = 0$ $\qquad$ $a = $ _____

10. $6x^2 + 12x - 3 = 0$ $\qquad$ $x = $ _____

11. $B^2 = 3(2B + 5)$ $\qquad$ $B = $ _____

12. $8I^2 = 4I(I - 3) + 16$ $\qquad$ $I = $ _____

The remaining exercises can be solved using either the factoring method or the quadratic formula.

13. The internal resistance of a 15 V power source is 0.75 Ω. When a load is connected to the source it dissipates 25 W. What circuit currents will satisfy the conditions of the load?

$I = $ _____

$I = $ _____

14. A source has an internal resistance of 1.2 Ω and a voltage of 14 V. Two resistors in parallel are placed across the source; one of them dissipates 12 W, while the other has a current of 1 A. What values of voltage can be measured across the parallel branch that will satisfy the circuit conditions?

$V = $ _____

$V = $ _____

15. Three resistors are in series across a 24 V supply. The value of one resistor is 22 Ω; the voltage measured across another is 10 V. The third resistor dissipates 0.5 W. What currents will satisfy these circuit conditions?

$I = $ _____

$I = $ _____

Part 3
Computer Mathematics, Logarithms, and Quadratic Equations

16. Two resistors are in series across a 50 V source. Their total resistance is 300 Ω. A meter with a resistance of 60 Ω is placed across one of the resistances, and the current through the meter is 200 mA. What is the resistance of the other resistor?

$R = $ _____

Appendix

TABLE A-1 Standard Annealed Copper Wire Solid American Wire Gage (Brown and Sharpe) (20°C)

Gage	Diameter, mm	Cross Section, sq mm	Ohms per Kilometer	Meters per Ohm	Kilograms per Kilometer
0000	11.68	107.2	0.160 8	6.219	953.2
000	10.40	85.01	0.202 8	4.931	755.8
00	9.266	67.43	0.255 7	3.911	599.5
0	8.252	53.49	0.322 3	3.102	475.5
1	7.348	42.41	0.406 5	2.460	377.0
2	6.543	33.62	0.512 8	1.950	298.9
3	5.827	26.67	0.646 6	1.547	237.1
4	5.189	21.19	0.815 2	1.227	188.0
5	4.620	16.77	1.028	0.972 4	149.0
6	4.115	13.30	1.297	0.771 3	118.2
7	3.665	10.55	1.634	0.612 0	93.80
8	3.264	8.367	2.061	0.485 3	74.38
9	2.906	6.631	2.600	0.384 6	58.95
10	2.588	5.261	3.277	0.305 2	46.77
11	2.30	4.17	4.14	0.242	37.1
12	2.05	3.31	5.21	0.192	29.4
13	1.83	2.63	6.56	0.152	23.4
14	1.63	2.08	8.28	0.121	18.5
15	1.45	1.65	10.4	0.095 8	14.7
16	1.29	1.31	13.2	0.075 8	11.6
17	1.15	1.04	16.6	0.060 3	9.24
18	1.02	0.823	21.0	0.047 7	7.32
19	0.912	0.653	26.4	0.037 9	5.81
20	0.813	0.519	33.2	0.030 1	4.61
21	0.724	0.412	41.9	0.023 9	3.66
22	0.643	0.324	53.2	0.018 8	2.88
23	0.574	0.259	66.6	0.015 0	2.30
24	0.511	0.205	84.2	0.011 9	1.82
25	0.455	0.162	106	0.009 42	1.44
26	0.404	0.128	135	0.007 43	1.14
27	0.361	0.102	169	0.005 93	0.908
28	0.320	0.080 4	214	0.004 67	0.715
29	0.287	0.064 7	266	0.003 75	0.575
30	0.254	0.050 7	340	0.002 94	0.450
31	0.226	0.040 1	430	0.002 33	0.357
32	0.203	0.032 4	532	0.001 88	0.288
33	0.180	0.025 5	675	0.001 48	0.227
34	0.160	0.020 1	857	0.001 17	0.179
35	0.142	0.015 9	1 090	0.000 922	0.141
36	0.127	0.012 7	1 360	0.735	0.113
37	0.114	0.010 3	1 680	0.595	0.091 2
38	0.102	0.008 11	2 130	0.470	0.072 1
39	0.089	0.006 21	2 780	0.360	0.055 2
40	0.079	0.004 87	3 540	0.282	0.043 3
41	0.071	0.003 97	4 340	0.230	0.035 3
42	0.064	0.003 17	5 440	0.184	0.028 2
43	0.056	0.002 45	7 030	0.142	0.021 8

44	0.051	0.002 03	8 510	0.118	0.018 0
45	0.0047	0.001 57	11 000	0.0910	0.014 0
46	0.0399	0.001 25	13 800	0.0724	0.011 1
47	0.0356	0.000 993	17 400	0.0576	0.008 83
48	0.0315	0.000 779	22 100	0.0452	0.006 93
49	0.0282	0.000 624	27 600	0.0362	0.005 55
50	0.0251	0.000 497	34 700	0.0288	0.004 41
51	0.0224	0.000 392	43 900	0.0228	0.003 49
52	0.0198	0.000 308	55 900	0.0179	0.002 74
53	0.0178	0.000 248	69 400	0.0144	0.002 21
54	0.0157	0.000 195	88 500	0.0113	0.001 7
55	0.0140	0.000 153	112 000	0.008 89	0.001 36
56	0.0124	0.000 122	142 000	0.007 06	0.001 08

TABLE A–2   Metric Units

Value	Prefix	Symbol	Example
$1\,000\,000\,000\,000 = 10^{12}$	tera	T	$THz = 10^{12}\ Hz$
$1\,000\,000\,000 = 10^9$	giga	G	$GHz = 10^9\ Hz$
$1\,000\,000 = 10^6$	mega	M	$MHz = 10^6\ Hz$
$1\,000 = 10^3$	kilo	k	$kV = 10^3\ V$
$100 = 10^2$	hecto	h	$hm = 10^2\ m$
$10 = 10$	deka	da	$dam = 10\ m$
$0.1 = 10^{-1}$	deci	d	$dm = 10^{-1}\ m$
$0.01 = 10^{-2}$	centi	c	$cm = 10^{-2}\ m$
$0.001 = 10^{-3}$	milli	m	$mA = 10^{-3}\ A$
$0.000\,001 = 10^{-6}$	micro	μ	$\mu V = 10^{-6}\ V$
$0.000\,000\,001 = 10^{-9}$	nano	n	$ns = 10^{-9}\ s$
$0.000\,000\,000\,001 = 10^{-12}$	pico	p	$pF = 10^{-12}\ F$

Additional prefixes are exa = 10^{18}, peta = 10^{15}, femto = 10^{-15}, and atto = 10^{-18}.

TABLE A-3 Trigonometric Functions for Decimal-Degrees

TRIGONOMETRIC FUNCTIONS FOR DECIMAL-DEGREES

Deg	Sin	Cos	Tan	Cot	Sec	Csc	Deg
0.0	0.000 00	1.000 0	0.000 00	Infin	1.000 0	Infin	90.0
.1	0.001 75	1.000 0	0.001 75	573.0	1.000 0	572.96	.9
.2	0.003 49	1.000 0	0.003 49	286.5	1.000 0	286.48	.8
.3	0.005 24	1.000 0	0.005 24	191.0	1.000 0	190.99	.7
.4	0.006 98	1.000 0	0.006 98	143.24	1.000 0	143.24	.6
.5	0.008 73	1.000 0	0.008 73	114.59	1.000 0	114.59	.5
.6	0.010 47	0.999 9	0.010 47	95.49	1.000 1	95.495	.4
.7	0.012 22	0.999 9	0.012 22	81.85	1.000 1	81.853	.3
.8	0.013 96	0.999 9	0.013 96	71.62	1.000 1	71.622	.2
.9	0.015 71	0.999 9	0.015 71	63.66	1.000 1	63.665	.1
1.0	0.017 45	0.999 8	0.017 46	57.29	1.000 2	57.299	89.0
.1	0.019 20	0.999 8	0.019 20	52.08	1.000 2	52.090	.9
.2	0.020 94	0.999 8	0.020 95	47.74	1.000 2	47.750	.8
.3	0.022 69	0.999 7	0.022 69	44.07	1.000 3	44.077	.7
.4	0.024 43	0.999 7	0.024 44	40.92	1.000 3	40.930	.6
.5	0.026 18	0.999 7	0.026 19	38.19	1.000 3	38.202	.5
.6	0.027 92	0.999 6	0.027 93	35.80	1.000 4	38.815	.4
.7	0.029 67	0.999 6	0.029 68	33.69	1.000 4	33.708	.3
.8	0.031 41	0.999 5	0.031 43	31.82	1.000 5	31.836	.2
.9	0.033 16	0.999 5	0.033 17	30.14	1.000 6	30.161	.1
2.0	0.034 90	0.999 4	0.034 92	28.64	1.000 6	28.654	88.0
.1	0.036 64	0.999 3	0.036 67	27.27	1.000 7	27.290	.9
.2	0.038 39	0.999 3	0.038 42	26.03	1.000 7	26.050	.8
.3	0.040 13	0.999 2	0.040 16	24.90	1.000 8	24.918	.7
.4	0.041 88	0.999 1	0.041 91	23.86	1.000 9	23.880	.6
.5	0.043 62	0.999 0	0.043 66	22.90	1.001 0	22.926	.5
.6	0.045 36	0.999 0	0.045 41	22.02	1.001 0	22.044	.4
.7	0.047 11	0.998 9	0.047 16	21.20	1.001 1	21.229	.3
.8	0.048 85	0.998 8	0.048 91	20.45	1.001 2	20.471	.2
.9	0.050 59	0.998 7	0.050 66	19.74	1.001 3	19.766	.1
3.0	0.052 34	0.998 6	0.052 41	19.081	1.001 4	19.107	87.0
.1	0.054 08	0.998 5	0.054 16	18.464	1.001 5	18.492	.9
.2	0.055 82	0.998 4	0.055 91	17.886	1.001 6	17.914	.8
.3	0.057 56	0.998 3	0.057 66	17.343	1.001 7	17.372	.7
.4	0.059 31	0.998 2	0.059 41	16.832	1.001 8	16.862	.6
.5	0.061 05	0.998 1	0.061 16	16.350	1.001 9	16.380	.5
.6	0.062 79	0.998 0	0.062 91	15.895	1.002 0	15.926	.4
.7	0.064 53	0.997 9	0.064 67	15.464	1.002 1	15.496	.3
.8	0.066 27	0.997 8	0.066 42	15.056	1.002 2	15.089	.2
.9	0.068 02	0.997 7	0.068 17	14.669	1.002 3	14.703	.1
4.0	0.069 76	0.997 6	0.069 93	14.301	1.002 4	14.336	86.0
.1	0.071 50	0.997 4	0.071 68	13.951	1.002 6	13.987	.9
.2	0.073 24	0.997 3	0.073 44	13.617	1.002 7	13.654	.8
.3	0.074 98	0.997 2	0.075 19	13.300	1.002 8	13.337	.7
.4	0.076 27	0.997 1	0.076 95	12.996	1.003 0	13.035	.6
.5	0.078 46	0.996 9	0.078 70	12.706	1.003 1	12.745	.5
.6	0.080 20	0.996 8	0.080 46	12.429	1.003 2	12.469	.4
.7	0.081 94	0.996 6	0.082 21	12.163	1.003 4	12.204	.3
.8	0.083 68	0.996 5	0.083 97	11.909	1.003 5	11.951	.2
.9	0.085 42	0.996 3	0.085 73	11.664	1.003 7	11.707	.1
5.0	0.087 16	0.996 2	0.087 49	11.430	1.003 8	11.474	85.0
.1	0.088 89	0.996 0	0.089 25	11.205	1.004 0	11.249	.9
.2	0.090 63	0.995 9	0.091 01	10.988	1.004 1	11.034	.8
.3	0.092 37	0.995 7	0.092 77	10.780	1.004 3	10.826	.7
.4	0.094 11	0.995 6	0.094 53	10.579	1.004 5	10.626	.6
.5	0.095 85	0.995 4	0.096 29	10.385	1.004 6	10.433	.5
.6	0.097 58	0.995 2	0.098 05	10.199	1.004 8	10.248	.4
.7	0.099 32	0.995 1	0.099 81	10.019	1.005 0	10.068	.3
.8	0.101 06	0.994 9	0.101 58	9.845	1.005 1	9.895 5	.2
.9	0.102 79	0.994 7	0.103 34	9.677	1.005 3	9.728 3	.1
6.0	0.104 53	0.994 5	0.105 10	9.514	1.005 5	9.566 8	84.0
.1	0.106 26	0.994 3	0.106 87	9.357	1.005 7	9.410 5	.9
.2	0.108 00	0.994 2	0.108 63	9.205	1.005 9	9.259 3	.8
.3	0.109 73	0.994 0	0.110 40	9.058	1.006 1	9.112 9	.7
.4	0.111 47	0.993 8	0.112 17	8.915	1.006 3	8.971 1	.6
.5	0.113 20	0.993 6	0.113 94	8.777	1.006 5	8.833 7	.5
.6	0.114 94	0.993 4	0.115 70	8.643	1.006 7	8.700 4	.4
.7	0.116 67	0.993 2	0.117 47	8.513	1.006 9	8.571 1	.3
.8	0.118 40	0.993 0	0.119 24	8.386	1.007 1	8.445 7	.2
.9	0.120 14	0.992 8	0.121 01	8.264	1.007 3	8.323 8	.1
7.0	0.121 87	0.992 5	0.122 78	8.144	1.007 5	8.205 5	83.0
.1	0.123 60	0.992 3	0.124 56	8.028	1.007 7	8.090 5	.9
.2	0.125 33	0.992 1	0.126 33	7.916	1.007 9	7.978 7	.8
.3	0.127 06	0.991 9	0.128 10	7.806	1.008 2	7.870 0	.7
.4	0.128 80	0.991 7	0.129 88	7.700	1.008 4	7.764 2	.6
.5	0.130 53	0.991 4	0.131 65	7.596	1.008 6	7.661 3	.5
.6	0.132 26	0.991 2	0.133 43	7.495	1.008 9	7.561 1	.4
.7	0.133 99	0.991 0	0.135 21	7.396	1.009 1	7.463 5	.3
.8	0.135 72	0.990 7	0.136 98	7.300	1.009 3	7.368 4	.2
.9	0.137 44	0.990 5	0.138 76	7.207	1.009 6	7.275 7	.1
Deg	Cos	Sin	Cot	Tan	Csc	Sec	Deg

Deg	Sin	Cos	Tan	Cot	Sec	Csc	Deg
8.0	0.139 17	0.990 3	0.140 54	7.115	1.009 8	7.185 3	82.0
.1	0.140 90	0.990 0	0.142 32	7.026	1.010 1	7.097 2	.9
.2	0.142 63	0.989 8	0.144 10	6.940	1.010 3	7.011 2	.8
.3	0.144 36	0.989 5	0.145 88	6.855	1.010 6	6.927 3	.7
.4	0.146 08	0.989 3	0.147 67	6.772	1.010 8	6.845 4	.6
.5	0.147 81	0.989 0	0.149 45	6.691	1.011 1	6.765 5	.5
.6	0.149 54	0.988 8	0.151 24	6.612	1.011 4	6.687 4	.4
.7	0.151 26	0.988 5	0.153 02	6.535	1.011 6	6.611 1	.3
.8	0.152 99	0.988 2	0.154 81	6.460	1.011 9	6.536 6	.2
.9	0.154 71	0.988 0	0.156 60	6.386	1.012 2	6.463 7	.1
9.0	0.156 43	0.987 7	0.158 38	6.314	1.012 5	6.392 5	81.0
.1	0.158 16	0.987 4	0.160 17	6.243	1.012 7	6.322 8	.9
.2	0.159 88	0.987 1	0.161 96	6.174	1.013 0	6.254 6	.8
.3	0.161 60	0.986 9	0.163 76	6.107	1.013 3	6.188 0	.7
.4	0.163 33	0.986 6	0.165 55	6.041	1.013 6	6.122 7	.6
.5	0.165 05	0.986 3	0.167 34	5.976	1.013 9	6.058 9	.5
.6	0.166 77	0.986 0	0.169 14	5.912	1.014 2	5.996 3	.4
.7	0.168 49	0.985 7	0.170 93	5.850	1.014 5	5.935 1	.3
.8	0.170 21	0.985 4	0.172 73	5.789	1.014 8	5.875 1	.2
.9	0.171 93	0.985 1	0.174 53	5.730	1.015 1	5.816 4	.1
10.0	0.173 6	0.984 8	0.176 3	5.671	1.015 4	5.758 8	80.0
.1	0.175 4	0.984 5	0.178 1	5.614	1.015 7	5.702 3	.9
.2	0.177 1	0.984 2	0.179 9	5.558	1.016 1	5.647 0	.8
.3	0.178 8	0.983 9	0.181 7	5.503	1.016 4	5.592 8	.7
.4	0.180 5	0.983 6	0.183 5	5.449	1.016 7	5.539 6	.6
.5	0.182 2	0.983 3	0.185 3	5.396	1.017 0	5.487 4	.5
.6	0.184 0	0.982 9	0.187 1	5.343	1.017 4	5.436 2	.4
.7	0.185 7	0.982 6	0.189 0	5.292	1.017 7	5.386 0	.3
.8	0.187 4	0.982 3	0.190 8	5.242	1.018 0	5.336 7	.2
.9	0.189 1	0.982 0	0.192 6	5.193	1.018 4	5.288 3	.1
11.0	0.190 8	0.981 6	0.194 4	5.145	1.018 7	5.240 8	79.0
.1	0.192 5	0.981 3	0.196 2	5.097	1.019 1	5.194 2	.9
.2	0.194 2	0.981 0	0.198 0	5.050	1.019 4	5.148 4	.8
.3	0.195 9	0.980 6	0.199 8	5.005	1.019 8	5.103 4	.7
.4	0.197 7	0.980 3	0.201 6	4.959	1.020 1	5.059 3	.6
.5	0.199 4	0.979 9	0.203 5	4.915	1.020 5	5.015 9	.5
.6	0.201 1	0.979 6	0.205 3	4.872	1.020 9	4.973 2	.4
.7	0.202 8	0.979 2	0.207 1	4.829	1.021 2	4.931 3	.3
.8	0.204 5	0.978 9	0.208 9	4.787	1.021 6	4.890 1	.2
.9	0.206 2	0.978 5	0.210 7	4.745	1.022 0	4.849 6	.1
12.0	0.207 9	0.978 1	0.212 6	4.705	1.002 3	4.809 7	78.0
.1	0.209 6	0.977 8	0.214 4	4.665	1.022 7	4.770 6	.9
.2	0.211 3	0.977 4	0.216 2	4.625	1.023 1	4.732 1	.8
.3	0.213 0	0.977 0	0.218 0	4.586	1.023 5	4.694 2	.7
.4	0.214 7	0.976 7	0.219 9	4.548	1.023 9	4.656 9	.6
.5	0.216 4	0.976 3	0.221 7	4.511	1.024 3	4.620 2	.5
.6	0.218 1	0.975 9	0.223 5	4.474	1.024 7	4.584 1	.4
.7	0.219 8	0.975 5	0.225 4	4.437	1.025 1	4.548 6	.3
.8	0.221 5	0.975 1	0.227 2	4.402	1.025 5	4.513 7	.2
.9	0.223 3	0.974 8	0.229 0	4.366	1.025 9	4.479 3	.1
13.0	0.225 0	0.974 4	0.230 9	4.331	1.026 3	4.445 4	77.0
.1	0.226 7	0.974 0	0.232 7	4.297	1.026 7	4.412 1	.9
.2	0.228 4	0.973 6	0.234 5	4.264	1.027 1	4.379 2	.8
.3	0.230 0	0.973 2	0.236 4	4.230	1.027 6	4.346 9	.7
.4	0.231 7	0.972 8	0.238 2	4.198	1.028 0	4.315 0	.6
.5	0.233 4	0.972 4	0.240 1	4.165	1.028 4	4.283 7	.5
.6	0.235 1	0.972 0	0.241 9	4.134	1.028 8	4.252 7	.4
.7	0.236 8	0.971 5	0.243 8	4.102	1.029 3	4.222 3	.3
.8	0.238 5	0.971 1	0.245 6	4.071	1.029 7	4.192 3	.2
.9	0.240 2	0.970 7	0.247 5	4.041	1.030 2	4.162 7	.1
14.0	0.241 9	0.970 3	0.249 3	4.011	1.030 6	4.133 6	76.0
.1	0.243 6	0.969 9	0.251 2	3.981	1.031 1	4.104 8	.9
.2	0.245 3	0.969 4	0.253 0	3.952	1.031 5	4.076 5	.8
.3	0.247 0	0.969 0	0.254 9	3.923	1.032 0	4.048 6	.7
.4	0.248 7	0.968 6	0.256 8	3.895	1.032 4	4.021 1	.6
.5	0.250 4	0.968 1	0.258 6	3.867	1.032 9	3.993 9	.5
.6	0.252 1	0.967 7	0.260 5	3.839	1.033 4	3.967 2	.4
.7	0.253 8	0.967 3	0.262 3	3.812	1.033 8	3.940 8	.3
.8	0.255 4	0.966 8	0.264 2	3.785	1.034 3	3.914 7	.2
.9	0.257 1	0.966 4	0.266 1	3.758	1.034 8	3.889 0	.1
15.0	0.258 8	0.965 9	0.267 9	3.732	1.035 3	3.863 7	75.0
.1	0.260 5	0.965 5	0.269 8	3.706	1.035 8	3.838 7	.9
.2	0.262 2	0.965 0	0.271 7	3.681	1.036 3	3.814 0	.8
.3	0.263 9	0.964 6	0.273 6	3.655	1.036 7	3.789 7	.7
.4	0.265 6	0.964 1	0.275 4	3.630	1.037 2	3.765 7	.6
.5	0.267 2	0.963 6	0.277 3	3.606	1.037 7	3.742 0	.5
.6	0.268 9	0.963 2	0.279 2	3.582	1.038 2	3.718 6	.4
.7	0.270 6	0.962 7	0.281 1	3.558	1.038 8	3.695 5	.3
.8	0.272 3	0.962 2	0.283 0	3.534	1.039 3	3.672 7	.2
.9	0.274 0	0.961 7	0.284 9	3.511	1.039 8	3.650 2	.1
Deg	Cos	Sin	Cot	Tan	Csc	Sec	Deg

TABLE A–3 Trigonometric Functions for Decimal-Degrees (Cont'd)

TRIGONOMETRIC FUNCTIONS FOR DECIMAL-DEGREES (Cont'd)

Deg	Sin	Cos	Tan	Cot	Sec	Csc	Deg	Deg	Sin	Cos	Tan	Cot	Sec	Csc	Deg
16.0	0.2756	0.9613	0.2867	3.487	1.0403	3.6280	74.0	24.0	0.4067	0.9135	0.4452	2.246	1.0946	2.4586	66.0
.1	0.2773	0.9608	0.2886	3.465	1.0408	3.6060	.9	.1	0.4083	0.9128	0.4473	2.236	1.0955	2.4490	.9
.2	0.2790	0.9603	0.2905	3.442	1.0413	3.5843	.8	.2	0.4099	0.9121	0.4494	2.225	1.0963	2.4395	.8
.3	0.2807	0.9598	0.2924	3.420	1.0419	3.5629	.7	.3	0.4115	0.9114	0.4515	2.215	1.0972	2.4300	.7
.4	0.2823	0.9593	0.2943	3.398	1.0424	3.5418	.6	.4	0.4131	0.9107	0.4536	2.204	1.0981	2.4207	.6
.5	0.2840	0.9588	0.2962	3.376	1.0429	3.5209	.5	.5	0.4147	0.9100	0.4557	2.194	1.0989	2.4114	.5
.6	0.2857	0.9583	0.2981	3.354	1.0435	3.5003	.4	.6	0.4163	0.9092	0.4578	2.184	1.0999	2.4022	.4
.7	0.2874	0.9578	0.3000	3.333	1.0440	3.4799	.3	.7	0.4179	0.9085	0.4599	2.174	1.1007	2.3931	.3
.8	0.2890	0.9573	0.3019	3.312	1.0446	3.4598	.2	.8	0.4195	0.9078	0.4621	2.164	1.1016	2.3841	.2
.9	0.2907	0.9568	0.3038	3.291	1.0451	3.4399	.1	.9	0.4210	0.9070	0.4642	2.154	1.1025	2.3751	.1
17.0	0.2924	0.9563	0.3057	3.271	1.0457	3.4203	73.0	25.0	0.4226	0.9063	0.4663	2.145	1.1034	2.3662	65.0
.1	0.2940	0.9558	0.3076	3.251	1.0463	3.4009	.9	.1	0.4242	0.9056	0.4684	2.135	1.1043	2.3574	.9
.2	0.2957	0.9553	0.3096	3.230	1.0468	3.3817	.8	.2	0.4258	0.9048	0.4706	2.125	1.1052	2.3486	.8
.3	0.2974	0.9548	0.3115	3.211	1.0474	3.3628	.7	.3	0.4274	0.9041	0.4727	2.116	1.1061	2.3400	.7
.4	0.2990	0.9542	0.3134	3.191	1.0480	3.3440	.6	.4	0.4289	0.9033	0.4748	2.106	1.1070	2.3314	.6
.5	0.3007	0.9537	0.3153	3.172	1.0485	3.3255	.5	.5	0.4305	0.9026	0.4770	2.097	1.1079	2.3228	.5
.6	0.3024	0.9532	0.3172	3.152	1.0491	3.3072	.4	.6	0.4321	0.9018	0.4791	2.087	1.1089	2.3144	.4
.7	0.3040	0.9527	0.3191	3.133	1.0497	3.2891	.3	.7	0.4337	0.9011	0.4813	2.078	1.1098	2.3060	.3
.8	0.3057	0.9521	0.3211	3.115	1.0503	3.2712	.2	.8	0.4352	0.9003	0.4834	2.069	1.1107	2.2976	.2
.9	0.3074	0.9516	0.3230	3.096	1.0509	3.2535	.1	.9	0.4368	0.8996	0.4856	2.059	1.1117	2.2894	.1
18.0	0.3090	0.9511	0.3249	3.078	1.0515	3.2361	72.0	26.0	0.4384	0.8988	0.4877	2.050	1.1126	2.2812	64.0
.1	0.3107	0.9505	0.3269	3.060	1.0521	3.2188	.9	.1	0.4399	0.8980	0.4899	2.041	1.1136	2.2730	.9
.2	0.3123	0.9500	0.3288	3.042	1.0527	3.2017	.8	.2	0.4415	0.8973	0.4921	2.032	1.1145	2.2650	.8
.3	0.3140	0.9494	0.3307	3.024	1.0533	3.1848	.7	.3	0.4431	0.8965	0.4942	2.023	1.1155	2.2570	.7
.4	0.3156	0.9489	0.3327	3.006	1.0539	3.1681	.6	.4	0.4446	0.8957	0.4964	2.014	1.1164	2.2490	.6
.5	0.3173	0.9483	0.3346	2.989	1.0545	3.1515	.5	.5	0.4462	0.8949	0.4986	2.006	1.1174	2.2412	.5
.6	0.3190	0.9478	0.3365	2.971	1.0551	3.1352	.4	.6	0.4478	0.8942	0.5008	1.997	1.1184	2.2333	.4
.7	0.3206	0.9472	0.3385	2.954	1.0557	3.1190	.3	.7	0.4493	0.8934	0.5029	1.988	1.1194	2.2256	.3
.8	0.3223	0.9466	0.3404	2.937	1.0564	3.1030	.2	.8	0.4509	0.8926	0.5051	1.980	1.1203	2.2179	.2
.9	0.3239	0.9461	0.3424	2.921	1.0570	3.0872	.1	.9	0.4524	0.8918	0.5073	1.971	1.1213	2.2103	.1
19.0	0.3256	0.9455	0.3443	2.904	1.0576	3.0716	71.0	27.0	0.4540	0.8910	0.5095	1.963	1.1223	2.2027	63.0
.1	0.3272	0.9449	0.3463	2.888	1.0583	3.0561	.9	.1	0.4555	0.8902	0.5117	1.954	1.1233	2.1952	.9
.2	0.3289	0.9444	0.3482	2.872	1.0589	3.0407	.8	.2	0.4571	0.8894	0.5139	1.946	1.1243	2.1877	.8
.3	0.3305	0.9438	0.3502	2.856	1.0595	3.0256	.7	.3	0.4586	0.8886	0.5161	1.937	1.1253	2.1803	.7
.4	0.3322	0.9432	0.3522	2.840	1.0602	3.0106	.6	.4	0.4602	0.8878	0.5184	1.929	1.1264	2.1730	.6
.5	0.3338	0.9426	0.3541	2.824	1.0608	2.9957	.5	.5	0.4617	0.8870	0.5206	1.921	1.1274	2.1657	.5
.6	0.3355	0.9421	0.3561	2.808	1.0615	2.9811	.4	.6	0.4633	0.8862	0.5228	1.913	1.1284	2.1584	.4
.7	0.3371	0.9415	0.3581	2.793	1.0622	2.9665	.3	.7	0.4648	0.8854	0.5250	1.905	1.1294	2.1513	.3
.8	0.3387	0.9409	0.3600	2.778	1.0628	2.9521	.2	.8	0.4664	0.8846	0.5272	1.897	1.1305	2.1441	.2
.9	0.3404	0.9403	0.3620	2.762	1.0635	2.9379	.1	.9	0.4679	0.8838	0.5295	1.889	1.1315	2.1371	.1
20.0	0.3420	0.9397	0.3640	2.747	1.0642	2.9238	70.0	28.0	0.4695	0.8829	0.5317	1.881	1.1326	2.1301	62.0
.1	0.3437	0.9391	0.3659	2.733	1.0649	2.9099	.9	.1	0.4710	0.8821	0.5340	1.873	1.1336	2.1231	.9
.2	0.3453	0.9385	0.3679	2.718	1.0655	2.8960	.8	.2	0.4726	0.8813	0.5362	1.865	1.1347	2.1162	.8
.3	0.3469	0.9379	0.3699	2.703	1.0662	2.8824	.7	.3	0.4741	0.8805	0.5384	1.857	1.1357	2.1093	.7
.4	0.3486	0.9373	0.3719	2.689	1.0669	2.8688	.6	.4	0.4756	0.8796	0.5407	1.849	1.1368	2.1025	.6
.5	0.3502	0.9367	0.3739	2.675	1.0676	2.8555	.5	.5	0.4772	0.8788	0.5430	1.842	1.1379	2.0957	.5
.6	0.3518	0.9361	0.3759	2.660	1.0683	2.8422	.4	.6	0.4787	0.8780	0.5452	1.834	1.1390	2.0890	.4
.7	0.3535	0.9354	0.3779	2.646	1.0690	2.8291	.3	.7	0.4802	0.8771	0.5475	1.827	1.1401	2.0824	.3
.8	0.3551	0.9348	0.3799	2.633	1.0697	2.8161	.2	.8	0.4818	0.8763	0.5498	1.819	1.1412	2.0757	.2
.9	0.3567	0.9342	0.3819	2.619	1.0704	2.8032	.1	.9	0.4833	0.8755	0.5520	1.811	1.1423	2.0692	.1
21.0	0.3584	0.9336	0.3839	2.605	1.0711	2.7904	69.0	29.0	0.4848	0.8746	0.5543	1.804	1.1434	2.0627	61.0
.1	0.3600	0.9330	0.3859	2.592	1.0719	2.7778	.9	.1	0.4863	0.8738	0.5566	1.797	1.1445	2.0562	.9
.2	0.3616	0.9323	0.3879	2.578	1.0726	2.7653	.8	.2	0.4879	0.8729	0.5589	1.789	1.1456	2.0498	.8
.3	0.3633	0.9317	0.3899	2.565	1.0733	2.7529	.7	.3	0.4894	0.8721	0.5612	1.782	1.1467	2.0434	.7
.4	0.3649	0.9311	0.3919	2.552	1.0740	2.7407	.6	.4	0.4909	0.8712	0.5635	1.775	1.1478	2.0371	.6
.5	0.3665	0.9304	0.3939	2.539	1.0748	2.7285	.5	.5	0.4924	0.8704	0.5658	1.767	1.1490	2.0308	.5
.6	0.3681	0.9298	0.3959	2.526	1.0755	2.7165	.4	.6	0.4939	0.8695	0.5681	1.760	1.1501	2.0245	.4
.7	0.3697	0.9291	0.3979	2.513	1.0763	2.7046	.3	.7	0.4955	0.8686	0.5704	1.753	1.1512	2.0183	.3
.8	0.3714	0.9285	0.4000	2.500	1.0770	2.6927	.2	.8	0.4970	0.8678	0.5727	1.746	1.1524	2.0122	.2
.9	0.3730	0.9278	0.4020	2.488	1.0778	2.6811	.1	.9	0.4985	0.8669	0.5750	1.739	1.1535	2.0061	.1
22.0	0.3746	0.9272	0.4040	2.475	1.0785	2.6695	68.0	30.0	0.5000	0.8660	0.5774	1.7321	1.1547	2.0000	60.0
.1	0.3765	0.9265	0.4061	2.463	1.0793	2.6580	.9	.1	0.5015	0.8652	0.5797	1.7251	1.1559	1.9940	.9
.2	0.3778	0.9259	0.4081	2.450	1.0801	2.6466	.8	.2	0.5030	0.8643	0.5820	1.7182	1.1570	1.9880	.8
.3	0.3795	0.9252	0.4101	2.438	1.0808	2.6354	.7	.3	0.5045	0.8634	0.5844	1.7113	1.1582	1.9821	.7
.4	0.3811	0.9245	0.4122	2.426	1.0816	2.6242	.6	.4	0.5060	0.8625	0.5867	1.7045	1.1594	1.9762	.6
.5	0.3827	0.9239	0.4142	2.414	1.0824	2.6131	.5	.5	0.5075	0.8616	0.5890	1.6977	1.1606	1.9703	.5
.6	0.3843	0.9232	0.4163	2.402	1.0832	2.6022	.4	.6	0.5090	0.8607	0.5914	1.6909	1.1618	1.9645	.4
.7	0.3859	0.9225	0.4183	2.391	1.0840	2.5913	.3	.7	0.5105	0.8599	0.5938	1.6842	1.1630	1.9587	.3
.8	0.3875	0.9219	0.4204	2.379	1.0848	2.5805	.2	.8	0.5120	0.8590	0.5961	1.6775	1.1642	1.9530	.2
.9	0.3891	0.9212	0.4224	2.367	1.0856	2.5699	.1	.9	0.5135	0.8581	0.5985	1.6709	1.1654	1.9473	.1
23.0	0.3907	0.9205	0.4245	2.356	1.0864	2.5593	67.0	31.0	0.5150	0.8572	0.6009	1.6643	1.1666	1.9416	59.0
.1	0.3923	0.9198	0.4265	2.344	1.0872	2.5488	.9	.1	0.5165	0.8563	0.6032	1.6577	1.1679	1.9360	.9
.2	0.3939	0.9191	0.4286	2.333	1.0880	2.5384	.8	.2	0.5180	0.8554	0.6056	1.6512	1.1691	1.9304	.8
.3	0.3955	0.9184	0.4307	2.322	1.0888	2.5282	.7	.3	0.5195	0.8545	0.6080	1.6447	1.1703	1.9249	.7
.4	0.3971	0.9178	0.4327	2.311	1.0896	2.5180	.6	.4	0.5210	0.8536	0.6104	1.6383	1.1716	1.9194	.6
.5	0.3987	0.9171	0.4348	2.300	1.0904	2.5078	.5	.5	0.5225	0.8526	0.6128	1.6319	1.1728	1.9139	.5
.6	0.4003	0.9164	0.4369	2.289	1.0913	2.4978	.4	.6	0.5240	0.8517	0.6152	1.6255	1.1741	1.9084	.4
.7	0.4019	0.9157	0.4390	2.278	1.0921	2.4879	.3	.7	0.5255	0.8508	0.6176	1.6191	1.1753	1.9031	.3
.8	0.4035	0.9150	0.4411	2.267	1.0929	2.4780	.2	.8	0.5270	0.8499	0.6200	1.6128	1.1766	1.8977	.2
.9	0.4051	0.9143	0.4431	2.257	1.0938	2.4683	.1	.9	0.5284	0.8490	0.6224	1.6066	1.1779	1.8924	.1
Deg	Cos	Sin	Cot	Tan	Csc	Sec	Deg	Deg	Cos	Sin	Cot	Tan	Csc	Sec	Deg

TABLE A–3 Trigonometric Functions for Decimal-Degrees (Cont'd)

TRIGONOMETRIC FUNCTIONS FOR DECIMAL-DEGREES (Cont'd)

Deg	Sin	Cos	Tan	Cot	Sec	Csc	Deg	Deg	Sin	Cos	Tan	Cot	Sec	Csc	Deg
32.0	0.5299	0.8480	0.6249	1.6003	1.1792	1.8871	58.0	39.0	0.6293	0.7771	0.8098	1.2349	1.2868	1.5890	51.0
.1	0.5314	0.8471	0.6273	1.5941	1.1805	1.8818	.9	.1	0.6307	0.7760	0.8127	1.2305	1.2886	1.5856	.9
.2	0.5329	0.8462	0.6297	1.5880	1.1818	1.8766	.8	.2	0.6320	0.7749	0.8156	1.2261	1.2904	1.5822	.8
.3	0.5344	0.8453	0.6322	1.5818	1.1831	1.8714	.7	.3	0.6334	0.7738	0.8185	1.2218	1.2923	1.5788	.7
.4	0.5358	0.8443	0.6346	1.5757	1.1844	1.8663	.6	.4	0.6347	0.7727	0.8214	1.2174	1.2941	1.5755	.6
.5	0.5373	0.8434	0.6371	1.5697	1.1857	1.8612	.5	.5	0.6361	0.7716	0.8243	1.2131	1.2960	1.5721	.5
.6	0.5388	0.8425	0.6395	1.5637	1.1870	1.8561	.4	.6	0.6374	0.7705	0.8273	1.2088	1.2978	1.5688	.4
.7	0.5402	0.8415	0.6420	1.5577	1.1883	1.8510	.3	.7	0.6388	0.7694	0.8302	1.2045	1.2997	1.5655	.3
.8	0.5417	0.8406	0.6445	1.5517	1.1897	1.8460	.2	.8	0.6401	0.7683	0.8332	1.2002	1.3016	1.5622	.2
.9	0.5432	0.8396	0.6469	1.5458	1.1910	1.8410	.1	.9	0.6414	0.7672	0.8361	1.1960	1.3035	1.5590	.1
33.0	0.5446	0.8387	0.6494	1.5399	1.1924	1.8361	57.0	40.0	0.6428	0.7660	0.8391	1.1918	1.3054	1.5557	50.0
.1	0.5461	0.8377	0.6519	1.5340	1.1937	1.8312	.9	.1	0.6441	0.7649	0.8421	1.1875	1.3073	1.5525	.9
.2	0.5476	0.8368	0.6544	1.5282	1.1951	1.8263	.8	.2	0.6455	0.7638	0.8451	1.1833	1.3093	1.5493	.8
.3	0.5490	0.8358	0.6569	1.5224	1.1964	1.8214	.7	.3	0.6468	0.7627	0.8481	1.1792	1.3112	1.5461	.7
.4	0.5505	0.8348	0.6594	1.5166	1.1978	1.8166	.6	.4	0.6481	0.7615	0.8511	1.1750	1.3131	1.5429	.6
.5	0.5519	0.8339	0.6619	1.5108	1.1992	1.8118	.5	.5	0.6494	0.7604	0.8541	1.1708	1.3151	1.5398	.5
.6	0.5534	0.8329	0.6644	1.5051	1.2006	1.8070	.4	.6	0.6508	0.7593	0.8571	1.1667	1.3171	1.5366	.4
.7	0.5548	0.8320	0.6669	1.4994	1.2020	1.8023	.3	.7	0.6521	0.7581	0.8601	1.1626	1.3190	1.5335	.3
.8	0.5563	0.8310	0.6694	1.4938	1.2034	1.7976	.2	.8	0.6534	0.7570	0.8632	1.1585	1.3210	1.5304	.2
.9	0.5577	0.8300	0.6720	1.4882	1.2048	1.7929	.1	.9	0.6547	0.7559	0.8662	1.1544	1.3230	1.5273	.1
34.0	0.5592	0.8290	0.6745	1.4826	1.2062	1.7883	56.0	41.0	0.6561	0.7547	0.8693	1.1504	1.3250	1.5243	49.0
.1	0.5606	0.8281	0.6771	1.4770	1.2076	1.7837	.9	.1	0.6574	0.7536	0.8724	1.1463	1.3270	1.5212	.9
.2	0.5621	0.8271	0.6796	1.4715	1.2091	1.7791	.8	.2	0.6587	0.7524	0.8754	1.1423	1.3291	1.5182	.8
.3	0.5635	0.8261	0.6822	1.4659	1.2105	1.7745	.7	.3	0.6600	0.7513	0.8785	1.1383	1.3311	1.5151	.7
.4	0.5650	0.8251	0.6847	1.4605	1.2120	1.7700	.6	.4	0.6613	0.7501	0.8816	1.1343	1.3331	1.5121	.6
.5	0.5664	0.8241	0.6873	1.4550	1.2134	1.7655	.5	.5	0.6626	0.7490	0.8847	1.1303	1.3352	1.5092	.5
.6	0.5678	0.8231	0.6899	1.4496	1.2149	1.7610	.4	.6	0.6639	0.7478	0.8878	1.1263	1.3373	1.5062	.4
.7	0.5693	0.8221	0.6924	1.4442	1.2163	1.7566	.3	.7	0.6652	0.7466	0.8910	1.1224	1.3393	1.5032	.3
.8	0.5707	0.8211	0.6950	1.4388	1.2178	1.7522	.2	.8	0.6665	0.7455	0.8941	1.1185	1.3414	1.5003	.2
.9	0.5721	0.8202	0.6976	1.4335	1.2193	1.7478	.1	.9	0.6678	0.7443	0.8972	1.1145	1.3435	1.4974	.1
35.0	0..5736	0.8192	0.7002	1.4281	1.2208	1.7434	55.0	42.0	0.6691	0.7431	0.9004	1.1106	1.3456	1.4945	48.0
.1	0.5750	0.8181	0.7028	1.4229	1.2223	1.7391	.9	.1	0.6704	0.7420	0.9036	1.1067	1.3478	1.4916	.9
.2	0.5764	0.8171	0.7054	1.4176	1.2238	1.7348	.8	.2	0.6717	0.7408	0.9067	1.1028	1.3499	1.4887	.8
.3	0.5779	0.8161	0.7080	1.4124	1.2253	1.7305	.7	.3	0.6730	0.7396	0.9099	1.0990	1.3520	1.4859	.7
.4	0.5793	0.8151	0.7107	1.4071	1.2268	1.7263	.6	.4	0.6743	0.7385	0.9131	1.0951	1.3542	1.4830	.6
.5	0.5807	0.8141	0.7133	1.4019	1.2283	1.7221	.5	.5	0.6756	0.7373	0.9163	1.0913	1.3563	1.4802	.5
.6	0.5821	0.8131	0.7159	1.3968	1.2299	1.7179	.4	.6	0.6769	0.7361	0.9195	1.0875	1.3585	1.4774	.4
.7	0.5835	0.8121	0.7186	1.3916	1.2314	1.7137	.3	.7	0.6782	0.7349	0.9228	1.0837	1.3607	1.4746	.3
.8	0.5850	0.8111	0.7212	1.3865	1.2329	1.7095	.2	.8	0.6794	0.7337	0.9260	1.0799	1.3629	1.4718	.2
.9	0.5864	0.8100	0.7239	1.3814	1.2345	1.7054	.1	.9	0.6807	0.7325	0.9293	1.0761	1.3651	1.4690	.1
36.0	0.5878	0.8090	0.7265	1.3764	1.2361	1.7013	54.0	43.0	0.6820	0.7314	0.9325	1.0724	1.3673	1.4663	47.0
.1	0.5892	0.8080	0.7292	1.3713	1.2376	1.6972	.9	.1	0.6833	0.7302	0.9358	1.0686	1.3696	1.4635	.9
.2	0.5906	0.8070	0.7319	1.3663	1.2392	1.6932	.8	.2	0.6845	0.7290	0.9391	1.0649	1.3718	1.4608	.8
.3	0.5920	0.8059	0.7346	1.3613	1.2408	1.6892	.7	.3	0.6858	0.7278	0.9424	1.0612	1.3741	1.4581	.7
.4	0.5934	0.8049	0.7373	1.3564	1.2424	1.6852	.6	.4	0.6871	0.7266	0.9457	1.0575	1.3763	1.4554	.6
.5	0.5948	0.8039	0.7400	1.3514	1.2440	1.6812	.5	.5	0.6884	0.7254	0.9490	1.0538	1.3786	1.4527	.5
.6	0.5962	0.8028	0.7427	1.3465	1.2456	1.6772	.4	.6	0.6896	0.7242	0.9523	1.0501	1.3809	1.4501	.4
.7	0.5976	0.8018	0.7454	1.3416	1.2472	1.6733	.3	.7	0.6909	0.7230	0.9556	1.0464	1.3832	1.4474	.3
.8	0.5990	0.8007	0.7481	1.3367	1.2489	1.6694	.2	.8	0.6921	0.7218	0.9590	1.0428	1.3855	1.4448	.2
.9	0.6004	0.7997	0.7508	1.3319	1.2505	1.6655	.1	.9	0.6934	0.7206	0.9623	1.0392	1.3878	1.4422	.1
37.0	0.6018	0.7986	0.7536	1.3270	1.2521	1.6616	53.0	44.0	0.6947	0.7193	0.9657	1.0355	1.3902	1.4396	46.0
.1	0.6032	0.7976	0.7563	1.3222	1.2538	1.6578	.9	.1	0.6959	0.7181	0.9691	1.0319	1.3925	1.4370	.9
.2	0.6046	0.7965	0.7590	1.3175	1.2554	1.6540	.8	.2	0.6972	0.7169	0.9725	1.0283	1.3949	1.4344	.8
.3	0.6060	0.7955	0.7618	1.3127	1.2571	1.6502	.7	.3	0.6984	0.7157	0.9759	1.0247	1.3972	1.4318	.7
.4	0.6074	0.7944	0.7646	1.3079	1.2588	1.6464	.6	.4	0.6997	0.7145	0.9793	1.0212	1.3996	1.4293	.6
.5	0.6088	0.7934	0.7673	1.3032	1.2605	1.6427	.5	.5	0.7009	0.7133	0.9827	1.0176	1.4020	1.4267	.5
.6	0.6101	0.7923	0.7701	1.2985	1.2622	1.6390	.4	.6	0.7022	0.7120	0.9861	1.0141	1.4044	1.4242	.4
.7	0.6115	0.7912	0.7729	1.2938	1.2639	1.6353	.3	.7	0.7034	0.7108	0.9896	1.0105	1.4069	1.4217	.3
.8	0.6129	0.7902	0.7757	1.2892	1.2656	1.6316	.2	.8	0.7046	0.7096	0.9930	1.0070	1.4093	1.4192	.2
.9	0.6143	0.7891	0.7785	1.2846	1.2673	1.6279	.1	.9	0.7059	0.7083	0.9965	1.0035	1.4118	1.4167	.1
38.0	0.6157	0.7880	0.7813	1.2799	1.2690	1.6243	52.0	45.0	0.7071	0.7071	1.0000	1.0000	1.4142	1.4142	45.0
.1	0.6170	0.7869	0.7841	1.2753	1.2708	1.6207	.9								
.2	0.6184	0.7859	0.7869	1.2708	1.2725	1.6171	.8								
.3	0.6198	0.7848	0.7898	1.2662	1.2742	1.6135	.7								
.4	0.6211	0.7837	0.7926	1.2617	1.2760	1.6099	.6								
.5	0.6225	0.7826	0.7954	1.2572	1.2778	1.6064	.5								
.6	0.6239	0.7815	0.7983	1.2527	1.2796	1.6029	.4								
.7	0.6252	0.7804	0.8012	1.2482	1.2813	1.5994	.3								
.8	0.6266	0.7793	0.8040	1.2437	1.2831	1.5959	.2								
.9	0.6280	0.7782	0.8069	1.2393	1.2849	1.5925	.1								
Deg	Cos	Sin	Cot	Tan	Csc	Sec	Deg	Deg	Cos	Sin	Cot	Tan	Csc	Sec	Deg

TABLE A-4 Logarithmic Table

N	0	1	2	3	4	5	6	7	8	9
10	0000	0043	0086	0128	0170	0212	0253	0294	0334	0374
11	0414	0453	0492	0531	0569	0607	0645	0682	0719	0755
12	0792	0828	0864	0899	0934	0969	1004	1038	1072	1106
13	1139	1173	1206	1239	1271	1303	1335	1367	1399	1430
14	1461	1492	1523	1553	1584	1614	1644	1673	1703	1732
15	1761	1790	1818	1847	1875	1903	1931	1959	1987	2014
16	2041	2068	2095	2122	2148	2175	2201	2227	2253	2279
17	2304	2330	2355	2380	2405	2430	2455	2480	2504	2529
18	2553	2577	2601	2625	2648	2672	2695	2718	2742	2765
19	2788	2810	2833	2856	2878	2900	2923	2945	2967	2989
20	3010	3032	3054	3075	3096	3118	3139	3160	3181	3201
21	3222	3243	3263	3284	3304	3324	3345	3365	3385	3404
22	3424	3444	3464	3483	3502	3522	3541	3560	3579	3598
23	3617	3636	3655	3674	3692	3711	3729	3747	3766	3784
24	3802	3820	3838	3856	3874	3892	3909	3927	3945	3962
25	3979	3997	4014	4031	4048	4065	4082	4099	4116	4133
26	4150	4166	4183	4200	4216	4232	4249	4265	4281	4298
27	4314	4330	4346	4362	4378	4393	4409	4425	4440	4456
28	4472	4487	4502	4518	4533	4548	4564	4579	4594	4609
29	4624	4639	4654	4669	4683	4698	4713	4728	4742	4757
30	4771	4786	4800	4814	4829	4843	4857	4871	4886	4900
31	4914	4928	4942	4955	4969	4983	4997	5011	5024	5038
32	5051	5065	5079	5092	5105	5119	5132	5145	5159	5172
33	5185	5198	5211	5224	5237	5250	5263	5276	5289	5302
34	5315	5328	5340	5353	5366	5378	5391	5403	5416	5428
35	5441	5453	5465	5478	5490	5502	5514	5527	5539	5551
36	5563	5575	5587	5599	5611	5623	5635	5647	5658	5670
37	5682	5694	5705	5717	5729	5740	5752	5763	5775	5786
38	5798	5809	5821	5832	5843	5855	5866	5877	5888	5899
39	5911	5922	5933	5944	5955	5966	5977	5988	5999	6010
40	6021	6031	6042	6053	6064	6075	6085	6096	6107	6117
41	6128	6138	6149	6160	6170	6180	6191	6201	6212	6222
42	6232	6243	6253	6263	6274	6284	6294	6304	6314	6325
43	6335	6345	6355	6365	6375	6385	6395	6405	6415	6425
44	6435	6444	6454	6464	6474	6484	6493	6503	6513	6522
45	6532	6542	6551	6561	6571	6580	6590	6599	6609	6618
46	6628	6637	6646	6656	6665	6675	6684	6693	6702	6712
47	6721	6730	6739	6749	6758	6767	6776	6785	6794	6803
48	6812	6821	6830	6839	6848	6857	6866	6875	6884	6893
49	6902	6911	6920	6928	6937	6946	6955	6964	6972	6981
50	6990	6998	7007	7016	7024	7033	7042	7050	7059	7067
51	7076	7084	7093	7101	7110	7118	7126	7135	7143	7152
52	7160	7168	7177	7185	7193	7202	7210	7218	7226	7235
53	7243	7251	7259	7267	7275	7284	7292	7300	7308	7316
54	7324	7332	7340	7348	7356	7364	7372	7380	7388	7396
55	7404	7412	7419	7427	7435	7443	7451	7459	7466	7474
56	7482	7490	7497	7505	7513	7520	7528	7536	7543	7551
57	7559	7566	7574	7582	7589	7597	7604	7612	7619	7627
58	7634	7642	7649	7657	7664	7672	7679	7686	7694	7701
59	7709	7716	7723	7731	7738	7745	7752	7760	7767	7774
60	7782	7789	7796	7803	7810	7818	7825	7832	7839	7846
61	7853	7860	7868	7875	7882	7889	7896	7903	7910	7917
62	7924	7931	7938	7945	7952	7959	7966	7973	7980	7987
63	7993	8000	8007	8014	8021	8028	8035	8041	8048	8055
64	8062	8069	8075	8082	8089	8096	8102	8109	8116	8122
65	8129	8136	8142	8149	8156	8162	8169	8176	8182	8189
66	8195	8202	8209	8215	8222	8228	8235	8241	8248	8254
67	8261	8267	8274	8280	8287	8293	8299	8306	8312	8319

	0	1	2	3	4	5	6	7	8	9
68	8325	8331	8338	8344	8351	8357	8363	8370	8376	8382
69	8388	8395	8401	8407	8414	8420	8426	8432	8439	8445
70	8451	8457	8463	8470	8476	8482	8488	8494	8500	8506
71	8513	8519	8525	8531	8537	8543	8549	8555	8561	8567
72	8573	8579	8585	8591	8597	8603	8609	8615	8621	8627
73	8633	8639	8645	8651	8657	8663	8669	8675	8681	8686
74	8692	8698	8704	8710	8716	8722	8727	8733	8739	8745
75	8751	8756	8762	8768	8774	8779	8785	8791	8797	8802
76	8808	8814	8820	8825	8831	8837	8842	8848	8854	8859
77	8865	8871	8876	8882	8887	8893	8899	8904	8910	8915
78	8921	8927	8932	8938	8943	8949	8954	8960	8965	8971
79	8976	8982	8987	8993	8998	9004	9009	9015	9020	9025
80	9031	9036	9042	9047	9053	9058	9063	9069	9074	9079
81	9085	9090	9096	9101	9106	9112	9117	9122	9128	9133
82	9138	9143	9149	9154	9159	9165	9170	9175	9180	9186
83	9191	9196	9201	9206	9212	9217	9222	9227	9232	9238
84	9243	9248	9253	9258	9263	9269	9274	9279	9284	9289
85	9294	9299	9304	9309	9315	9320	9325	9330	9335	9340
86	9345	9350	9355	9360	9365	9370	9375	9380	9385	9390
87	9395	9400	9405	9410	9415	9420	9425	9430	9435	9440
88	9445	9450	9455	9460	9465	9469	9474	9479	9484	9489
89	9494	9499	9504	9509	9513	9518	9523	9528	9533	9538
90	9542	9547	9552	9557	9562	9566	9571	9576	9581	9586
91	9590	9595	9600	9605	9609	9614	9619	9624	9628	9633
92	9638	9643	9647	9652	9657	9661	9666	9671	9675	9680
93	9685	9689	9694	9699	9703	9708	9713	9717	9722	9727
94	9731	9736	9741	9745	9750	9754	9759	9763	9768	9773
95	9777	9782	9786	9791	9795	9800	9805	9809	9814	9818
96	9823	9827	9832	9836	9841	9845	9850	9854	9859	9863
97	9868	9872	9877	9881	9886	9890	9894	9899	9903	9908
98	9912	9917	9921	9926	9930	9934	9939	9943	9948	9952
99	9956	9961	9965	9969	9974	9978	9983	9987	9991	9996

*This table gives the mantissas of numbers with the decimal point omitted in each case. Characteristics are determined by inspection from the numbers.

Answers to Odd-Numbered Exercises

CHAPTER 1
Exercise 1-1
1. $\dfrac{a}{b} = d$
3. $A = \dfrac{ab}{2}$
5. $A + B = \dfrac{D}{2}$
7. $0.4\,V_m = v$
9. $t = \dfrac{1}{f}$

Exercise 1-2
1. 18
3. −27.5
5. 18
7. 181.3
9. −1660
11. −5.1
13. 0
15. −132
17. 17
19. 0
21. 25
23. 1860
25. 88.7
27. 49
29. 60

Exercise 1-3
1. 6 V
3. 0 V
5. 25°
7. 6.4 ft
9. 32°
11. 6 V
13. 8°
15. 0

Exercise 1-4
1. −42
3. 21
5. −24
7. −360
9. 7680
11. 4
13. 4
15. −32.8
17. −26.9
19. 0.72
21. −6
23. 10
25. −40
27. 1.14

CHAPTER 2
Exercise 2-1
ADD:
1. $47a^2$
3. $1860b^2$
5. $-41I^2$
7. $136a^2b + 36ab^2$
9. $25iR + 10Ir$
11. $-59IR$

SUBTRACT:
1. $12a$
3. $175I$
5. $18.8V$
7. $16I^2 - 4i^2$
9. $48A$
11. $212Z$

Exercise 2-2
1. $7a$
3. $41R - 2R^2$
5. $3IX^2 + 3X - 4IR - 6$
7. $-9I^2 - 13E + 24$
9. 0
11. $-8v^2 + 6v - 9$

Exercise 2-3
1. $48v^5V$
3. $72a^9b^4c$
5. $800R^3V^3$
7. $36I^5R$
9. a^{13}
11. 648

Exercise 2-4
1. $4a$
3. R^2
5. $\dfrac{4i}{r}$
7. $\dfrac{0.333H^2}{R^2}$
9. $16M^4$
11. $\dfrac{2b^4}{a^4}$

Exercise 2-5
1. $24v^3 + 18Rv$
3. $-48V^3v + 36V^2 - 18V^2IR$
5. $8a^4b^3c^2 + 8a^5b^2c + 8a^3b^3c + 8a^3b^2c^4$
7. $-bc - 7 + ab - a^4$
9. $9i + 8R^2 - 6E$
11. $c^2 - 18cV + 32c$

Exercise 2-6
1. $6b - \dfrac{c}{2a}$
3. $2IR \quad 0.5R$
5. $\dfrac{3R}{a} - \dfrac{7b^2}{R} + \dfrac{8}{aR}$
7. $\dfrac{12V^2}{I} + 7V - \dfrac{5I^2}{V}$
9. $v - 4 + \dfrac{6}{v^2}$
11. $\dfrac{3F^2}{f} - \dfrac{f^3}{F} + 9 - \dfrac{12}{F}$

Exercise 2-7
1. $6V - 10$
3. $-4E + 25$
5. $7a^2 - a - 10$
7. $24v^3 - 36v^2$

CHAPTER 3
Exercise 3-1
1. $a^2 + 10a + 16$
3. $18E^2 + 39E + 15$
5. $20a^2 - 43a + 21$
7. $6V^2 - V - 35$
9. $21E^2 - 23E + 6$
11. $8b^2 + 18bc - 35c^2$
13. $15V^2 + 12V - 36$
15. $12a^2c^2 + 29abc + 15b^2$
17. $10D^2 + 67D + 42$
19. $36 - 24a - 12a^2$

Exercise 3-2
1. $V^2 - 36$
3. $144E^2 - 144$
5. $81a^2b^2 - 9c^2$
7. $V^2R^2 - A^2$
9. $V^6 \times 9b^4$

Exercise 3-3
1. $B^2 + 12B + 36$
3. $16a^2 - 16a + 4$
5. $P^2 + 14P + 49$
7. $9E^2 + 24EI + 16I^2$
9. $\dfrac{a^2}{4} + 6a + 36$
11. $144c^2 - 144c + 36$
13. $16a^2b^2 + 24abc + 9c^2$
15. $I^2 - \dfrac{IR}{2} + \dfrac{R^2}{16}$

Exercise 3-4
1. $6B + 13B^2 - 12B - 12$
3. $3R^3 + 8R^2 - 7R - 12$
5. $2T^4 - 5T^3 + 7T^2 + 5T$
7. $4A^3 + 8A^2 - 5A - 12$

Exercise 3-5
1. $I + 2$
3. $R + 8$
5. $B - 6 + \dfrac{3}{B - 5}$
7. $3c - 7 + \dfrac{4}{2R + 3}$

CHAPTER 4
Exercise 4-1
1. $5(3)$
3. $2^2, 3^2, 5$
5. $2^4, 3^2$
7. $2^3, 5, b^4, c^5$

Exercise 4-2
1. $6a^2(a - 3)$
3. $2IR(4IR^2 + 6I^2R - 3)$
5. $3p(p^2 + 5pq - 2q)$
7. $12ab(2a^3b - 3a^2b^2 + 1)$
9. $3c^3d(1 - 4c^2d^3)$
11. $12I^2R$
13. $5c(c^2 - 1 + c)$
15. $a(10a^4 - 8a^3 - 6a^2 + 2a - 1)$

Exercise 4-3
1. $(I + 5R)^2$
3. $(V - 1)^2$
5. NPS
7. NPS
9. $4(2D + 5B)^2$
11. NPS
13. $(2a + 8)^2$
15. $2(B - 8)^2$
17. NPS

Exercise 4–4
1. $(I - 5)(I - 4)$
3. $(a - 7)(a - 4)$
5. $(E + 10)(E - 3)$
7. $(I - 4)(I - 3)$
9. NF
11. $(8I - 1)(I + 6)$
13. $(3R + 5)(R + 4)$
15. $(3a - 7)(a - 1)$
17. NF
19. $2(8P^2 - 33P - 4)$
21. $(3i + 2)(9i - 1)$
23. $(4c + 7)(c + 7)$

Exercise 4–5
1. $(5I + 3)(5I - 3)$
3. $(3I + 1)(3I - 1)$
5. $(a + 1)(a - 1)$
7. $(5AB + D)(5AB - D)$
9. $4(3p + 10)(3p - 10)$
11. $3e(e + 5)(e - 5)$
13. $(A^2 + 9)(A + 3)(A - 3)$

CHAPTER 5
Exercise 5–1
1. 8.675×10^3
3. 1.78×10^{-2}
5. 8.675×10^1
7. 1.00×10^{-5}
9. 6.28×10^6

Exercise 5–2
1. 8.65×10^7
3. 1.6×10^{-6}
5. 9.2×10^0
7. 1.6×10^{-5}
9. 5×10^1
11. 1.18×10^{-1}
13. 6.28×10^{-4}
15. 1.56×10^{10}

Exercise 5–3
1. 2   3. 1   5. 6   7. 4   9. 3   11. 3

Exercise 5–4
1. 179   3. 0.0632   5. 200   7. 291   9. 100

Exercise 5–5
1. 1.00×10^3
3. 2.45×10^7
5. 3.94×10^1
7. 9.00×10^{-13}
9. 4.00×10^{-1}

Exercise 5–6
1. 2.71×10^1
3. 2.00×10^2
5. 1.06×10^{-4}
7. 5.50×10^{-3}

Exercise 5–7
1. 61.8×10^4
3. 1.80×10^{-8}
5. 1.20×10^{14}
7. 1.81×10^{-7}
9. 4.10×10^3
11. 1.07×10^5
13. 3.16×10^{-5}
15. 8.25×10^{-3}

Exercise 5–8
1. 1.98×10^8
3. 5.593×10^{-4}
5. 5.5×10^{-3}
7. 9.9×10^5

CHAPTER 6
Exercise 6–1
1. 0.036
3. 1,050,000
5. 0.000002
7. 1.6
9. 0.013
11. 0.0000065
13. 55

Exercise 6–2
1. $150 \times 10^{-3}, 1.5 \times 10^{-1}$
3. $300 \times 10^{-3}, 3 \times 10^{-1}$
5. $36 \times 10^{-2}, 3.6 \times 10^{-1}$
7. $1000 \times 10^6, 10^9$
9. $150,000 \times 10^{-3}, 1.5 \times 10^2$
11. $0.004 \times 10^3, 4$
13. $0.250 \times 10^{-6}, 2.5 \times 10^{-7}$
15. $15000 \times 10^3, 1.5 \times 10^7$
17. $10^3, 10^3$
19. $6.3 \times 10^0, 6.3$

Exercise 6–3
1. $50 \times 10^{-6}, 50\ \mu A$
3. $12 \times 10^{-6}, 12\ \mu A$
5. $75 \times 10^{-9}, 75\ nA$
7. $1.75 \times 10^{-3}, 1.75\ mA$
9. $125 \times 10^3, 125\ A$

Exercise 6–4
1. $V = 10.8$ mV
3. $R = 200\ \Omega$
5. $I = 42$ nV
7. $V = 81.6$ mV
9. $V = 6$ V
11. $R = 141$ kΩ

Exercise 6-5
1. $V_1 = 7.14$ V, $V^2 = 8.60$ V, $V_3 = 2.20$ V
3. $P_1 = 1.5$ W, $P_2 = 2.54$ W, $P_T = 4.04$ W
5. 5.32 kΩ

CHAPTER 7
Exercise 7-1
1. -4
3. -5.5
5. 4
7. 7.33
9. 4.3
11. 3
13. 0
15. -1
17. -8.5
19. -9
21. -4.5
23. 1.22

Exercise 7-2
1. -12
3. 2.38
5. 1.5
7. -3
9. 0
11. -0.562

Exercise 7-3
1. $V = ZI$
3. $X_L = QR$
5. $f = \dfrac{X_L}{2\pi L}$
7. $R = \dfrac{V^2}{P}$
9. $g_m = \dfrac{\mu}{r_p}$
11. $I_C = \beta I_b$
13. $A_o = A_v A_i$
15. $f = \dfrac{1}{2\pi f X_C C}$
17. $V_{CC} = V_C + i_C R_L$
19. $V_m = \dfrac{V_{rms}}{0.707}$
21. $f = \dfrac{1}{t}$
23. $V_m = \dfrac{V_{ave}}{0.636}$
25. $E_s = \dfrac{N_S E_p}{N_p}$
27. $E = IR + e$

CHAPTER 8
Exercise 8-1
1. 24
3. ab^2
5. $2i^2 + 8i$
7. $R^2 - R$
9. $e^2 - 8e + 15$

Exercise 8-2
1. $\dfrac{1}{5}$
3. $\dfrac{1}{R}$
5. $\dfrac{E}{I}$
7. $\dfrac{a+b}{a-b}$
9. $\dfrac{I(I+6)}{I+1}$
11. $\dfrac{i+r}{i-r}$
13. $\dfrac{E+4}{E-4}$

Exercise 8-3
1. $\dfrac{4}{5}$
3. $\dfrac{R-E}{E^2}$
5. $\dfrac{(-a^2 b)}{a-b}$
7. $-\left(\dfrac{3}{5}\right)$
9. $-\left(\dfrac{i+r}{r-i}\right)$
11. $-\left(\dfrac{a}{b-a-c}\right)$
13. -1
15. R
17. 1

Exercise 8-4
1. $\dfrac{1}{16}$
3. $\dfrac{acD + b^2 D - c^2 b}{bcD}$
5. $\dfrac{6R + 4I + 3}{IR}$
7. $\dfrac{-4er}{(e+r)(e-r)}$
9. $\dfrac{R_2 R_3 + R_1 R_3 + R_1 R_2}{R_1 R_2 R_3}$
11. $\dfrac{3er + e - 4r}{(e-r)(e-2r)}$
13. $\dfrac{24 - 4I^2}{(I-2)(I+2)^2}$

Exercise 8-5
1. $3ab$
3. $\dfrac{(e+r)}{r}$
5. $\dfrac{4b(a+b)}{a}$
7. $\dfrac{3(I-1)}{I}$
9. $E^2 - 12IE + 12I^2$

Exercise 8-6
1. $\dfrac{I^3}{R^2}$
3. $\dfrac{R(E+1)}{I}$
5. $\dfrac{(P^2+P)}{(P-1)}$
7. $\dfrac{(e-3)(2i+3)}{(3i+4)(e+5)}$

CHAPTER 9
Exercise 9-1
1. 2.67 3. 1 5. 15 7. 6.43 9. −9.57

Exercise 9-2
1. 0.333 3. 2.07 5. 0.75 7. 5.5 9. 6.93

Exercise 9-3
1. $\dfrac{1}{2\pi X_C C}$
3. $\dfrac{X_L}{2\pi L}$
5. $\dfrac{R_2 R_T}{R_2 - R_T}$
7. $\dfrac{V_{bb} - V_C}{R_S}$
9. $\dfrac{C_2 C_T}{C_2 - C_T}$
11. $\dfrac{f_r}{f_2 - f_1}$
13. $\dfrac{A(R_L + r_p)}{R_L}$
15. $\left(\dfrac{V_T}{I}\right) - R_1$

Exercise 9-4
1. 31 Ω 3. 208 kΩ 5. 446 Ω 7. 1.24 μA, 1.79 V

Exercise 9-5
1. 1431 Ω 3. 27.4 Ω 5. 21.2 Ω 7. 28 Ω 9. 23.3 Ω 11. 2.17 kΩ 13. 820 Ω 15. 4.29 V

CHAPTER 10
Exercise 10-1
1. 51.9 Ω 3. 8.83 kΩ 5. 0.268 Ω

Exercise 10-2
1a. sin 62° = 0.8829, cos = 0.4695, tan = 1.881
1c. sin 8.6° = 0.1495, cos = 0.9888, tan = 0.1512
1e. sin 45° = 0.7071, cos = 0.7071, tan = 1.000
2a. 45.0°
2c. 49.5°
2e. 13.7°
3a. 45°
3c. 5.62°
3e. 18.9°
4a. 35.3°
4c. 86.5°
4e. 5.00°
5. 0.5934
7. 0.9291

Exercise 10-3
1. 617 Ω 3. 81.8 Ω 5. 0.997 Ω

Exercise 10-4
1. 416 Ω 3. 107 kΩ 5. 1958 MΩ

Exercise 10-5
1. 671 Ω 3. 156 Ω 5. 1000 kΩ

Exercise 10-6
1. 55.2° 3. 55.3° 5. 8.37°

CHAPTER 11
Exercise 11-1
1. 1.13 3. 4.19 5. 91.7° 7. 138° 9. 430°

Exercise 11-3

Angle	Algebraic sign	Ratio	Quadrant	θ
sin 118	+	0.8829	2	62°
cos 342.5	+	0.9537	4	17.5°
cos 2.5 rad	−	0.8013	2	36.8°
tan −5.35 rad	+	1.349	1	53.4°
cos −1525	+	0.0872	4	85°
sin 26.5	−	0.4462	4	26.5°
sin 45	+	0.7071	1	45°
cos −942	−	0.7431	3	42°

Exercise 11-4
1. $\theta' = 70.3°, \theta = -70.3°$
3. $\theta' = 55.6°, \theta = 124.4°$
5. $\theta' = 74.2°, \theta = 105.8°$
7. $\theta' = 70.5°, \theta = -70.5°$
9. $\theta' = 34.3°, \theta = -34.3°$

CHAPTER 12
Exercise 12-1
1. $A = 7, B = 6$
3. $v = 2, \mu = -1$
5. $I = 2, I_2 = 5$
7. $x = 0.5, y = 0.75$
9. $E = -20.5, e = 17$

Exercise 12-2
1. $I = -1.25, i = -0.5$
3. $B = -1.71, b = -0.625$
5. $B = -7, b = -5$
7. $E = 0.462, e = -1.23$

Exercise 12-3
1. −7
3. −112
5. −20
7. 468

Exercise 12-4
1. $a = 3, y = 2$
3. $e = 6, E = 5$
5. $B = 1.88, A = 2.5$
7. $a = 0.323, b = 2.42$

Exercise 12-5
1. 12
3. 471
5. 260

Exercise 12-6
1. $x = -2, y = 4, z = 1$
3. $I = 2.09, i = -4.15, e = -9.47$
5. $M = 1.68, m = -0.559, I = -3.10$

CHAPTER 13
Exercise 13-1
1. $I_a = 387$ mA, $I_b = -1.92$ A, $I_L = 2.31$ A
3. $I_a = 57$ mA, $I_b = -24.2$ mA, $I_L = 81.2$ mA
5. $I_a = 3.85$ A, $I_b = 3.75$ A, $I_L = 100$ mA
7. $I_a = 375$ mA, $I_b = -1.78$ A

Exercise 13-2
1. $I_a = 652$ mA, $I_b = 304$ mA
3. $I_a = 783$ mA, $I_b = 96.0$ mA, $I_c = 921$ mA
5. $I_a = 320$ mA, $I_b = 39.5$ mA, $I_c = 368$ mA

CHAPTER 14
Exercise 14-1
1. $I = 385$ mA, $I_b = 1.92$ A, $I_L = 2.31$ A
3. $I_a = 57$ mA, $I_b = 24.2$ mA, $I_L = 81.2$ mA
5. $I_a = 3.85$ A, $I_b = 3.75$ A, $I_L = 100$ mA

Exercise 14-2
1. 94.2 mA
3. 14.5 mA
5. 379 mA

CHAPTER 15
Exercise 15-1
1. balanced   3. balanced   5. balanced

Exercise 15-2
1. $I_m = 33$ mA, $I_1 = 108$ mA, $I_3 = 256$ mA   3. 7.95 mA

Exercise 15-3
1. 17.4 mA   3. 117 mA

CHAPTER 16
Exercise 16-2
1. $x = 6.43, y = 41.5$ 5. $x = 0, y = 36$ 9. $x = -45.9, y = -40.5$
3. $x = -72, y = 0$ 7. $x = -454, y = 1534$ 11. $x = 0.57, y = -0.32$

Exercise 16-3
1. $76.3/\underline{31.6°}$ 3. $184/\underline{40.6°}$ 5. $41.5/\underline{220°}$ 7. $1/\underline{90°}$ 9. $48/\underline{180°}$

Exercise 16-4
1. $120/\underline{45°}$ 3. $90.7/\underline{151°}$ 5. $56/\underline{5.32°}$ 7. $140/\underline{134°}$ 9. $41.6/\underline{49.9°}$

CHAPTER 17
Exercise 17-1
1. 28.7°   3. 591 V   5. -15.6 μA   7. -26.5 mA   9. 295 μV   11. 0

Exercise 17-2
1. 29 mA   3. -73.1 V   5. -629 μA   7. 80 V   9. 918 μA

Exercise 17-3

1.
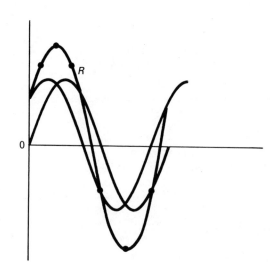

Figure Ex. 17-3(1)

3.
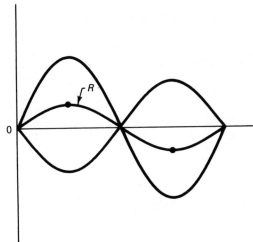
Figure Ex. 17-3(3)

5.
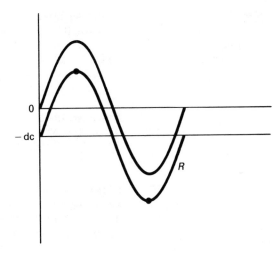
Figure Ex. 17-3(5)

Exercise 17–4
1. -0.584 V 3. 741 mV 5. 10.3 mA 7. -572 μV 9. 150 mV

CHAPTER 18
Exercise 18–1
1. $150\underline{/-20°}$, $x = 141$, $y = -51.3$
3. $30\underline{/180°}$, $x = 30$, $y = 0$
5. $75\underline{/-145°}$, $x = -61.4$, $y = -43$
7. $350\underline{/-90°}$, $x = 0$, $y = -350$
9. $117\underline{/70°}$, $x = 40$, $y = 110$

Exercise 18–2
1. $V_R = 75.5 \sin(\omega t + 14.8°)$ V
3. $V_R = 38 \sin(\omega t + 23.2°$ μA$)$
5. $V_R = 307 \sin(\omega t + 38.5°)$
7. $V_R = 21 \sin(\omega t + 26.2°$ mV$)$

Exercise 18–3
1. $j7$ 3. $j6.71$ 5. $j4z$ 7. $j5.66i$ 9. $j\dfrac{V}{I}$

Exercise 18–4
1. $50\underline{/30°}$, $43.3 + j25$
3. $130\underline{/-60°}$, $65 - j113$
5. $7.5\underline{/-140°}$, $-5.74 + j4.82$
7. $165\underline{/240°}$, $-82.5 - j143$
9. $440\underline{/-90°}$, $0 - j440$

Exercise 18–5
1. j 3. -8 5. 18 7. -4 9. 12 11. $j0.5$

CHAPTER 19
Exercise 19–1
1. $9 + j10$
3. $166 - j63$
5. $-0.35 + j10$
7. $28 - j5$
9. $47 + j10$
11. $1000 + j800$
13. $20 - j16$
15. $0 - j15$ or $-j15$
17. $6 - j1$
19. $7 + j4$
21. $-9 + j16$
23. $18 - j11$
25. $15 - j87$

Exercise 19–2
1. $24 - j8$ 3. $3 + j6$ 5. $a^2 + b^2$ 7. $-j20$ 9. 40 11. $-14 + j60$

Exercise 19–3
1. $0.769 - j1.15$ 3. $0.559 + j0.265$ 5. $0.692 - j1.46$ 7. $-j5$ 9. 1

527

Exercise 19-4
1. 240/107°   3. 64.8/106°   5. 252/−41°   7. 2550/−140°   9. 16.9/0°

Exercise 19-5
1. 5.67/25°   5. 2.4/186°   9. 0.286/−440°
3. 0.578/−37°   7. 0.333/65°

Exercise 19-6
1. 15.8/18.4°   5. 1768/135°   9. 61.4 + j43   13. −27.6 + j23
3. 154/13.1°   7. 69.9/−20.8°   11. −0.3 − j0.52   15. 23.9 + j9.86

CHAPTER 20
Exercise 20-1
1. $0 + j2512\ \Omega$, $2512/90°\ \Omega$ 5. $2700 + j0\ \Omega$, $2700/0°\ \Omega$
3. $0 − j7.96\ \Omega$, $7.96/−90°\ \Omega$ 7. $0 + j471\ \Omega$, $471/90°\ \Omega$

Exercise 20-2
1. $Z_T = 1.53/38.1°$ kΩ, $I_T = 76.5/−38.1°$ mA, $V_R = 92/−38.1°$ V, $V_L = 72.3/51.9°$ V
3. $Z_T = 115/79°$ kΩ, $I_T = 461/−79°$ μA, $V_R = 10.1/−79°$ V, $V_L = 52.1/11°$ V
5. $V_R = 48.8/−21°$ V, $V_L = 18.5/69°$ V, $v_R = 73.8 \sin \omega t$ V

Exercise 20-3
1. $Z_T = 64.4/−29.6°\ \Omega$, $I_T = 3.42/29.6°$ A, $V_R = 191/29.6°$ V, $V_C = 109/−60.4°$ V
3. $V_T = 32.9/0°$ V, $V_C = 27.3/−37.4°$ V
5. $V_R = 53.2/−35°$ V, $V_C = 37.3/−65°$ V

Exercise 20-4
1. $Z_T = 107/9.64°\ \Omega$, $I_T = 793/−9.64°$ mA
3.

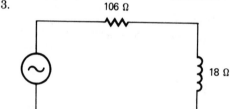

Figure Ex. 20-4(3)

5. $V_1 = 97.7/−42.9°$ V, $V_2 = 66.6/−47.9°$ V, $V_3 = 120/105°$ V
7. $Z_T = 252/−7.97°\ \Omega$, $I_T = 337/7.97°$ mA
9.

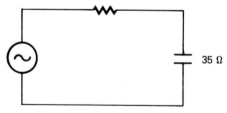

Figure Ex. 20-4(9)

11. $Z_T = 347/−23.8°\ \Omega$, $I_T = 46.1/23.8°$ mA

Exercise 20-5
1. $PF = 73.3\%$, $P_T = 25.8$ W   7. $P_A = 4.2$ VA, $P_R = 4.06$ VAR, $P_T = 1.08$ W
3. $\theta = 34.9°$, $Q = 0.698$ 9. $PF = 73.3\%$, $P_T = 28.3$ μW
5. $P_T = 1853$ μW

CHAPTER 21
Exercise 21-1
1. $Z_T = 218\underline{/29.1°}$ kΩ, $I_T = 185\underline{/-29.1°}$ μA
3. $Z_T = 22.4\underline{/26.6°}$ kΩ, $I_T = 400\underline{/-26.6°}$ mA
5. $Z_T = 53.6\underline{/-45.3°}$ Ω, $I_T = 857\underline{/45.3°}$ mA
7. $Z_T = 98.3\underline{/-4.7°}$ kΩ, $I_T = 509\underline{/4.7°}$ μA

Exercise 21-2
1. $Z_T = 69.2\underline{/4.72°}$ Ω 3. $I_2 = 286\underline{/-30°}$ mA, $333\underline{/-10°}$ mA   5. $Q = 7.4$
7.

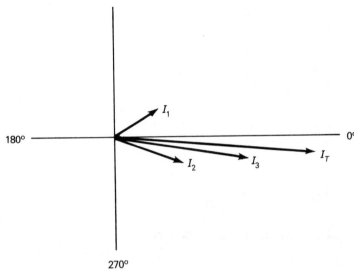

Figure Ex. 21-2(7)

9. $P_T = 27$ W   11. $I_T = 291\underline{/-3.46°}$ mA

Exercise 21-3
1. $P_R = 4526$ VAR, $P_A = 9192$ VA   3. 29.3 μF   5. 25.3 μF

CHAPTER 22
Exrcise 22-1
1. $Z_T = 39.8\underline{/-15.7°}$ Ω
3. 246 μF
5. $P_L = 13.2$ W, $P_S = 727$ mW
7. $13.8\underline{/-14.1°}$ V
9. $P_L = 12.7$ W, $P_S = 4.25$ W
11. $233\underline{/8.02°}$ μA
13. 3.53 μW
15. 0.178

Exercise 22-2
1. 10.6 kHz
3. $V_R = 50.0$ mV, $V_L = 333$ mV, $V_C = 333$ mV
5. $V_R = 277$ μV, $V_C = 277$ μV, $V_R = 300$ μV
7. $I_{line} = 355$ nA, $Z_{cir} = 183$ kΩ
9. $Z_{cir} = 1.25$ MΩ, $I_{line} = 96$ nA

CHAPTER 23
Exercise 23-1
(answers are given in clockwise rotation)
1. 14.4 Ω, 19.3 Ω, 8.89 Ω
3. 35 Ω, 23.3 Ω, 43.8 Ω
5. $2.92\underline{/74.8°}$ Ω, $1.96\underline{/-41.8°}$ Ω, $2.48\underline{/-7.1°}$ Ω
7. $10.4 + j15.2$ Ω, $10.4 + j15.2$ Ω, $21.1 - j3.53$ Ω

Exercise 23-2
1. 6.8 mA   3. $13.1\underline{/9.25°}$ Ω 5. $173\underline{/-58°}$ mA

Exercise 23-3
1. 10.7 mA   3. 536/74.7 µA

Exercise 23-4

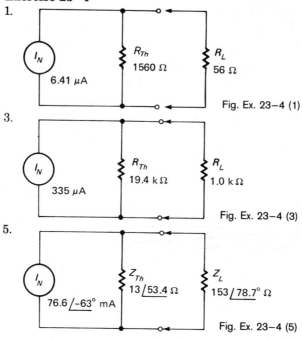

Fig. Ex. 23-4 (1)

Fig. Ex. 23-4 (3)

Fig. Ex. 23-4 (5)

CHAPTER 24
Exercise 24-1
1. 111_2 3. 10011_2 5. 11001_2 7. 1110011_2

Exercise 24-2
1. 130_8 3. 710_8 5. 124_8 7. 617_8

Exercise 24-3
1. 58_{16} 3. 163_{16} 5. $1E9C_{16}$ 7. $87FC_{16}$

Exercise 24-4
1. 7   3. 4   5. 57   7. 493

Exercise 24-5
1. 63   3. 86   5. 255   7. 343

Exercise 24-6
1. 308   3. 1251   5. 11026   7. 4012

Exercise 24-7
1. 265_8 5. 2336_8 9. 110110_2 13. 1011010_2
3. 33264_8 7. 30515_8 11. 1101100_2 15. 1001100100_2

Exercise 24-8
1. $147AC_{16}$ 5. $1FC04_{16}$ 9. 10001001_2 13. 11000010001_2
3. $1EABF_{16}$ 7. $1FF0_{16}$ 11. 1011010101_2 15. 1000010000110001_2

Exercise 24-9
1. 0.001010_2, 0.121_8 5. 0.1101_2, 0.6525_8
3. 0.110010_2, 0.620_8 7. 0.0011_2, 0.1400_8

Exercise 24-10
1. 0.9375   3. 0.59375   5. 0.336   7. 0.2147

Exercise 24–11
1. 10100
3. 101010
5. 100001
7. 11000000.001
9. 101011
11. 1101.001

Exercise 24–12
1. 111
3. 101101
5. 1101010
7. 1010
9. 1100

CHAPTER 25
Exercise 25–1
1. 1   3. 1   5. 0   7. 0   9. 0

Exercise 25–2

1.
A	B	AB
0	0	0
0	1	1
1	0	0
1	1	0

3.
A	B	AB
0	0	1
0	1	0
1	0	0
1	1	0

5.
A	B	C	ABC
0	0	0	0
0	0	1	0
0	1	0	0
0	1	1	0
1	0	0	0
1	0	1	0
1	1	0	0
1	1	1	1

Exercise 25–3

1.
A	B	AB
F	F	F
F	T	F
T	F	F
T	T	T

3.
A	B	A + B
F	F	F
F	T	T
T	F	T
T	T	T

5.
A	B	A + B
F	F	F
F	T	T
T	F	T
T	T	T

Figure Ex. 25-3(1)

Figure Ex. 25-3(3)

Figure Ex. 25-3(5)

7.
A	B	C	D	A + B + C + D
F	F	F	F	F
F	F	F	T	T
F	F	T	F	T
F	F	T	T	T
F	T	F	F	T
F	T	F	T	T
F	T	T	F	T
F	T	T	T	T
T	F	F	F	T
T	F	F	T	T
T	F	T	F	T
T	F	T	T	T
T	T	F	F	T
T	T	F	T	T
T	T	T	F	T
T	T	T	T	T

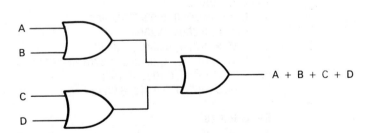

Figure Ex. 25-3(7)

Exercise 25-4
1. A(B + C)

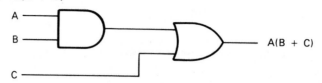

Figure Ex. 25-4(1)

3. $\overline{AB}$

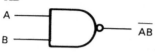

Figure Ex. 25-4(3)

5. $\overline{A} + \overline{B}$

Figure Ex. 25-4(5)

7. $\overline{\overline{AB} + \overline{C}}$

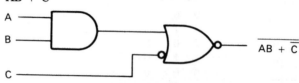

Figure Ex. 25-4(7)

Exercise 25-5
1. A + B

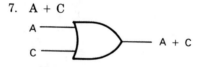

Figure Ex. 25-5(1)

3. 1
5. DE

Figure Ex. 25-5(5)

7. A + C

Figure Ex. 25-5(7)

9. DE

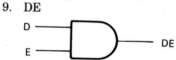

Figure Ex. 25-5(9)

11. AB   13. DEF   15. $\overline{AB}$ 17. Y   19. $\overline{A} + \overline{BC}$ 21. BC

CHAPTER 26
Exercise 26-1
1. 100,000   3. NT   5. 34.6   7. 0.000012   9. NT

Exercise 26-2
1. 1.246×10^3, 3, 0.0955, 3.0955
3. 1.78, 0, 0.2504, 0.2504
5. 2.25×10^4, 4, 0.3522, 4.3522
7. 1.66×10^{-1}, -1, 0.2201, 1.2201 or -7.7989
9. 1.28×10^5, 5, 0.1072, 5.1072
11. 9.00×10^6, 6, 0.9542, 6.9542

Exercise 26-3
1. 1.33×10^1 5. 10^{-3} 9. 3.00
3. 2.75×10^{-3} 7. 7.73×10^8 11. 1.47×10^{-8}

Exercise 26-4
1. 3.798, 8.746   5. 3.208, 7.386   9. 2.718
3. 4.653, 10.7   7. 6.278   11. 0.3679

Exercise 26-5
1. $\log_2 32 = 4$
3. $\log_x R = y$
5. $\log_e Y = X$
7. $3^3 = 1000$
9. $M^P = Y$
11. $e^{2.079} = 8$

Exercise 26-6
1. $\log 5 + \log x$
3. $\log I + \log R$
5. $\log V - \log R$
7. $\dfrac{\log a + \log b}{2}$
9. $\log pv$
11. $\log 2.71a$
13. $\log \dfrac{P_2}{P_1}$
15. $\log \dfrac{X^3 Y^2}{6}$

Exercise 26-7
1. 22.9   3. 215   5. 10^3 7. 1.96   9. 3.8

CHAPTER 27
Exercise 27-1
1. 534 Ω   3. 220 Ω   5. #10   7. 0.748 in   9. #19

Exercise 27-2
1. 10.3 mA
3. $V_C = 48V$, $V_R = 102$ V
5. 6.85 msec
7. 88.8 mA
9. 48 mA
11. 25.5 msec

Exercise 27-3
1. 3.01 dB, 6.02 dB, 9.03 dB, 12.04 dB, 15.05 dB
3. 14 dB, 20 dB, 23.5 dB, 26 dB, 28 dB
5. 72.6 dB
7. −1.4 dB
9. 5.62 W
11. 65.6 dB
13. 1.75 mW

Exercise 27-4
1. 10 mW, 100 mW, 316 mW, 5623 mW, 1778 W, 1000 kW
3. 707 mV = 0 dB
5. −11.9 dBm
7. 116 mV

Exercise 27-5
1. 63.1 W   3. 3.40 dB   5. 157 W   7. −16 dB   9. 39.8 W

CHAPTER 28
Exercise 28-1
1. ±10
3. ±3
5. 31, 19
7. $\pm\sqrt{c - b}$
9. $\pm\sqrt{\dfrac{P}{R}}$
11. ±9.49

Exercise 28-2
1. 3, 2   3. 0.667, −1   5. 0.667b, 0.5b   7. 1, −2.67   9. 1.5, 0.75

Exercise 28-3
1. −4, 2   3. 2, 6   5. 0.920, −4.92   7. 2.73, −0.73   9. 4.79, −3.13

Exercise 28-4
1. −3, −4   3. 0.571, −1.57   5. 0.25, 0

Exercise 28-5
1. 45 A, 3 A   3. 3.42 V, 1.92 V   5. 22.8 Ω

Index

Abbreviations, 514
Absolute value, 5
Ac bridge, 385
Ac parallel circuits, 331
    current, 332, 333
    current ratio, 335
    equivalent series circuit, 339
    impedance, 332, 333
    phasor diagram, 332, 334
    power dissipation, 322, 339
    Quality, Q, 340
    R-C circuits, 333
    R-L circuits, 331
Addition
    binary numbers, 417
    complex number, 293
    fractions, 96
    monomials, 11, 12
    powers of ten, 60
    signed numbers, 3
    vectors–phasors, 234–236
Algebraic expression, 9
    classification of expressions, 10
    simplifying algebraic expressions, 18
Alternating current
    average value, 257
    defined, 254
    effective value, 258
    instantaneous value, 258
    maximum value, 257
    peak-to-peak value, 257
    peak value, 257
    RMS, 258
"And" function, 427
Angles
    converting radians and degrees, 147
    decimal angles, 148
    degrees, 146
    degrees-minutes-seconds, 148
    generation of, 151
    measurement, 146
    negative generation, 151
    radians, 146
Angular velocity, 267
Antilogarithms, 460
Apparent power, 322

Bases, 403
    binary to decimal, 407
    binary to hexadecimal, 411
    binary to octal, 410
    decimal to binary, 404
    decimal to hexadecimal, 406
    decimal to octal, 405
    hexadecimal to decimal, 409
    octal to decimal, 408

Bel, 481
Binary fractions, 413–415
Binary numbers
    addition, 417
    conversion to decimal, 407
    conversion to hexadecimal, 411
    conversion to octal, 410
    fractions, 415
    subtraction by complements, 419
Binomials
    binomial times binomial, 25
    binomial times polynomial, 30
    defined, 10
    special cases of binomial multiplication, 28
    squaring binomials, 29
Boolean algebra
    associative laws, 441
    commutative laws, 441
    defined, 426
    distributive laws, 441
    double negatives, 443
    equivalent functions, 438
    logic statements, 426, 436
    truth tables, 433
Braces, 84
Brackets, 84
Bridge circuit, 213
    ac bridge, 385
    balanced, 213
    solving bridges with determinants, 217
    Thevenizing a bridge, 220
    unbalanced, 215

Capacitive reactance, 307
Cartesian coordinates, 148
Coefficient, 10
Complex numbers, 286
    adding and subtracting, 293
    division, 296
    j operators, 287, 295, 296
    multiplication, 295
Cosine curve
    lagging curve, 261
    leading curve, 261
Current gain, 482

Dc parallel circuits, 114
    branch currents, 116
    current, 115
    power dissipated by the circuit, 117
    power dissipated by each component, 116
    resistance, 115
    voltage, 115
Decibel, 481
    dBm (defined), 485
    to overall gain, 488
    reference levels, 485

Index

Delta (Δ) circuit, 374
    analyzing the bridge, 383
    delta-to-wye conversions, 374
    delta-to-wye conversions using impedance, 378
    wye-to-delta conversions, 377
Determinants
    equations with third-order determinant, 179
    evaluating a matrix, 172
    matrix for third-order determinant, 177
    second-order determinants, 172
    third-order determinants, 177
Division
    binomials, 31
    complex numbers, 296
    by conjugate, 297
    fractions, 102
    monomials, 14
    phasors in polar form, 300
    polynomials, 16, 31
    powers of ten, 57

Engineering notation, 68
Epsilon (e), 461
Equations, 79
    axioms, 79
    defined, 2
    fractional, 106
    graphing linear equations, 164
    with literal numbers, 86, 111
    literal numbers defined, 37
    with numerical denominators, 106
    quadratic, 495
    regime of grouping, 84
    transposing factors, 86
    transposing terms, 80
    with unknowns in the denominators, 108
Equivalent circuits
    delta-to-wye, 374
    Norton's, 206, 392
    Thevenin's, 196, 388
    wye-to-delta, 377
Exponent, 10

Factoring
    the difference of two squares, 46
    by removing the common expression, 39
    trinomial expressions, 43
    trinomial squares, 41
Factors
    defined, 10
    literal, 37
    of numerals, 37
    prime, 37
Fractions, 92
    adding, 96
    binary, 415
    division, 102
    equations, 106
    multiplication, 99
    octal, 415
    reducing to lowest terms, 93
    signs of fractions, 94
    subtracting, 96
Frequency, 255

Harmonic distortion, 271
Harmonics, 270

Imaginary numbers, 283, 284
Impedance, 129, 309
Inductance
    inductive reactance, 307
    parallel circuits, 331, 332
    series circuits, 309, 310
Inverse trigonometric functions, 158

J operator, 283

Kirchhoff's Law, 185

Literal numbers, 9
Logarithmic table, 518–519
Logarithms, 456
    applications, 471
    antilog, 460
    base epsilon, 461
    characteristic, 457
    characteristic is negative, 458
    equations, 462–465
    mantissa, 457
    natural log, 461
    on the calculator, 457
    operation of a log, 456
Logic statements, 426, 436
Loop equations, 184
    three loops, 192
    two loops, 184–191

Metric system, 64
    conversion by using powers of ten, 66
    converting units within the metric system, 64
    SI (Systeme International), 64
Metric units table, 514
Monomials, 10
    addition, 11
    division, 14
    multiplication, 13
    subtraction, 11
Multiplication
    binomials, 25, 28, 29, 30
    complex numbers, 295
    monomials, 13, 15
    phasors in polar form, 298
    polynomials, 30
    using j operator, 295

Natural logs, 461
Networks, 184
Norton's Theorem
    ac, 392
    dc, 206
"Not" function, 428
Numbering systems, 403

Ohm's Law, 69
"Or" function, 428
    exclusive "or", 428
    inclusive "or", 428

Parallel circuits
    alternating current, 331
    direct current, 114
Parentheses, 84
Period, 267
Phase shift, 260
Phasors, 277, 284
    converting polar and rectangular, 302
    dividing, 300
    multiplying, 298
Phi (ϕ), 129
Pi (π) circuit, 374
Polar coordinates, 150
Polynomials, 10
    division, 16, 31
    multiplication, 15, 30
Power dissipation
    in an ac circuit, 322, 339
    in a dc circuit, 116, 117
Power factor, 322
Power factor correction, 347

Power of ten
    addition and subtraction, 60
    division, 57
    engineering notation, 68
    multiplication, 56
    squares and square roots, 59
Pythagorean Theorem, 129

Quadratic equations, 495
    application, 504
    pure quadratics, 495
    quadratic formula, 501
    solution by completing the square, 499
    solution by factoring, 497
Quality, or Q, 325

R-C circuits
    parallel, 333
    series, 314
    time constant, 475
Reactance
    capacitive, 307
    defined, 129
    inductive, 307
Rectangular coordinates, 148
Resistance, 72, 129
    dc parallel circuits, 114–118
    dc series circuits, 72–75
    power in resistive circuits, 75
Resonance, 363
    frequency, 366
    parallel, 364
    quality, Q, 366
    series, 363
    tank circuit, 365
Right triangle, 128
    altitude, 128
    base, 128
    in electronics, 129
    hypotenuse, 128
R-L circuits
    parallel, 331
    series, 309
    time constant, 477
Rotating radius, 256

Scientific notation, 52
Series circuits
    ac circuit, 308
    dc resistance circuits, 72
    equivalent circuit, 317
    power dissipation, 75
    R-C circuits, 314, 475
    R-L circuits, 309, 477
Series-parallel ac circuits, 356
Series-parallel dc circuits, 120
Signed numbers, 3
    addition, 3
    multiplication and division, 6
    subtraction, 4
Significant figures
    defined, 54
    rounding, 55
    rules for determining, 54

Simultaneous equations
    application, 184
    solution by addition, 165
    solution by substitution, 167
    solution by subtraction, 165
Sine waves, 254
    addition, 264, 279
    average value, 257
    current curves, 257
    graphical generation, 256
    instantaneous values, 258
    mechanical generation, 254
    peak-to-peak value, 257
    peak value, 257
    RMS, 258
    voltage curves, 257
Square wave, 273

Tank circuit, 365
Tee circuit, 374
Terms, 9
Theta (θ), 129
Thevenin's Theorem (ac), 387
    equivalent resistance, 387
    internal impedance, 389
    open-circuit voltage, 387
    solution of impedance bridge, 388
Thevenin's Theorem (dc), 196
    equivalent resistance (R), 197
    one power source, 196
    open-circuit voltage, 197
    two power source, 199
Time constant (R-C), 475
Time constant (R-L), 477
Transient response, 476, 478
    R-C circuit, 475
    R-L circuit, 477
Transmission lines, 471
    coaxial cable, 471
    open wire, 471
Trigonometric functions, 131
    "arc" function, 134
    cosine ratio, 132, 137
    inverse functions, 134
    sine ratio, 132, 135
    tangent ratio, 132, 139
Trigonometric tables, 515–517
Trinomials, 10

Vector algebra, 229
    adding at other than 90°, 234
    adding at right triangles, 232
    adding more than two vectors, 236
    computing the resultant, 242–247
    computing resultant vectors at
        other than right angles, 247
    multiple vectors, 231
    single vector force, 230
    vector representation, 229
    x and y components, 240
Voltage gain, 482

Wire gauge table, 513
Wye circuit, 374